THE MOLECULAR SWITCH

THE MOLECULAR SWITCH

Signaling and Allostery

Rob Phillips

Illustrated by Nigel Orme

PRINCETON UNIVERSITY PRESS

Princeton and Oxford

Published by Princeton University Press
41 William Street, Princeton, New Jersey 08540
6 Oxford Street, Woodstock, Oxfordshire OX20 1TR

press.princeton.edu

ISBN 978-0-691-20024-8
ISBN (e-book) 978-0-691-20025-5

Preface epigraph translation courtesy of Carolina Tropini. Chapter 2
epigraph from Cohen, J. E. (2004), "Mathematics is biology's next
microscope, only better; biology is mathematics' next physics, only
better," *PLoS Biol.* 2(12):e439, with permission (CC BY-NC 4.0).
Chapter 3 epigraph from Mermin, N. D. (April 1989), "What's wrong
with this pillow?" *Physics Today*, with permission of AIP Publishing.
Chapter 5 epigraph from Stanislaw Ulam, quoted in Campbell, D. K.,
"Nonlinear science from paradigms to practicalities," *Los Alamos Science
Number 15: Special Issue on Stanislaw Ulam 1909–1984*, p. 259; used
with permission of the Estate of Stanislaw Ulam.

Library of Congress Control Number: 2020936993

British Library Cataloging-in-Publication Data is available

Editorial: Ingrid Gnerlich and Arthur Werneck
Production Editorial: Karen Carter
Text Design: Lorraine Doneker
Cover Design: Lorraine Doneker
Production: Jacqueline Poirier
Publicity: Katie Lewis and Matthew Taylor
Copyeditor: Barbara Liguori
Cover Art: Antique switchboard / yogesh_more / iStock

This book has been composed in Minion Pro for text
Helvetica Neue for display

Printed on acid-free paper. ∞

Printed in China

TO AMY

FOR EVERYTHING

CONTENTS

PREFACE

The universe cannot be understood without first learning the language, and being conversant with the characters with which it is written. It is written in mathematical language, and the characters are triangles, circles and other geometrical figures, without which means it is humanly impossible to comprehend a single word.
                    —Galileo Galileo, *Opere: Il Saggiatore*

Biology is sometimes disparaged as a subject without far-reaching conceptual ideas. I think this is often the result of the way our understanding of the living world is presented in introductory courses rather than of any flaw in the subject itself. Indeed, for me, biology is a subject that harbors perhaps the most beautiful of phenomena we are aware of in the natural world and the most interesting scientific puzzles of our time. Further, belying the biology portrayed in many survey courses and fact-filled textbooks, it is also a subject with many overarching conceptual ideas, and this claim has been made for decades (see the references at the end of the preface for examples). One of these ideas, namely, that of allostery, has just recently passed its 50th anniversary and serves as an intuitive explanation of how molecules function across all domains of biology, from physiology to neuroscience, from the processes of the central dogma to those of quorum sensing.

My own fascination with the subject was solidified at a meeting at the Institut Pasteur to celebrate the 50th anniversary of the allostery concept, with many of the central fields of biology on display. Neuroscience had its advocates on hand to talk about ion channels from an allostery perspective. The processes of the central dogma were represented on many fronts, with researchers discussing everything from how genes are regulated by allosteric transcription factors to the structural conformations of the ribosome during translation. Human physiology was represented in the form of studies on hemoglobin. G protein–coupled receptors took center stage as an example of how the allostery phenomenon impacts cell signaling. Other examples abounded as well, including conformational changes in viral capsids, the workings of bacterial chemotaxis and molecular motors such as myosin. See the special volume of the *Journal of Molecular Biology*, "Allosteric interactions and biological regulation (Part I)" (Kalodimos and Edelstein 2013) for the impressive breadth of topics described at that meeting. From my own point of view, the aspect of these talks I found most interesting was that all these seemingly disparate examples were to a first approximation described by the same fundamental equation that serves as the central equation of allostery, namely,

$$p_{active}(c) = \frac{(1 + \frac{c}{K_A})^n}{(1 + \frac{c}{K_A})^n + e^{-\beta \varepsilon}(1 + \frac{c}{K_I})^n}. \tag{1}$$

This equation tells us the probability of a molecule with n binding sites for ligands being in its active state is a function of ligand concentration c in terms of the energy difference between the inactive and active conformations ε, and the affinity K_A and K_I of these ligands for the active and inactive states, respectively.

Often during the years over which this book was under construction, I have given a talk entitled "The Other Bohr and Biology's Greatest Model," in some sense as a provocation to see if I could get anyone to name for me a model they think has broader reach in helping explain more biology than does the two-state allostery model of Monod, Wyman, and Changeux (and its model cousins) celebrated at that wonderful Institut Pasteur meeting. I didn't hear many persuasive alternatives.

This book is one in a collection of books entitled *Studies in Physical Biology* with one central aim, namely, to celebrate the deep, quantitative principles that preside over the subject of biology. This series aims to complement an earlier book, *Physical Biology of the Cell*, by taking particular themes laid out there and probing them more deeply, more in research monograph format than in textbook format. The audience for this book is the same as that for *Physical Biology of the Cell* or *Cell Biology by the Numbers*. That list includes biologists interested in seeing how the language of physics and mathematics can sharpen our biological thinking, physicists who want an introduction to some of the most exciting themes of modern biology, and engineers who want to see the scientific underpinnings of their subject. Though it is way too lofty an ideal to ever live up to, I always had the model of the classic series of Landau and Lifshitz in mind as an inspiration for how entire domains of a quantitative field could be revealed and, more recently, the more accessible (at least to me) work of Blandford and Thorne. The idea of this first book in the series is to bring together in one place the rudiments of how to describe the allostery phenomenon using ideas from statistical physics and to showcase some of the most intriguing usage of these models across a broad spectrum of applications. The book is a caricature of a subject (allostery) that itself is full of caricatures of many important biological problems. I adopt this approach of simplification and abstraction without apology. My argument is that simplified models are often nothing more than a restatement of many of the key cartoons and schematics that dominate the biological literature, recast in a mathematical form that permits us to more rigorously subject those models to corroboration. One of my favorite scientific essays by my friend Jeremy Gunawardena aptly channels the great James Black in noting that our models are "not meant to be descriptions, pathetic descriptions, of nature; they are designed to be accurate descriptions of our pathetic thinking about nature." That in a nutshell is the philosophy that animates the entirety of this book, namely, that the history of science shows us over and over that our impressions of subjects are often naive or just plain wrong, and the idea behind formulating our views in mathematical terms is that we can sharpen those views, making it easier to subject them to scrutiny. To that end, we try to make our pathetic thinking as clear as we possibly can and then use mathematical reasoning to explicate the experimental implications of that thinking.

It is important to be clear from the outset, this is a book about models more than it is a book about biological phenomena. We knowingly make simplifications of many phenomena as a vehicle to introduce the conceptual framework. Of course, this leaves the question of how to best think about the phenomena themselves. To go even further, our ambitions are even more limited because for the most part, the entire book is restricted to one special class of models as described using equilibrium statistical mechanics, though we will make several forays into the ways in which these ideas have to be generalized when thinking about the dynamical situations that arise out of equilibrium.

I have been particularly inspired by two books that attempt to bring unity to problems of signaling and regulation in biology, namely, *Genes and Signals*, by Mark Ptashne and Alex Gann, and *Cell Signaling*, by Wendell Lim, Bruce Mayer, and Tony Pawson. These excellent books attempt to find overarching principles that describe how molecules can mediate signaling processes. This book tackles some of the same questions but with a change in philosophy in the sense that it argues that by using mathematics to sharpen our thinking, we can sometimes go further in our understanding of the underlying phenomena, gaining precise insights into both the physiological and evolutionary parameters that characterize the phenotypes of the allosteric molecules of the cell.

The book is organized around a series of vignettes, each of which tells a little story about a biological problem which is a caricature of the real biology. With the biological problem in hand, we then show how statistical mechanical models of allostery can be erected for that problem as an explanatory and predictive framework. Signaling has many levels of description, ranging from the ecology and physiology of organisms to the natures of the networks to the molecules that make up those networks. In this book we will tell stories about the higher levels, but the rigorous quantitative discussion will focus on the molecular implementations of these signaling concepts with specific reference to allosteric interactions. The formulation might appear repetitive at times. Why? Because the statistical mechanical states are conceptually the same, and the resulting mathematics ends up often being identical, with the nouns changing but the verbs staying the same. Alfred Hershey once introduced the concept of a perfect experiment. "When asked what his idea of happiness would be, [Hershey] replied, 'to have an experiment that works, and do it over and over again.'" (Hodgkin 2001), In some sense, the statistical mechanical theory of allostery is the perfect theory for the same reason, because we can do it over and over again to shed light on a huge variety of different biological phenomena.

I was surprised to find how many fierce and personal debates and feelings I would uncover during the decade I spent digging deeply into models of allostery. There are still exposed nerves from the 1960s over topics such as the relative merits of the MWC and KNF models. But further, discussions of these ideas in the scientific public and in the reviewing process have exposed the widely divergent views about what it actually means to know something. Opinions on the status of theoretical models of allostery range from the idea that the success of the approach is "well known" to the fact that such models "don't work," with both opinions held with great conviction and stated as fact. MWC seems to be a world fraught with opinions not based on explicitly having a dialogue between theory and experiment, where theory says which knobs to tune, and then experiments are designed to tune those knobs. There is much received wisdom about the applicability or lack thereof of this framework, and part of the goal here is to make that conversation more clear. On that note, because in fact many of the applications of models of allostery to real-world problems have not been subjected to a rigorous theory-experiment dialogue, in some cases the data I appeal to will not be characterized by the same attention to error bars and fits that more modern treatments would. The reader is urged to overcome those shortcomings of my presentation by participating in the next phase of development of the study of allostery.

A subject I am very sensitive to is the enormous contributions made in all corners of science by a veritable army of researchers young and old, of different races and genders. These contributions should be honored, and I am grateful to be the beneficiary of these contributions every day. That said, this book makes no attempt at a scholarly report on the original literature on the many topics that appear throughout the text. As already noted, the book is built around intentional simplifications of the enormous atomic-level complexity found in the macromolecules of life. My main focus is to provide a coherent and pedagogical discussion of how one class of models can answer to diverse biological phenomena. My referencing is idiosyncratic and primarily focused in two directions: those papers that were the basis of some particular case study I am introducing or those papers I found to be particularly accessible for describing some topic.

Acknowledgments

I have always found there are two great and very distinct pleasures in writing a book. The first is personal and solitary and is a pleasure born of sitting and thinking and calculating and trying to come to a deep and rigorous understanding of the topic by myself. The second is quite the opposite and is fundamentally social and involves invoking the generosity of colleagues who are happy to talk about their ideas, to offer critiques of their own, and to provide an education in the subject. Though I have benefited from many such conversations now for decades, for this book I am particularly grateful to Frances Arnold, Tony Auerbach, Rachel Banks, Stephanie Barnes, Nathan Belliveau, Howard Berg, Bill Bialek, Shelby Blythe, Jean-Pierre Changeux, Griffin Chure, Eric Davidson, Julia Duque, Bill Eaton, Tal Einav, Luke Funk, Vahe Galstyan, Hernan Garcia, John Gerhart, Lea Goentoro, Chandana Gopalakrishnappa, Shura Grosberg, Jeremy Gunawardena, Bob Haselkorn, Liz Haswell, Dan Herschlag, Hopi Hoekstra, Christina Hueschen, Kabir Husain, Oleg Igoshin, Zofii Kaczmarek, Marc Kirschner, Tolya Kolomeisky, Jane Kondev, John Kuriyan, Michael Laub, Thomas Lecuit, Heun Jin Lee, Henry Lester, Mitch Lewis, Wendell Lim, Rod MacKinnon, Sanjoy Mahajan, Rodrigo Maillard, Madhav Mani, Sarah Marzen, Linas Mazutis, Chris Miller, Ron Milo, Leonid Mirny, Tim Mitchison, Richard Murray, Arvind Murugan, Brigitte Naughton, Phil Nelson, Elad Noor, Noah Olsman, Arthur Pardee, Isabelle Peter, Niles Pierce, Steve Quake, Rama Ranganathan, Manuel Razo, Doug Rees, Greg Reinhart, Kimberly Reynolds, Tyler Ross, Helmut Schiessel, Clarissa Scholes, Mikhail Shapiro, Boris Shraiman, Parijat Sil, David Sivak, Ed Taylor, Matt Thomson, Denis Titov, Reza Vafabakhsh, Navish Wadhwa, Jon Widom (sorely missed), Ned Wingreen, Shahrzad Yazdi, and Zechen Zhang, Special thanks for Manuel Razo, Griffin Chure, Stephanie Barnes, and Nathan Belliveau, who undertook turning our theoretical musings into a rigorous experimental endeavor. I am deeply grateful to the Benjamin and Donna Rosen Center for Bioengineering, which has supported my work at every turn, and to the NIH for generous support for curiosity-driven research.

I especially need to single out Jean-Pierre Changeux, Leonid Mirny, Peter Swain, Julie Theriot, and Ned Wingreen as the five people who have most inspired this book. Through their papers, generous conversations, and a long string of email queries and responses, they have all been models of scientific depth and friendliness. Much of this book has also been a joint venture with Tal Einav, who spent five years of his PhD thesis research accompanying me, and often leading me, on this adventure. Some of the sections of several chapters are lifted directly from papers he and I wrote together and from Tal's unpublished work. Similarly, over the last few years, Vahe Galstyan has taken up the baton and driven several chapters described here. I also need to give a special nod to Christina Hueschen, Phil Nelson, and Clarissa Scholes, who have thoughtfully and constructively shaped my thinking and writing. Suzy Beeler, Tal Einav, Jeremy Gunawardena, Brigitte Naughton, and Yuhai Tu read every page of the book and gave constant constructive feedback. Most everyone ends their acknowledgments by noting that any faults that remain are strictly the responsibility of the author and that is certainly true here. I can hope only that my immature efforts do not disappoint these many generous and distinguished colleagues, and I apologize in advance for the many places I surely could have done better. I end with loving thanks to Amy, who has been there in every way.

REFERENCES

Bonner, J. T. (2002) *The Ideas of Biology*. New York: Dover.

Dobzhansky, T. (1973) "*Nothing in biology makes sense Except in the light of evolution.*" Am. Bio. Teach. 35:125–129.

Hodgkin, J. (2001) "Hershey and his heaven." *Nat. Cell Biol.* 3:E77.

Kalodimos, C., and S. Edelstein, eds. (2013) "Allosteric interactions and biological regulation (Part I)." *J. Mol. Biol.* 425 (9): 1391–1592.

Nurse, P. (2003) "The great ideas of biology." *Clin. Med.* 3:560–568.

Thorne, K. S., and R. D. Blandford (2018) *Modern Classical Physics.* Princeton, NJ: Princeton University Press.

PART I

THE MAKING OF
MOLECULAR SWITCHES

IT'S AN ALLOSTERIC WORLD

1

What's in a name? That which we call a rose by any other word would smell as sweet.

—William Shakespeare

1.1 The Second Secret of Life

The 1953 discovery of the structure of DNA ushered in the molecular era in biology with a vengeance. As with many other great discoveries, the determination of the molecular basis of heredity spawned a host of new questions. One of the dominant mysteries was the nature of regulation. How are the many molecules of the cell (including DNA itself) regulated so that they carry out their functions both when and where they are needed and not otherwise? Such questions arise in all corners of biology. In the context of metabolism, it was clear that bacterial cells rank-order their preferences for different carbon sources, raising the question of how the cell acts on these preferences. The study of enzymes revealed that some enzymes are active in the absence of some inhibitor and are shut down in its presence. Animal body plans are set up by particular spatiotemporal patterns of gene expression, making it clear that whole batteries of genes are switching between different states. These examples and many others reveal that regulation is one of the most widespread molecular processes in all of biology.

Ten short years after the secrets of the great molecule of heredity were uncovered, a second molecular discovery of central importance to biology was announced. That discovery, the formulation of the allostery concept, is relevant to thinking about the function of molecules across the entire domain of biological inquiry. Stated simply, the allostery concept harkens back to the Roman deity Janus, shown in Figure 1.1(A), symbolized by his two faces and noted for presiding over all kinds of transitions. In this book, we will take a broad view of allostery as the phenomenon in which a molecule has more than one state of activity, as shown in Figure 1.1(B), with the relative probabilities of those different states controlled by some effector(s).

One of the discoverers of the allostery concept, Jacques Monod, referred to this discovery as the "second secret of life." But what exactly was this secret? We will refine our answer to that question through the various case studies that make up the chapters that follow, though here we give a brief qualitative sketch

Figure 1.1

Allostery defined. (A) The Roman god Janus. (B) Molecules such as transcriptional repressors have a Janus-like existence as they switch between active and inactive conformations.

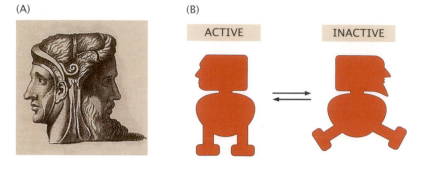

(A)   (B)

ACTIVE   INACTIVE

Figure 1.2

The molecular switch. Different classes of molecules exemplify the allostery phenomenon, highlighting the inactive and active conformational states of these molecules. Ion channels can be either closed or open, with the binding of a ligand favoring the open state. An enzyme can be in an inactive state, in which it is unable to cleave a substrate, or in an active state, in which it is competent to perform such a cleavage reaction. The presence of an effector (red) favors the active state. A membrane-bound receptor can be in either an inactive state or an active state when it is bound to a ligand, in which it can perform a phosphorylation reaction leading to a subsequent signaling cascade. For all the examples shown here, the ligands shown in red tip the balance in favor of the active conformation over the inactive conformation.

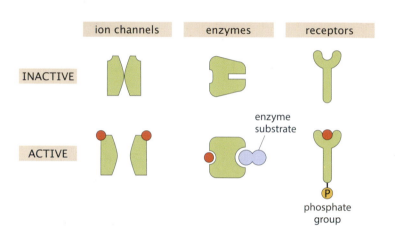

ion channels   enzymes   receptors

INACTIVE

ACTIVE

enzyme substrate

phosphate group

of the key elements of the concept. Stated simply, many biological molecules behave as molecular switches. In its most basic incarnation, the idea is that many biological molecules have two distinct conformations, which we will often think of as inactive and active, as shown in Figure 1.2. As a result of the exchange of energy with the molecules of the surrounding solution (i.e., thermal energy), these molecules are constantly flipping back and forth between these two conformational states. In equilibrium, the relative proportion of these two states is fixed by the energy difference between them. However, the interesting regulatory behavior of these molecules is that the binding of a ligand can change the relative probabilities of the inactive and active states. Specifically, the binding affinity of the ligand for each state is different, resulting in a shift in the relative probabilities of the inactive and the active conformations when the ligand concentration is changed. The outcome is that a ligand can serve to regulate when molecules like those shown in Figure 1.2 are active. Our task is to explore different biological phenomena that are controlled by such molecules and to examine what physical models of these molecules have to say about their function.

1.2   The Broad Reach of the Allostery Concept

Biology is a science full of beautiful and fascinating exceptions. But the ease with which we can find such exceptions is not a proof of the absence of broad and overarching ideas. One such motif that has captured my imagination and which serves as the basis of this book is that there is a unifying mathematical description of the way that many of the macromolecules of life can exist in several distinct states (see Figure 1.2, p. 4). For example, as we have already remarked, ion channels can be open or closed. Proteins can be phosphorylated or not. Receptors can be active or not. Often, which of these two different states is more likely depends in turn upon whether or not a particular ligand is bound to the molecule of interest. In such cases, by titrating the amount of ligand competing for the attentions of our molecule of interest, we can shift the balance between these two states. But only when viewed using equations rather than words and cartoons is the full impact of the allostery idea made clear. To foreshadow the kinds of phenomena that fall within the purview of the allostery concept, here we consider several illustrative examples.

Figure 1.3 shows how allosteric processes are central to many signaling pathways. As seen in the figure, both the membrane-bound receptors that receive signals and the soluble proteins that mediate the processes at the end of the signaling cascade are often themselves allosteric. For example, in the context of metabolism, as shown in the first panel of the figure, many enzymatic reactions are catalyzed by proteins that are subject to feedback such that the reaction occurs only when it is needed. This case illustrates a metabolic enzyme that is activated only in the presence of some effector molecule (shown in red). We will explore these processes more deeply in the next section.

Further, the processes of the central dogma of molecular biology such as transcription involve regulatory proteins. Transcription factors can be localized to the nucleus and bind DNA depending upon the presence or absence of some ligand, as shown in the middle panel of Figure 1.3. In the presence of inducer for the regulatory architecture shown here, the repressor protein is released from its binding site on the DNA, allowing for the expression of the regulated gene.

As indicated schematically in the final panel of the figure, activation of cytoskeletal growth and remodeling makes processes such as eukaryotic chemotaxis possible. Indeed, one of the most inspiring microscopy time-lapse sequences of all time shows a neutrophil engaged in the process of tracking down a bacterium. Several snapshots from this process are shown in Figure 1.4. In order for the cell to change direction, the leading edge has to be remodeled with new actin filaments synthesized in the correct direction of motion. To that end, these motile cells have an impressive signaling pathway that allows them to detect extracellular ligands, resulting in a subsequent molecular cascade within the cell that ends with the construction of actin filaments at the leading edge. A schematic representation of this pathway is given in Figure 1.5 and is mediated by molecules that can exist in several conformational states (i.e., allosteric) with different abilities to catalyze reactions.

1.2.1   Sculpting Biochemistry via Allostery

We can get a higher-resolution view of the ubiquitous nature of allosteric regulation by turning to one of the best-understood biochemical pathways, namely, that associated with carbon metabolism. Figure 1.6 gives a depiction of the key enzymes that mediate glycolysis, as well as their various substrates. What is not at all evident from this figure is that many of the enzymes in this pathway are

Figure 1.3

Signaling pathways and allostery. Each panel shows a schematic representation of inactive and active signaling pathways. In each case, an extracellular ligand binds to a receptor resulting in a cascade involving intracellular signaling proteins. These proteins in turn influence a variety of other proteins that can carry out specific biological processes, including activating metabolic enzymes (top panel), turning on the transcription of key genes (middle panel), or turning on cytoskeletal polymerization in particular regions within the cell (bottom panel).

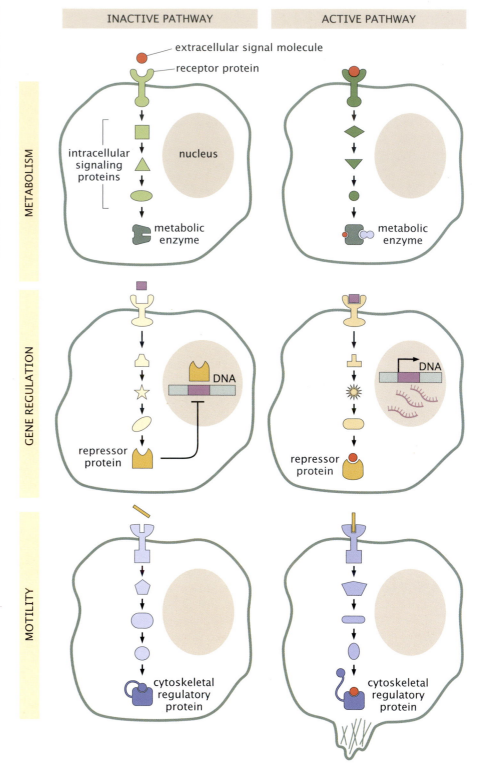

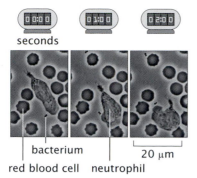

seconds

bacterium   20 μm

red blood cell   neutrophil

Figure 1.4

Eukaryotic chemotaxis. Snapshots from a video taken by David Rogers of the dynamics of a neutrophil hunting down a bacterium. This classic video raises myriad questions about the mechanisms of signaling and motility in cell biology. We use snapshots from this video as a reminder of the ubiquitous nature of cell signaling, exemplified here by the way the neutrophil is "aware" of its environment. From Phillips, R., J. Kondev, J. Theriot, and H. Garcia (2013), *Physical Biology of the Cell*, 2nd ed. Reproduced by permission of Taylor & Francis LLC, a division of Informa plc. Adapted from from a video by David Rogers.

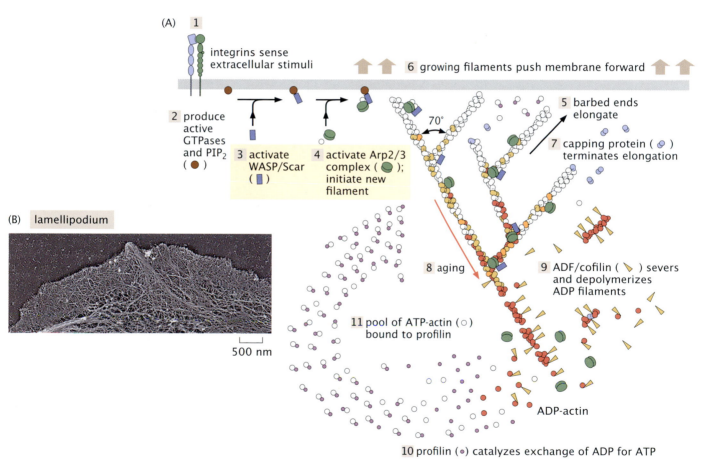

Figure 1.5

Cell signaling and the dendritic nucleation model. (A) The nucleation of new actin filaments such as those involved in the famous video of a neutrophil chasing down a bacterium (see Figure 1.4) require the activation of proteins such as N-WASP and Arp2/3 (highlighted in yellow). Protein activation can be viewed through the prism of the allosteric models this book discusses. (B) Leading edge of a motile cell as viewed using an electron microscope. The positioning of the actin network shown here is controlled by the signaling cascade shown in part (A). (A) Adapted from Pollard, T. D., and G. G. Borisy (2003) "Cellular motility driven by assembly and disassembly of actin filaments," *Cell* 112:456. with permission of Elsevier. (B) Adapted from Phillips, R., R. Milo (2016) *Cell Biology by the Numbers* (Garland Science), Estimate 3–7. Adapted from Svitkina, T. M., A. B. Verkhovsky, K. M. McQuade, and G. G. Borisy (1997) *J. Cell Biol* 139:397–415.

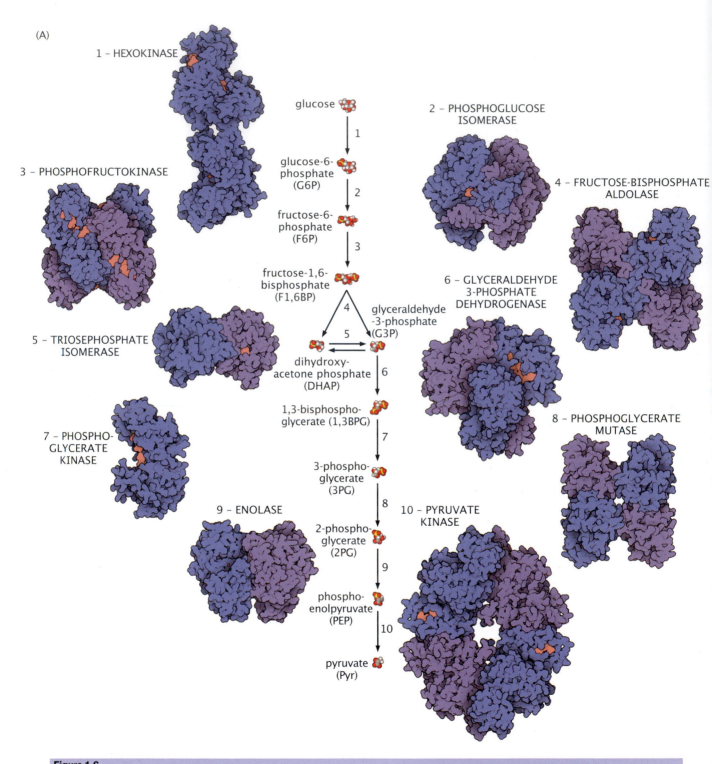

(A)

1 – HEXOKINASE

2 – PHOSPHOGLUCOSE ISOMERASE

3 – PHOSPHOFRUCTOKINASE

4 – FRUCTOSE-BISPHOSPHATE ALDOLASE

6 – GLYCERALDEHYDE 3-PHOSPHATE DEHYDROGENASE

5 – TRIOSEPHOSPHATE ISOMERASE

8 – PHOSPHOGLYCERATE MUTASE

7 – PHOSPHO-GLYCERATE KINASE

9 – ENOLASE

10 – PYRUVATE KINASE

glucose

glucose-6-phosphate (G6P)

fructose-6-phosphate (F6P)

fructose-1,6-bisphosphate (F1,6BP)

glyceraldehyde-3-phosphate (G3P)

dihydroxy-acetone phosphate (DHAP)

1,3-bisphospho-glycerate (1,3BPG)

3-phospho-glycerate (3PG)

2-phospho-glycerate (2PG)

phospho-enolpyruvate (PEP)

pyruvate (Pyr)

Figure 1.6

Enzymes in the glycolysis pathway. The molecules running down the center of the figure reveal the life history of a glucose molecule once it enters the glycolysis pathway. The large protein structures mediate these molecular transformations as glucose is turned into pyruvate. As shown in Figure 1.7, many of these proteins are modulated by the binding of small molecules. Courtesy of David Goodsell.

activated and inhibited by a suite of small molecules. Figure 1.7 gives an impression of this small-molecule regulatory landscape by presenting the complement of known inhibitors and activators. For example, if we look at phosphofructo-kinase, we see that it is inhibited by four distinct molecules and is activated by three others, giving the cell a suite of regulatory knobs with which to regulate this one step of this complex and important pathway.

As seen in Figure 1.7, there are a number of small molecules that play a role in repeated signaling and regulation. That qualitative impression has been made more concrete by counting up the number of regulatory interactions these small molecules participate in. A systematic analysis of the panoply of small molecules that preside over the regulation of proteins is shown in Figure 1.8 which counts the number of times that a given small molecule is known to participate in either inhibitory or activating interactions. It is clear that potassium ions are one of the most critical players in regulating the activity of proteins. Perhaps even more interesting is the role that ATP plays both as inhibitor and activator, including in the glycolysis pathway.

Sometimes biochemical reactions are mediated by ligands that are tethered to their receptor, giving rise to biochemistry on a leash. An example of this generalized allostery is offered in Figure 1.9. For example, as seen in Figure 1.9(A), when the tethered ligand is bound to the tethered receptor, the protein is in an inactive state. As the concentration of the free ligand is increased, at some point that concentration exceeds the "effective concentration" (estimated as $c_{eff} = 1/\frac{4}{3}\pi R^3$, where R is approximately the tether length), and hence the soluble ligands bind the tethered receptor, thus opening up the protein to its active conformation. The second example (N-WASP) shown in Figure 1.9(B) features a multidomain protein with a tethered linker. The relative equilibrium of active and inactive states is modulated by several binding partners (Cdc42 and PIP_2).

The example in Figure 1.9(B) teaches us another lesson as well. In particular, this is our first encounter with the concept of combinatorial control, as illustrated schematically in Figure 1.10. Here the idea is that a given cellular action, whether the activation of transcription or the activity of an enzyme, is dependent upon the status of several inputs simultaneously. For example, in the genetic network shown in Figure 1.10(A), the gene at the bottom of the diagram is regulated by two distinct activators. For this particular construct, the genetic circuit behaves as an AND gate, with high expression occurring only when both activators are present. As we will explore in detail in chapter 9, (p. 303), allosteric molecules can themselves behave as logical elements, as indicated schematically in Figure 1.10(B). Only in the presence of both inputs will the allosteric molecule be active, as already shown in the case of Cdc42 and PIP_2 in the context of Figure 1.9(B).

1.2.2   One- and Two-Component Signal Transduction and the Two-State Philosophy

So far, we have focused on specific molecular pathways that feature allosteric molecules. These case studies naturally lead us to wonder about the broader reach of the allostery concept. Despite the amazing advances of the high-throughput era, it remains a daunting challenge to identify genome-wide which

Figure 1.7

Small-molecule interactions in *E. coli* central carbon metabolism. The enzymes that mediate various steps in the pathway are denoted in black. Activators of those enzymes are denoted in green, and inhibitors are denoted in red. The structure of the enzyme phosphofructokinase (pfk) is shown, as well as its substrate F6P and product FDP. Adapted from Reznick et al. (2017).

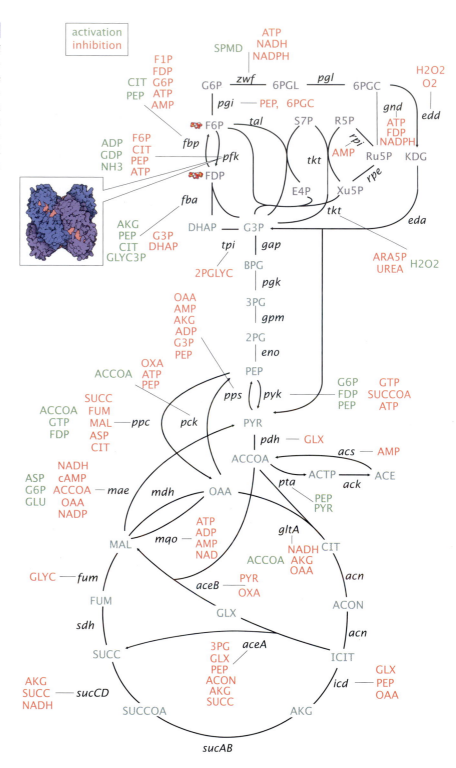

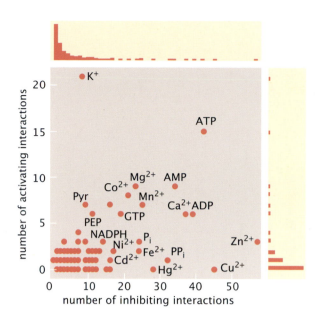

Figure 1.8

Small molecules with broad regulatory reach. These plots show the number of activating and inhibiting interactions that different small molecules engage in. Clearly, ATP is especially important both as an activator and inhibitor. Adapted from Reznick et al. (2017).

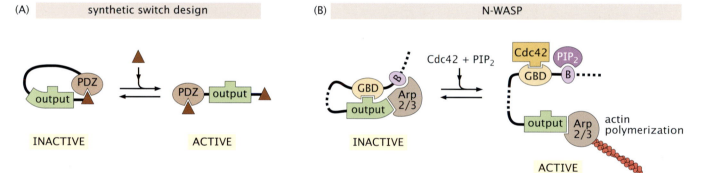

Figure 1.9

Biochemistry on a leash. Allostery can be based upon tethering motifs. (A) Tethering a receptor to a protein allows the state of activity of the protein to be controlled by the binding of soluble ligands which can outcompete the tethered ligand. The active state is characterized by exposure of the output domain. (B) Tether motif in the context of activation of actin polymerization. In the presence of Cdc42 and PIP_2, N-WASP is activated, which in turn activates Arp2/3. Adapted from Dueber et al. (2003).

proteins are allosterically regulated. Even more tricky is identifying what small molecules regulate them.

One context in which attempts have been made to survey the allosteric landscape is signaling in bacteria, specifically in the context of the two-component signal transduction systems in bacteria. The idea broadly is that the cell membrane is occupied by a wide variety of different receptors which flip between inactive and active states, as indicated schematically in Figure 1.11. In particular, as a result of the presence or absence of some external ligand, these receptors then switch between states in which they are either active or inactive for phosphorylating their cytoplasmic response regulator. These soluble proteins are now able to perform cellular functions such as changing the frequency

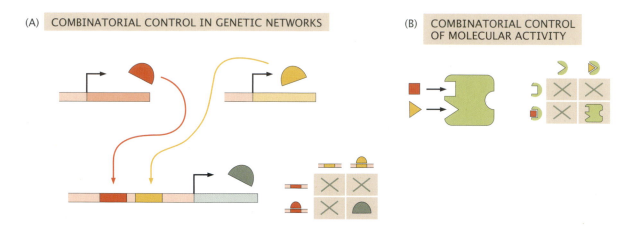

(A) COMBINATORIAL CONTROL IN GENETIC NETWORKS

(B) COMBINATORIAL CONTROL OF MOLECULAR ACTIVITY

Figure 1.10

Introducing combinatorial control. (A) Combinatorial control in genetic networks. The gene in blue at the bottom of the schematic has binding sites for two distinct transcription factors. The logical truth table shows that the gene is on only when both transcription factors are bound. (B) Combinatorial control of allosteric molecules. The enzyme or signaling molecule (green) requires the presence of both ligands to be activated.

Figure 1.11

Schematic showing reactions of the sensor histidine kinase and response regulator in a typical bacterial two-component system. In the presence of ligand, the kinase has a higher probability of being in the active state where it is competent to carry out the phosphorylation reaction.

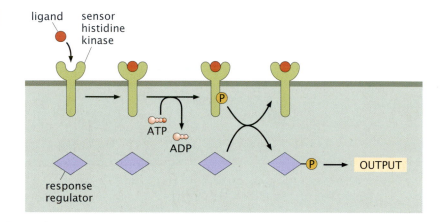

of tumble motions during bacterial chemotaxis, as will be taken up in detail in chapter 4 (p. 124).

To give a sense of the diversity of such two-component signaling systems in *E. coli*, Figure 1.12 shows the sensor histidine kinases and their corresponding response regulators. We see that there are a wide variety of inputs into these signaling systems that then lead to changes in cellular physiology and behavior. Figure 1.13 goes further by providing some insight into the distribution of one- and two-component signaling systems in bacteria by examining the databases of sequenced bacterial and archaeal genomes as of 2005 (it would be great to see these studies modernized). Specifically, the deeply interesting question that was examined in that work was the nature of the signaling systems in bacteria with special reference to whether those are one-component or two-component signaling systems. In the one-component signaling systems, the input and output domains are present on the same protein. Two classic

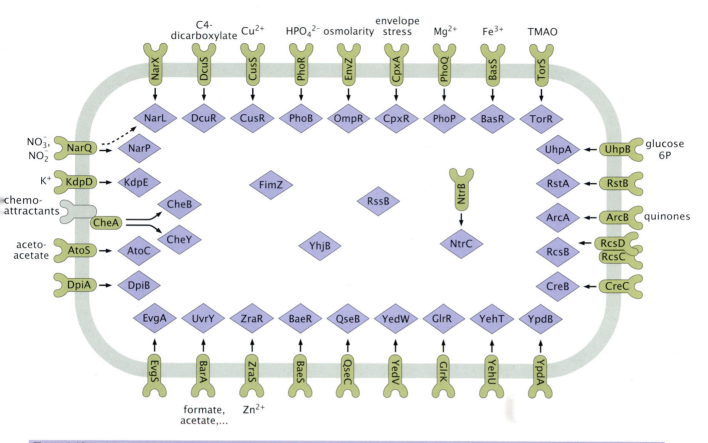

Figure 1.12

The vast array of sensor histidine kinases and response regulators found in *E. coli*. Schematic showing all sensor histidine kinases (green) and response regulators (blue) as identified by sequence in the *E. coli* genome. Both the sensor and the response regulator can be allosteric. Courtesy of Mark Goulian and Michael Laub.

examples of one-component signaling systems in bacterial transcription will be considered in chapter 8 (p. 272) when we discuss repression by the Lac repressor and activation by CRP. By way of contrast (see Figure 1.11), two-component signaling systems are characterized by having the input domain on one protein (usually a membrane-bound receptor) and the output domain (usually a cytoplasmic protein) on another protein. As we see in Figure 1.13, one-component signaling systems far outnumber their two-component counterparts, suggesting a rich proteomic reservoir of possible allosteric signaling molecules for further investigation.

Though data like those described here are not a proof that those molecules are allosteric, it is at least a tantalizing hint that the proteins that serve as one-component signaling systems may well be allosteric. Several key observations were made on the basis of this bioinformatic analysis of sequenced genomes and their putative signaling systems. First, that work found that roughly 85% of the output domains on these one-component signaling molecules are DNA-binding helix-turn-helix domains, indicating that often these signaling pathways appear

Figure 1.13

Number of one-component and two-component signal transduction systems as a function of genome size. Adapted from Ulrich, Koonin, and Zhulin (2005).

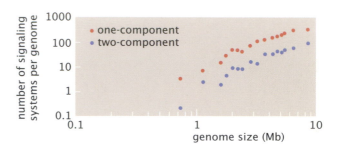

to be tied to transcriptional regulation, the example of which we will take up in detail in chapter 8 (p. 272). Similarly, this work found systematic trends in the input domains of these one-component systems, with more than 90% of them involving small-molecule binding domains, again providing a tantalizing hint of the possible allosteric control of these proteins.

1.3   Reasoning about Feedback: The Rise of Allostery

1.3.1   The Puzzle

Feedback is one of the greatest of ideas. A visit to a science museum such as the Musée des Arts et Métiers in Paris reveals century-old machines with their spinning "governors" that served to prevent them from running out of control, as shown in Figure 1.14. As hinted at in the figure, as the system rotates ever more quickly, the two balls will lift higher and in so doing will let some of the pressure bleed off, thus reducing the driving force that increases the rate of rotation.

In a fascinating reminder of the way that science and technology have always gone hand in hand, Figure 1.15 shows how the very same James Clerk Maxwell of Maxwell's equations fame worked on governors, noting that they are machines "by means of which the velocity of the machine is kept nearly uniform, notwithstanding variations in the driving power or the resistance." The modern world depends upon mechanical governors as well. In fact, we need look no farther than our toilets to see feedback in action, as also shown in Figure 1.14.

Like their macroscopic counterparts, the macromolecules of life are replete with examples of molecular governors which inhibit or enhance key biochemical reactions in response to either too much or too little of some substrate of interest. The bottom panel of Figure 1.14 shows regulatory feedback in the context of transcriptional autorepression, where the gene product of the gene of interest "governs" its own production. This will be the subject of chapter 8 (p. 272), where we will consider transcriptional regulation and its connection to allostery in detail.

But more generally, how do molecular governors work? One of the original ideas in the context of enzymes was that there are inhibitory molecules which compete for the attentions of the active site of some enzyme (as a concrete molecular example), thereby slowing down the reaction of interest. To be specific, people envisioned that the inhibitor molecule could actually bind to

mechanical feedback

ACTIVE   REPRESSED

ACTIVE   REPRESSED

genetic feedback

ACTIVE   REPRESSED

R

Figure 1.14

Control by feedback. (Top) Mechanical governors for feedback on a steam engine and in a toilet tank. (Bottom) Negative feedback in a gene regulatory circuit. The gene produces a protein that represses itself.

The following communications were read:—

I. "On Governors." By J. CLERK MAXWELL, M.A., F.R.SS.L. & E. Received Feb. 20, 1868.

A Governor is a part of a machine by means of which the velocity of the machine is kept nearly uniform, notwithstanding variations in the driving-power or the resistance.

Figure 1.15

James Clerk Maxwell and the mechanics of governors. (*Lower left*) Lowes Cato Dickinson, portrait of James Clerk-Maxwell (1891). Photo courtesy of Master and Fellows of Trinity College, Cambridge. (*Lower right*) Courtesy of Ray Tomes of Auckland, New Zealand. Taken at MOTAT (Museum of Transport and Technology) in Auckland.

Figure 1.16

Regulation of an enzyme by an inhibitor. (A) The inhibitor fits in the active site occluding the site from possible binding by the correct substrate. (B) Not all inhibitors have the correct shape to fit into the active site of the enzyme. In the early 1960s, this phenomenon led to the question of how such inhibitors regulate their target enzymes.

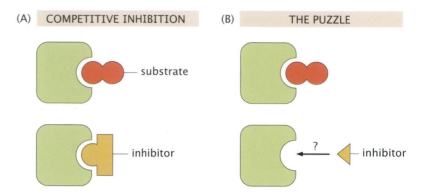

the enzyme in such as way as to literally block access of the real substrate to the active site, as indicated schematically in Figure 1.16(A). However, for such a mechanism to work, it would seem that the inhibitory molecule would need to have the same shape and size as the substrate molecule whose enzymatic modification was being inhibited. The question of how molecules would activate an enzyme was even more puzzling to contemplate.

One of the early molecules that focused the attention of scientists on questions of enzymatic regulation was aspartate transcarbamoylase, one of the key enzymes in pyrimidine biosynthesis. Recall that the DNA double helix is made up of the repeated pairing of pyrimidine-purine pairs, with cytosine, thymine, and uracil making up the pyrimidine derivatives. The *E. coli* version of aspartate transcarbamoylase is made up of 12 subunits with half coming from two trimers and the other half coming from three dimers, the two classes serving as the enzymatic and regulatory parts of the complex (see Figure 6.1 for more details). This enzyme functions in pyrimidine biosynthesis by mediating the interaction of aspartate and carbamyl phosphate to form *N*-carbamyl-L-aspartate and inorganic phosphate. For our purposes, the reason this enzyme was and is so interesting is because the rate with which it carries out the reaction is modulated by the levels of both pyrimidines and purines. Specifically, the final product of the pyrimidine pathway, namely, CTP, feeds back into the reaction and slows it down. By way of contrast, ATP, the final product in the purine pathway, speeds up the pyrimidine synthesis reactions. Thus, the system is subject to both negative and positive feedback in order to tune the quantities of the pyrimidine substrate. The kinds of questions that arose in light of these observations centered on how molecules such as CTP and ATP could interact with the enzyme itself in such a way as to tune the reaction rate.

The puzzle faced by early investigators of proteins that were subject to inhibition and activation was how a battery of regulatory molecules could be fine-tuned to fit into the active sites of their binding partners, as shown in Figure 1.16(B). The simple answer is that often they don't. Rather, groups in Paris and Berkeley realized that a different regulatory strategy could be "action at a distance," in which the binding of a regulatory ligand in one part of a macromolecule could lead to a conformational change elsewhere in the molecule such that the activity of the enzyme was changed. This thinking has been codified in the so-called Monod-Wyman-Changeux (MWC) model. We now turn to this allosteric resolution of the regulatory puzzle.

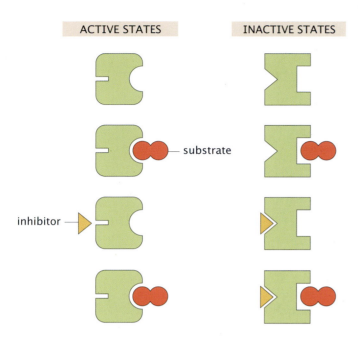

ACTIVE STATES   INACTIVE STATES

substrate

inhibitor

Figure 1.17

Allosteric regulation of an enzyme by an inhibitor. The enzyme can exist in both active (left column) and inactive (right column) states. For each conformation, there are four states of occupancy—empty, bound by substrate, bound by inhibitor, and bound by inhibitor and substrate. The binding affinities for both inhibitor and substrate are different in the two states, with the binding of the inhibitor favored in the inactive state.

1.3.2   The Resolution of the Molecular Feedback Puzzle

Figure 1.17 gives a schematic view of the extremely clever hypothesis that was formulated as the allosteric alternative to the kind of direct regulation posited originally and schematized in Figure 1.16. The essence of the cartoon is that there is a regulatory site where a ligand binds that tunes the relative probability of the active and inactive states. Note however that there is a nuance that is captured in our cartoon. Specifically, the regulatory ligand does *not* have the same binding energy when bound to the active and the inactive states, as indicated by the difference in the shape of the binding site in the two conformations. This critical mechanistic feature is the entire basis of the allostery framework, as we will show in equation 1.2 (p. 26). Note further that the allosteric strategy is noncommittal with respect to the question of whether the regulatory ligand leads to inhibition or activation of its binding partner. If the effector molecule favors binding the active state, then it will serve as an activator. If the effector molecule favors binding the inactive state, then it will serve as an inhibitor. By way of contrast, for the strategy highlighted in Figure 1.16, there is no obvious mechanism for activation.

Though we provided a caricature of some of the classes of MWC molecules we will consider here, in Figure 1.2 (p. 4), in fact, the detailed atomic structures of some of these molecules are known for a variety of different conformational states both in the absence and presence of their substrates and effectors. Structure has become one of the most powerful tools in modern biology. The conceptual argument associated with the great push for structural insights into biological problems is a deep confidence in the structure-function paradigm that holds that function follows structure. Several examples of structures of key allosteric molecules that will occupy our attention throughout the book are shown in Figure 1.18. For example, our first concrete case study in quantitative allosteric thinking will focus in chapter 3 (p. 77) on the ligand-gated ion

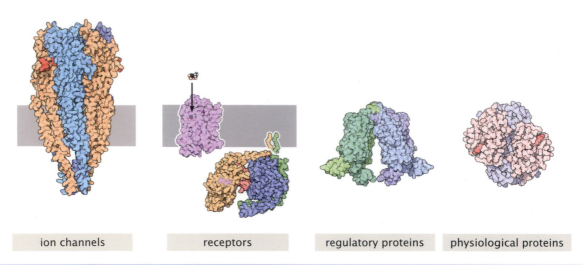

| ion channels | receptors | regulatory proteins | physiological proteins |

Figure 1.18

Structures of MWC molecules. Many different macromolecules can exist in distinct conformational states including ion channels, such as the nicotinic acetylcholine receptor shown here, G protein–coupled receptors such as the adrenergic receptor shown here, transcription factors such as the Lac repressor shown here, and proteins relevant to human physiology such as hemoglobin. Illustrations courtesy of David Goodsell.

channel known as the nicotinic acetylcholine receptor and shown in the left of Figure 1.18.

Chapters 4 (p. 124) and 5 (p. 170) take up the topic of allosteric membrane receptors like those shown in Figure 1.18. First, we tackle the behavior of the bacterial receptors responsible for chemotaxis and quorum sensing, followed by an in-depth examination of G protein–coupled receptors.

Another classic example that dates all the way back to the inception of the allostery concept itself is the Lac repressor molecule (third structure shown in Figure 1.18) that binds DNA, thus shutting down transcription of genes associated with lactose usage. This molecule can be thought of as an MWC molecule because in the presence of allolactose it undergoes a conformational change that reduces its binding affinity for DNA, thus permitting the transcription of the genes for β-galactosidase that make it possible to metabolize this alternative carbon source. There are also numerous examples of transcriptional activators.

The final example shown in Figure 1.18 is hemoglobin, the critical oxygen carrier. We devote chapter 7 (p. 231) to the fascinating allosteric mechanisms of hemoglobin and the basis for its physiological and evolutionary adaptation.

A zoomed-out view of the secondary structure of some representative allosteric proteins is shown in Figure 1.19. For example, in our discussion of gene regulation in chapter 8 (p. 272), we will consider both repression and activation, using classic examples from bacteria as our critical case studies. One of the most beloved and well-studied examples of activation is offered by the bacterial protein CRP, shown in the upper left panel of Figure 1.19, bound to its effector cAMP. The remaining structures in Figure 1.19 give other examples of allosteric proteins, revealing transcription factors, macromolecular assemblies, and enzymes.

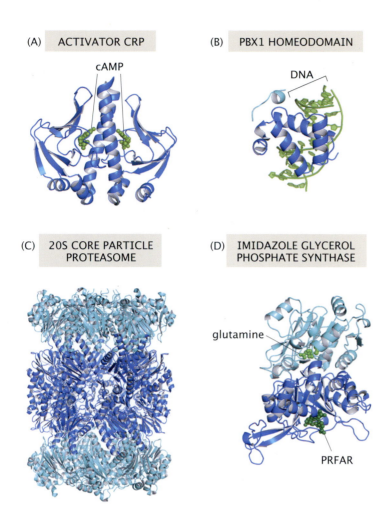

(A)   **ACTIVATOR CRP**

cAMP

(B)   **PBX1 HOMEODOMAIN**

DNA

(C)   **20S CORE PARTICLE PROTEASOME**

(D)   **IMIDAZOLE GLYCEROL PHOSPHATE SYNTHASE**

glutamine

PRFAR

Figure 1.19

Structures of MWC molecules in complex with effectors and substrates. (A) The bacterial transcriptional activator CRP bound to cAMP (green). (B) PBX1 homeodomain bound to DNA. (C) Structure of the protein degradation apparatus from the archaeon *Thermoplasma acidophilum*. (D) Key metabolic enzyme relevant to both amino acid (histidine) and nucleotide (purine) biosynthesis. The structure shows both the substrate (glutamine) and the allosteric effector (PRFAR). Adapted from Grutsch, Bruschweiler, and Tollinger (2016), CC BY-NC 4.0.

Though it is clear that structural insights are a critical cornerstone of modern biology, the approach advocated here is quite different. Indeed, in many ways the entire goal of the kinds of models to be described throughout the book is to telescope out from the atomic-level mindset to more coarse-grained perspectives which make predictions about how MWC molecules will respond quantitatively in new situations. In some sense, the way that structure is internalized at the level of the models described here is through a very small set of parameters. As will be shown in the next section, there is one overarching conceptual framework for describing MWC molecules that in its simplest incarnation features three key parameters, namely, the difference in energy between the active and inactive states in the absence of ligand, and the dissociation constants (K_A and K_I) for ligand binding in the active and inactive states, respectively. In the context of the MWC model, the structural details for MWC molecules found in the Protein Data Bank influence only these three parameters. Our aim is to show how one can talk about the huge swaths of biology reflected in the case studies shown in Figure 1.18 in terms of abstract models without appealing to detailed atomic-level positions.

A fundamental pillar upon which the entire book is constructed is the idea of the power, the rigor, and the intuition that can be developed by self-consciously suppressing features of a system. As will be highlighted repeatedly throughout the book, statistical mechanics teaches us how to "integrate out" degrees of freedom. This does not mean that we approximate the system by ignoring some feature (such as the existence of intrinsically disordered domains in a protein) but rather that we formally and mathematically compute the implications of those hidden degrees of freedom for the rest of the system.

1.3.3   Finding the Allosterome

A puzzle that remains in the field of allostery in this high-throughput era is that we have had very limited tools that allow us to answer the general question of which proteins in the proteome are allosteric and who their binding partners are. Despite Monod's characterization of the allostery phenomenon as the second secret of life, because of this important knowledge gap, as a field we are often flying blind because of our ignorance of how the key molecular players in signaling pathways have their activity modified by other chemical agents, and because of our ignorance of the identity of those chemical agents themselves.

To that end, the emergence of mass spectrometry has provided an exciting opportunity to query not only the posttranslational modifications suffered by a given signaling molecule but also, because of recent innovations, when signaling molecules have bound a given small molecule. The idea of one such method is shown in Figure 1.20. We see that by lysing cells in the absence and in the presence of some small-molecule allosteric effector candidate, some proteins will bind that small molecule and, as a result, be resistant to limited proteolysis by proteinase K. This means that when the proteins are denatured and trypsin digested, the pattern of cuts in the polypeptide chain will be different for any protein that was bound to the candidate small molecule, as indicated in Figure 1.20(B). Approaches such as this hold the promise of systematic identification of the allosterome for any organism and will be a critical part of our resolution of the puzzle of how the macromolecules of the cell are controlled by a battery of small molecules.

1.4   Mathematicizing the Two-State Paradigm

By peering at allostery through a mathematical lens we learn there are many common biophysical features shared by these molecules, as shown in Figure 1.21. This figure focuses our attention on the function of these molecules rather than their structure. One important feature is how much activity they exhibit even in the absence of ligand, a quantity we will call the *leakiness*. Just as we interest ourselves in the activity of allosteric molecules in the absence of ligand, their behavior in "saturating" concentrations is also critical to their function. Another parameter of great physiological and evolutionary significance is the critical concentration at which the activity reaches the midpoint between the inactive and active states, sometimes denoted as the EC_{50}. We will also be deeply interested in how sharp the switching events are between the two

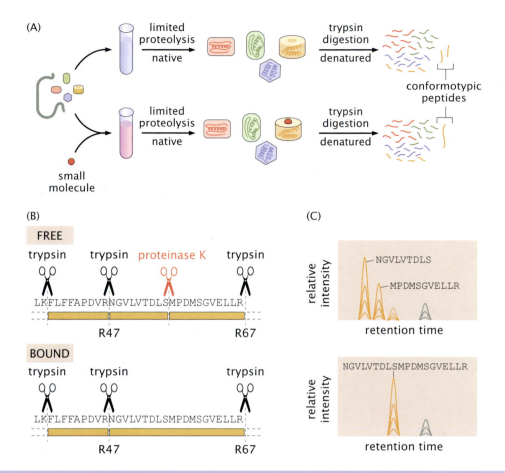

Figure 1.20

Finding allosteric proteins and the molecules that regulate them. (A) Incubation of cell lysate with and without some candidate small molecule leads some proteins to have a different pattern of peptide fragmentation because the binding of the small molecule protects some parts of the polypeptide. (B) Partial proteolysis is performed with proteinase K, followed by a trypsin digestion of denatured protein. This leads to peptide fragments that can be measured using mass spectrometry. In the example shown here, the measurement is made on the protein FixJ bound to aspartyl phosphate (PDB: 1DBW) and its ligand-free form (PDB: 1D5W) (C) Differences in spectrum resulting from mass spectrometry for the two different situations, revealing which parts of the molecule has been protected. Adapted from Piazza et al. (2018).

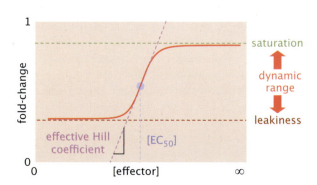

Figure 1.21

Macromolecular activity as a function of ligand concentration. This curve shows key phenotypic properties of allosteric molecules including leakiness, dynamic range, the EC_{50}, and the effective Hill coefficient, which gives a measure of the sharpness of the regulatory response.

conformational states of interest as a function of the ligand concentration that drives this shift. The examination of such sharpness in molecular responses led to one of the most preeminent ideas in biology, namely, cooperativity, an idea that falls very naturally within the purview of the statistical mechanical models of allostery to be described throughout the book.

The MWC model is the mathematical framework that was introduced to describe molecules with two states of activity and modulated by the binding of regulatory ligands. Monod, Wyman, and Changeux saw how to enumerate the various microscopic states of the system and to compute their relative probabilities. As any lover of statistical mechanics knows, the two-state paradigm is one of the centerpieces of statistical physics and has had enormous reach in the form of the Ising model and its generalizations. In the physics setting, the two states referred to in the statistical mechanical setting of Ising models can refer to the orientations of magnetic "spins," for example. The kinds of questions that people were interested in addressing with such two-state models centered on phase transitions between the low-temperature magnetic state and a high-temperature nonmagnetic state of some materials. Monod, Wyman, and Changeux, without knowing it at the time, were introducing another overwhelmingly important statistical mechanical model that could have impact in biology similar to that of the Ising model in physics. To get a better idea of how specific models can have such broad reach, we consider examples of such transcendent concepts in physics.

1.4.1   Transcendent Concepts in Physics

How can we describe the behavior of allosteric transitions mathematically? The answer to that question is the subject of this book! The goal will be to show in many different biological contexts like those described earlier in the chapter how the allostery phenomenon can be described in mathematical terms. In particular, we aim to reveal how to enumerate both the various microscopic states that are available to an MWC molecule and the probabilities of these different states as a function of the concentration of various ligands. The simplest version of these ideas will unfold here, and then in the remaining chapters, we will see how those ideas can be generalized to include features such as oligomerization, cooperativity, and applications to a variety of distinct biological situations.

Certain scientific concepts like the MWC model have very broad reach. To clarify what I mean by that, let's explore some examples of scientific broad reach in physics. Young scientists and engineers of all stripes are subjected to a first indoctrination in physics during their early years in university. Shortly after beginning a foray into mechanics, these students are exposed to the seemingly sterile world of masses and springs. After drawing a free-body diagram to reckon how all the forces act on the mass, they obtain an equation of motion that gives the position of that mass as a function of time. Little do they expect that in talking about the abstract behavior of blocks and springs, they have opened a vista onto one of the most far reaching of ideas: periodic motion around an equilibrium point. If they are lucky, these same students will later see that, in fact, the mass-spring problem lays the groundwork for thinking about very different problems such as the pendulum and electrical circuits built up of

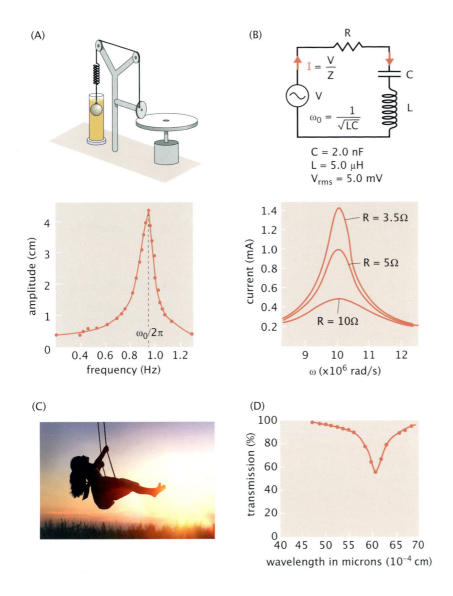

Figure 1.22

Resonance as a concept that transcends any particular example. (A) Mass-spring system. Forced oscillator with damping due to motion in a fluid. The graph shows the amplitude of the oscillations as a function of the frequency of the driving force. (B) Resonance in an *RLC* circuit. The circuit is composed of a resistor (R), a capacitor (C), and an inductor (L). Z is the impedance. The graph shows the current as a function of frequency of the voltage for several choices of resistor. (C) Child pushed on a swing. (D) Transmission of infrared radiation passing through a thin film of sodium chloride as a function of the wavelength of the incident radiation. (A) adapted from French (1965); (B) adapted from Hyperphysics website (C.R. Nave); (C) © Konstantin Yuganov, Dreamstime.com / ID 75499691; and (D) adapted from Feynman, Leighton, and Sands (1963), see Further Reading.

resistors, capacitors, and inductors. All are surprisingly described by the same equation,

$$\ddot{x} + \gamma\dot{x} + \omega^2 x = F(t), \qquad (1.1)$$

where x is the displacement from equilibrium, $\dot{x} = dx/dt$ is the velocity, $\ddot{x} = d^2x/dt^2$ is the acceleration, γ provides a measure of the damping, and ω is the vibrational frequency. Furthermore, depending upon the behavior of the forcing function $F(t)$, the periodic motions can give rise to the general phenomenon of resonance, as shown in Figure 1.22. Here we see that for certain driving frequencies, the amplitude of the vibrations become very large—the phenomenon of resonance familiar to anyone who has pushed a child on a swing.

This resonance idea is so far-reaching as to be astonishing. The mechanics of a pendulum, represented by a child on a swing set in the figure, can be

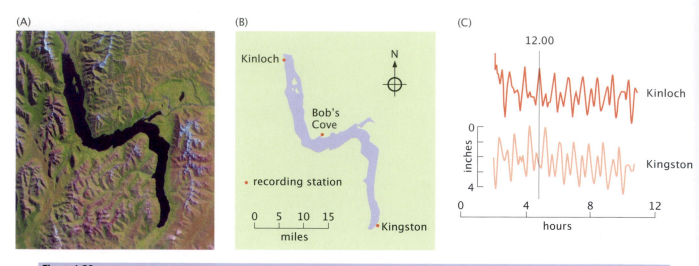

Figure 1.23

Seiche in Lake Wakatipu. A large-scale example of resonance in an unexpected place. (A) Satellite image of the lake. (B) Scale map showing the sites of measurement. (C) Lake height at different points in the lake as a function of time. Adapted from Bottomley (1956).

mapped onto the problem of a mass-spring system, and hence all the things we learned about resonance in the context of the mass-spring system apply just as well to the pendulum (as long as the amplitude of the swinging is not too large). Things become increasingly surprising when we learn that precisely the same mathematics describes the dynamics of charge flow in simple electrical circuits featuring capacitors and inductors. These insights become even more impressive when we find that the very same thinking helps us understand the sloshing motion of a giant lake such as Lake Wakatipu in New Zealand, home to a famed seiche, as shown in Figure 1.23. My point here is to demonstrate a fundamental principle in physics that is now ready for prime time in biology, too: the mathematical unity of apparently disparate phenomena.

Another example of this kind of surprising deep connection between apparently quite different phenomena is offered by the ubiquitous random-walk concept, shown in Figure 1.24. The key point here is that an idea so simple as rolls of a die or flips of a coin can be repurposed to help us understand phenomena as diverse as the diffusion of molecules in solutions or cells, or the statistical conformations of polymer molecules such as DNA. But how? The middle panel of the figure shows how we can think of a random walker as being able to march in any one of six directions each step: east, west, north, south, and up or down. The outcome of the roll of our die tells the walker which one of those steps to make. Importantly, the outcome of this analysis is a *statistical* description of the molecular configurations.

A final physical example of the transcendence of certain physical concepts is given by the all-important wave phenomenon of interference, one of the fruits of Thomas Young's interconnected thinking on physiology and physics (see Figure 1.25 and the article by Mollon (2002) referenced in the Further Reading section). Young was the first to see the phenomenon of interference in all of its sameness, applying it not only to the well-known example of light but also to auditory beats and to the seemingly obscure phenomenon of the tides in the Gulf of Tonkin, which don't exhibit the usual twice-daily tides we are

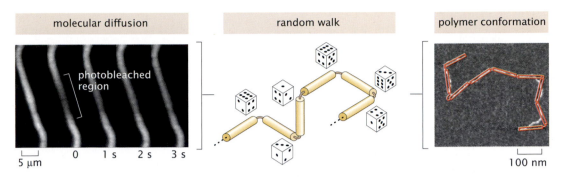

Figure 1.24

The broad reach of the random-walk concept. The center panel of the figure shows how successive rolls of a die determine the nature of the walk. Molecular diffusion, shown on the left, can be analyzed using nothing more than this simplified die rolling. In this case, a bacterium has had the fluorescence in its middle destroyed through photobleaching, and over time, because of diffusion, the fluorescence is restored in the photobleached region. Similarly, the conformations of a polymer such as DNA can be thought of using the same ideas. (*Left and right*) From Phillips, R., J. Kondev, J. Theriot and H. Garcia (2013), *Physical Biology of the Cell*, 2nd ed. Reproduced by permission of Taylor & Francis LLC. (*Left*) Adapted from Mullineaux, C. W., A. Nenninger, N. Ray, and C. Robinson (2006), *J. Bacteriol.* 188:3442, fig. 5. Amended with permission from American Society for Microbiology. (*Right*) Adapted from Wiggins, P. A., et al. (2006), *Nat. Nanotech* 1:37, fig. 1a.

accustomed to at most beaches. Most of us have experienced interference first hand in the form of beautifully colored oil slicks on our driveway after a rain. Isaac Newton mapped out how the color depends upon the thickness of the air layer between two glass plates (one of which was curved; see Figure 1.25(A)), and Thomas Young saw how to compute those colors on the basis of the simple idea of waves either reinforcing one another or canceling each other out, as shown in Figure 1.25(B). The critical idea of sameness is that in all these cases, the interference phenomenon can be simply expressed as the result of several waves adding up either constructively or destructively in a way that is relatively indifferent to whether those waves are sea waves, sound waves, or light waves. This idea was described by Young in what might be thought of as *The Feynman Lectures on Physics* of his time, and his explanation is shown in Figure 1.25(D). As Young responded to critics of his idea, "I was so forcibly impressed with the resemblance of the phenomena that I saw, to those of the colours of thin plates, with which I was already acquainted, that I began to suspect the existence of a closer analogy between them than I could before have easily believed." It is just such a resemblance of phenomena that the allostery concept allows us to understand, as we will see in the pages that follow.

1.4.2   One Equation to Rule Them All

But what does this have to do with biology in general, and this book in particular? A superficial appearance that phenomena are different may mask a very subtle but deep connection between those phenomena or the theory used to describe them. The argument of this book is that the allostery phenomenon is such an example that has the same status as ideas such as resonance or random walks, but unlike those cases, the allostery universality is one that animates the subject of biology.

   One of the deep appeals of the kind of universality offered by the allostery concept and its statistical mechanical implementation via the MWC model

Figure 1.25

The broad impact of the interference concept. (A) Newton examined the colors in a layer of air between two glass surfaces. (B) Using his theory of interference, Thomas Young predicted the observed wavelengths as a function of film thickness. The final column is the addition of Mollon (2002), who gave the wavelengths in nanometers. (C) Figure from Thomas Young's *Lectures on Natural Philosophy* (1807) that shows the interference in water "obtained by throwing two stones of equal size into a pond at the same instant." (D) Page from Thomas Young's *Lectures on Natural Philosophy* that show how waves interfere. The dark lines correspond to the waves being superposed, and the broken line shows their composition (though Young used a different scale.) Adapted from Mollon (2002), see Further Reading.

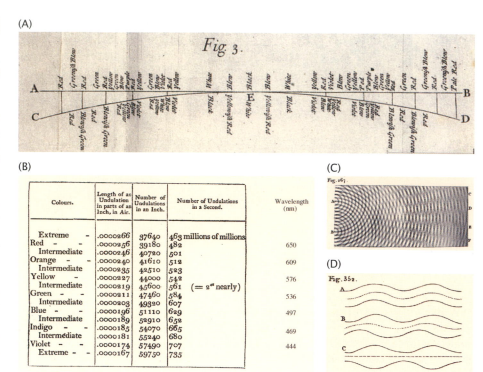

(A)

(B)

Colours.	Length of an Undulation in parts of an Inch, in Air.	Number of Undulations in an Inch.	Number of Undulations in a Second.	Wavelength (nm)
Extreme	.0000266	37640	463 millions of millions	
Red	.0000256	39180	482	650
Intermediate	.0000246	40720	501	
Orange	.0000240	41610	512	609
Intermediate	.0000235	42510	523	
Yellow	.0000227	44000	542	576
Intermediate	.0000219	45600	561 ($= 2^a$ nearly)	
Green	.0000211	47460	584	536
Intermediate	.0000203	49320	607	
Blue	.0000196	51110	629	497
Intermediate	.0000189	52910	652	
Indigo	.0000185	54070	665	469
Intermediate	.0000181	55240	680	
Violet	.0000174	57490	707	444
Extreme	.0000167	59750	735	

(C)

(D)

and its generalizations is that it provides a different way of connecting biological phenomena. Thus, we can imagine organizing biological phenomena on the basis of their biological proximity. For example, we might talk about ion channels and their role in muscle contraction in a physiology course and talk about transcription factors and their induction in a systems biology course. Alternatively, as suggested by Figure 1.26, we can organize biological phenomena according to their physical proximity. In this case, the ion channels and the transcription factors can be seen as the "same" phenomenon, despite how apparently different the biological phenomena they explain mechanistically may be. Figure 1.26 attempts to make this point by showing the MWC model as an intellectual node that links many disparate biological phenomena.

To be specific, Figure 1.26 puts forward the suggestion that phenomena as diverse as the packing of DNA in nucleosomes and the binding of oxygen to hemoglobin are related through physical proximity. But the relatedness of these different problems becomes really clear only when formulated mathematically, in precisely the same way that a mass-spring system and an *LC* circuit are the "same" thing is revealed by the underlying mathematics. The allostery phenomenon as embodied in the MWC model can be stated through the idea that there is one equation to rule them all, namely,

$$p_{active}(c) = \frac{\left(1 + \frac{c}{K_A}\right)^n}{\left(1 + \frac{c}{K_A}\right)^n + L\left(1 + \frac{c}{K_I}\right)^n}. \tag{1.2}$$

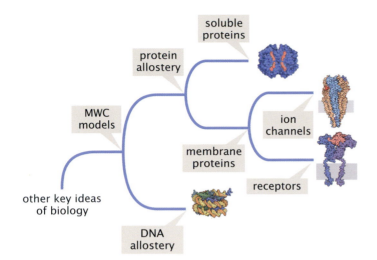

Figure 1.26

Physical proximity of diverse biological phenomena. The MWC model links hemoglobin, ligand-gated ion channels, G protein–coupled receptors, and nucleosomes all through one physical/mathematical framework. Courtesy of David Goodsell.

This equation tells us the probability that an MWC molecule with n ligand binding sites will be in the active state as a function of ligand concentration c. The model features three key parameters L, K_A and K_I. The parameter L is the equilibrium constant between the active and inactive states in the absence of ligand, K_A and K_I are the dissociation constants for ligand binding to the active and inactive states, respectively.

A recent exciting development in evolutionary cell biology is the realization that it is possible to explore the biophysical basis of the parameters that yield the different phenotypes shown in Figure 1.21 such as leakiness, EC_{50}, and effective Hill coefficient explored by evolution. In the context of the MWC model, the entire space of phenotypes is determined by the three molecular parameters introduced in the context of equation 1.2: L, K_A and K_I. That is to say, the only way molecular structure reflects on function in the context of the MWC model is through the value of these three parameters which set key characteristics such as the leakiness, the dynamic range, the EC_{50}, and the effective Hill coefficient which determine the sensitivity of the molecule to ligand concentration. Of course, this is a very naive view of molecules and their evolution, but it will serve as the jumping off point for our thinking.

In many ways, the task of the book is to show where equation 1.2 comes from, why it is the same for so many distinct biological problems, and what its implications are for thinking about biological phenomena ranging from quorum sensing to enzyme feedback. An array of examples of the way this equation can be used to describe different biological phenomena is highlighted in Figure 1.27. This figure is intended to whet the reader's appetite for statistical mechanical modeling of a host of different biological phenomena. There we see examples as diverse as the activity of enzymes of glycolysis such as phosphofructokinase (see chap. 6), oxygen binding to hemoglobin (see chap. 7), ligand-gated ion channels (see chap. 3), the activity of chemotaxis receptors (see chap. 4), and the activity of G protein–coupled receptors (see chap. 5). For now, we content ourselves with admiring these various activity curves for the generality of the allosteric phenomena that they reveal.

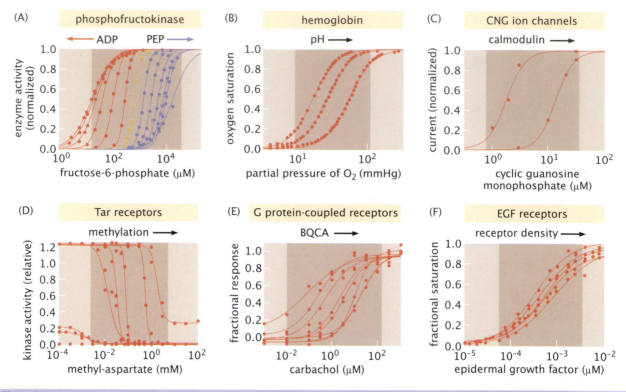

Figure 1.27

Diversity of activity curves of key allosteric molecules. Each dose-response curve shows the activity as a function of an effector molecule. In each case, there is a family of curves reflecting the fact that the activity curve can be tuned by other molecules (shown by the arrows above each graph) in the process of adaptation. Adapted from Olsman and Goentoro (2016).

1.5   Beyond the MWC Two-State Concept

From the outset, I want to make clear that our subject has a long and rich history filled with subtle phenomena, deep and creative models, and colorful personalities. What this means to those that participated in the creation of the subject is that there are many nuances that represent years of work tied to fierce intellectual battles. Though it may seem to miss some of the nuance, my plan is largely to fly below the radar of these debates and to provide a warm statistical mechanical embrace to many different variations of the same basic theme. Specifically, we will argue that these different categories of models all fall within the same statistical mechanical fold because they are based upon discrete state spaces and because ligand binding tunes the relative free energies of inactive and active configurations. We provide a first view of these ideas now.

1.5.1   Molecular Agnosticism: MWC versus KNF versus Eigen

A central theme of the book is the power and beauty of coarse-grained descriptions that intentionally suppress reference to the microscopic degrees of freedom. A corollary of this point of view will be that sometimes we will have a measured indifference to the molecular particulars of a given problem. As such,

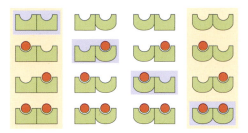

Figure 1.28

Compact representation of the MWC (yellow states), KNF (blue states), and Eigen models (all states) for a two-site allosteric molecule. The active state has the rectangular shape, and the inactive monomers have a rounded shape.

we will pass freely between descriptions based on either all-or-nothing (MWC) or sequential (KNF) or hybrid models of molecular conformations.

To get a flavor for the different conformational states and states of occupancy that allosteric molecules can sustain, Figure 1.28 shows some of the different ways that one can imagine assigning microscopic states to allosteric molecules. The MWC model will be our reference model, culminating in equation 1.2. However, an alternative picture is offered by the sequential model of Koshland, Némethy, and Filmer developed in 1966. This model imagines obligate conformational changes whenever a subunit is bound by a ligand, as seen in the blue states of Figure 1.28. An even broader generalization introduced by Manfred Eigen (1968) allows for the possibility that the different subunits independently change between the inactive and active states and that both sets of states allow for ligand binding, albeit with different dissociation constants. As already mentioned, the debates surrounding these different molecular possibilities have engendered passionate debates. For our purposes, we will view them much more liberally as comprising different collections of allowed states but with the key unifying property of the presence of inactive and active states that have different binding constants for ligands.

One concern justifiably put forth by those critical of the two-state concept is that real biological macromolecules sometimes have more than two dominant conformational states. Part of the reason that I find debates about the nature of allostery dated is that such arguments miss for me the statistical mechanical essences which are simpler and can be flexibly altered to account for the more complex situations, going all the way to the case of intrinsically disordered proteins in which there is a continuum of different states, as will be discussed in detail in chapter 11 (p. 347).

To be specific, we can easily accommodate generalizations to cases involving more substates, as shown in Figure 1.29, where we consider a molecule with three conformational states. In this case we show the "three-state MWC molecule" in which there are three conformational states, each of which is subject to ligand binding or not, resulting in a total of six possible states. The statistical mechanical protocol we will develop in the next chapter will allow us to write the energies of each of these states, as well as their statistical weights, resulting in expressions for their different probabilities, though we need to be alert to the proliferation of parameters as we accept more and more states. Similarly, we can also accommodate the even further generalizations of these multistate models to the full level of the Eigen model by allowing for all states of occupancy and all partial conformational states.

We can explore this proliferation of parameters more systematically by reflecting on what we have done thus far and by generalizing to cases involving N states and M binding sites. For a molecule with only one conformational state

Figure 1.29

Generalizations of the two-state concept. In this example we consider a three-state, one-site model in which there are three conformational states, each of which can be empty or bound by ligand.

but M binding sites, the number of distinct states is 2^M, since each site can be either empty or occupied by ligand. The general rule for the generalized MWC model is that if we have N states and M binding sites, then the total number of distinct configurations that we need to consider in our states-and-weights diagram is

$$\sum_{i=1}^{N} 2^M = N2^M. \tag{1.3}$$

Each of the N conformational states will have its own ϵ_i characterizing the energy of that conformation. Further, for each of those states, there will be a distinct binding energy (or K_d).

The two-state paradigm can be generalized even further to the case in which there is a continuum of allowed states, as we will discuss in chapter 11 (p. 347). As early as the 1980s, Cooper and Dryden explored generalizations of the allostery concept in which there was not a strict conformational change, but, nevertheless, the inactive and active conformations had different free energies of binding because such binding altered the vibrational free energy of the system. An exciting recent development that builds on this kind of thinking has been the emergence of the paradigm of intrinsically disordered proteins, which forces us to think more broadly about the different states allowed to allosteric proteins.

1.6   On Being Wrong

As already highlighted in the preface, the aim of this book is to explore a highly idealized and abstracted view of models of allosteric molecules. There are many opinions on the value of models in the context of biology, some of which are negative and focus on how such models are wrong by "missing" some key element of the system. In that vein, it has become a cliché to quote George Box's refrain that "all models are wrong, but some are useful", at this point, however, I think that this quote itself has outlived its usefulness. In his great essay *Common Sense*, Thomas Paine notes "A long habit of not thinking a thing wrong gives it a superficial appearance of being right." I also worry that a long habit of not thinking a thing right gives it a superficial appearance of being wrong. In the context of the subject of this book, I have been amazed at the diversity of unsubstantiated opinions starting with the words "we all know that" that then go on to assert that either the MWC model has already been shown either to work or to fail, with both opinions held with equal conviction.

It is hard to escape the feeling after studying the literature that often the wrongness is not with the model but with a lack of a truly rigorous dialogue between theory and experiment or an incomplete generalization of the model when faced with new circumstances. For example, there are some that continue to use equation 1.2 (p. 26) as "the MWC model" even when they are considering cases with more than one molecular species. However, in the case in which there is some competitor ligand, equation 1.2 needs to be generalized to

$$p_{active}(c_1, c_2) = \frac{(1 + \frac{c_1}{K_{A,1}} + \frac{c_2}{K_{A,2}})^n}{(1 + \frac{c_1}{K_{A,1}} + \frac{c_2}{K_{A,2}})^n + L(1 + \frac{c_1}{K_{I,1}} + \frac{c_2}{K_{I,2}})^n} \tag{1.4}$$

where c_1 is the concentration of species 1, and c_2 is the concentration of species 2, the competitor.

A second way in which the model of equation 1.2 must be generalized is in the case of important situations like those shown in Figure 1.7, (p. 10) in which there are multiple inputs to the same allosteric molecule. The one equation that rules them all (the traditional MWC model) must account for these multiple inputs. For example, if there are two ligands that interact with the MWC molecule, equation 1.2 must be generalized as

$$p_{active}(c_1, c_2) = \frac{(1 + \frac{c_1}{K_{A,1}})^n (1 + \frac{c_2}{K_{A,2}})^n}{(1 + \frac{c_1}{K_{A,1}})^n (1 + \frac{c_2}{K_{A,2}})^n + L(1 + \frac{c_1}{K_{I,1}})^n (1 + \frac{c_2}{K_{I,2}})^n}. \tag{1.5}$$

For the case in which there are M distinct sites subject to binding by different ligands labeled by the index i, the MWC equation introduced earlier as eqn. 1.2 must be generalized even further to

$$p_{active}(\{c_i\}) = \frac{\prod_i^M (1 + \frac{c_i}{K_{A,i}})^n}{\prod_i^M (1 + \frac{c_i}{K_{A,i}})^n + L \prod_i^M (1 + \frac{c_i}{K_{I,i}})^n}. \tag{1.6}$$

The point here is that these kinds of generalizations are really necessary in order to broaden the scope of the original MWC formulation. Each scenario will have a slightly different equation, and as far as this book is concerned, they all fall within the bailiwick of the MWC framework. Failure to use the right equation will invariably lead to a superficial appearance of wrongness that may not be justified at all.

1.7 Summary

Our first chapter set the stage for all that will follow. It began with one of the most important *facts* of biology (many of the macromolecules of life are allosteric) and one of the most important *concepts* in biology (the MWC model of allostery). I heard of a very distinguished and well-known biologist who refused to talk to physicists entering biology until they could properly define allostery, and the aim of this chapter was to give the reader enough background to pass that test. To that end, I showed how allosteric molecules are found in all corners of biology, whether neuroscience or physiology or evolution. The mathematical implementation of the allostery concept leads to a transcendent biological framework in many ways analogous to the way that concepts such as resonance or interference are transcendent in physics. The chapter ended by openly acknowledging the many oversimplifications and caricatures inherent in this class of models but argued that despite these shortcomings, the framework is extremely potent.

1.8 Further Reading

Alon, U. (2007) *An Introduction to Systems Biology*. Boca Raton, FL: Chapman and Hall/CRC. This excellent book is visionary in showing the interplay between careful theoretical thinking and well-designed experiments to attempt to deeply understand biological problems.

Ben-Naim, A. (2001) *Cooperativity and Regulation in Biochemical Processes*. New York: Kluwer Academic/Plenum. This excellent book channels earlier work from Terrell Hill that demonstrates the naturalness of Gibbs's grand partition function for describing the binding and activity of allosteric molecules.

Cantor, C. R., and P. R. Schimmel (1980) *Biophysical Chemistry*. New York: W. H. Freeman. This series of books digs deeply into many aspects of allostery. The cover of my edition of the book pays homage to the concept of allostery.

Feynman, R. P., R. B. Leighton, and M. Sands (1963) *The Feynman Lectures on Physics*. Reading, MA: Addison-Wesley. Everything here is worth reading, but the chapter on resonance is particularly delightful.

Lim, W. B. Mayer, and T. Pawson (2014) *Cell Signaling*. New York: Garland Science. Over a happy and challenging six months I read every page of this excellent book. My book is an attempt to see what happens when one tries to mathematicize the ideas in this work.

Lindsley, J. E., and J. Rutter (2006) "Whence cometh the allosterome?" *Proc. Natl. Acad. Sci.* 103:10533–10535. This very important paper lays down the gauntlet by noting the challenge of figuring out which proteins are allosteric and if they are, what molecules control that allostery.

Martins, B. M. C., and P. S. Swain (2011) "Trade-offs and constraints in allosteric sensing." *PLoS Comput. Biol.* 7(11):e1002261. This article is a must-read for anyone truly interested in allostery. The authors explore many important facets of the statistical mechanics of models of allostery.

Mollon, J. D. (2002) "The origins of the concept of interference." *Phil. Trans. R. Soc. London* A360, 807–819. This fascinating article describes how Thomas Young brought unity to our understanding of the phenomenon of wave interference. For those interested in delving more deeply into the topic and especially how it touches on biology, see Nelson, P. (2017) *From Photon to Neuron*: *Light, Imaging, Vision*. Princeton, NJ: Princeton University Press. And any interested scientific reader should study Feynman, R. P. (2014) *QED: The Strange Theory of Light and Matter*. Princeton, NJ: Princeton University Press.

Motlagh, H. N., J. O. Wrab, J. Li, and V. J. Hilser (2014) "The ensemble nature of allostery." *Nature* 508:331. This paper reflects on the evolution of our understanding of the allostery phenomenon with the emergence of new experimental techniques such as NMR which have substantially generalized the mechanistic underpinnings of how molecules can behave allosterically.

Nussinov, R., C.-J. Tsai, and B. Ma (2013) "The underappreciated role of allostery in the cellular network." *Annu. Rev. Biophys.* 42:169–89. Nussinov's work provides an inspiring enlargement of the concept of allostery.

Ptashne, M. (2004) *A Genetic Switch.* Cold Spring Harbor, NY: Cold Spring Harbor Laboratory Press. This book retains its position easily on the top 10 of scientific books I have read over my life as an enthusiastic reader. Obviously, the title of my book is inspired by Ptashne's, though I should count myself lucky to attain half the success of that book.

Ptashne, M., and A. Gann (2002) *Genes and Signals.* Cold Spring Harbor, NY: Cold Spring Harbor Laboratory Press. Another book that was a huge inspiration for the present book. Searching my e-reader version of this book revealed 43 mentions of the word "allostery," many of which correspond to thoughtful insights into the topic of the present book.

Raman, A. S., K. I. White, and R. Ranganathan (2016) "Origins of allostery and evolvability in proteins: A case study." *Cell* 166:468–480. A shortcoming of my book is that I have been unable to rise to the occasion of properly describing the evolution of allostery. The work of Ranganathan and collaborators is providing deep insights into these questions, and this paper gives an indication of where that work is going.

Thirmulai, D., C. Hyeon, P. I. Zhuravlev, and G. H. Lorimer (2019) "Symmetry, rigidity and allosteric signaling: From monomeric proteins to molecular machines." *Chem. Rev.* 119:6788–6821. One of the exciting frontiers of allostery that is underdeveloped in the present book is the connection between what has been learned from structural biology and how to reconcile it with statistical physics. This paper showcases how these worldviews complement each other and how one can be used to inform the other.

Wyman, J., and S. J. Gill (1990) *Binding and Linkage: Functional Chemistry of Biological Macromolecules.* Mill Valley, CA: University Science Books. This excellent book gives a definitive statement of the key ideas about binding in cooperative models as understood in the era when the allostery concept was in full development.

1.9 REFERENCES

Bottomley, G. A. (1956) "Seiches on Lake Wakatipu, New Zealand." *Trans. Proc. R. Soc. N. Z.* 83:579–587.

Cooper, A. and D.T.F. Dryden (1984) "Allostery without conformational change: A plausible model." *Eur. Biophys. J.* 11:103–109.

Dueber, J. E., B. J. Yeh, K. Chak, and W. A. Lim (2003) "Reprogramming control of an allosteric signaling switch through modular recombination." *Science* 301:1904.

Eigen, M. (1968) "New looks and outlooks on physical enzymology." *Quart. Rev. Biophys.* 1:3–33.

French, A. P. (1965) *Vibrations and Waves.* Boca Raton, FL: CRC Press.

Grutsch, S., S. Bruschweiler, and M. Tollinger (2016) "NMR methods to study dynamic allostery." *PLoS Comp. Biol.* 12:e1004620.

Koshland, D. E., G. Némethy, and D. Filmer (1966) "Comparison of experimental binding data and theoretical models in proteins containing subunits." *Biochemistry* 5:365–385.

Olsman, N. and L. Goentoro (2016) "Allosteric proteins as logarithmic sensors." *Proc. Natl. Acad. Sci.* 113:E4423–E4430.

Piazza, I., K. Kochanowski, V. Cappelletti, T. Fuhrer, E. Noor, U. Sauer, and P. Picotti (2018) "A map of protein-metabolite interactions reveals principles of chemical communication." *Cell* 172:1–15.

Pollard, T. D. and G. G. Borisy (2003) "Cellular motility driven by assembly and disassembly of actin filaments." *Cell* 112:453–465.

Steven, A. C., W. Baumeister, L. N. Johnson, and R. N. Perham (2016) *Molecular Biology of Assemblies and Machines*. New York: Garland Science.

Reznick, E., D. Christodoulou, J. E. Goldford, E. Briars, U. Sauer, D. Segre, and E. Noor (2017) "Genome-scale architecture of small molecule regulatory networks and the fundamental trade-off between regulation and enzymatic activity." *Cell Rep.* 20:2666–2677.

Ulrich, L. E., E. V. Koonin, and I. B. Zhulin (2005) "One-component systems dominate signal transduction in prokaryotes." *Trends Microbiol.* 13:52–56.

THE ALLOSTERICIAN'S TOOLKIT

2

Mathematics is biology's next microscope, only better; biology is mathematics' next physics, only better.

—Joel Cohen

2.1 A Mathematical Microscope: Statistical Mechanics Preliminaries

Ultimately, this is a book about calculations. The reason for this underlying philosophy is a belief that when we state our biological hypotheses in mathematical language, they are less slippery. Using mathematical arguments, we can sharpen the statements and questions we have about biological systems by forcing ourselves to be semantically and quantitatively precise. The MWC model is but one example of many in which the act of mathematicizing our intuition about molecular mechanism has implications for the way we understand many phenotypic properties of the system of interest.

The goal of many of the calculations central to our story is to find probabilities. For example, we will want to know the probability of a particular allosteric molecule being active as a function of ligand concentration. The primary calculational engine we will be using throughout the book to find these probabilities is equilibrium statistical mechanics, though we will also explore kinetic descriptions of these probabilities using chemical master equations. This is not the place to provide a detailed exposition of statistical mechanics, but in this chapter, we review a few of the high points of the subject as a backdrop for the treatment later in this chapter of the statistical mechanical description of generic MWC molecules. For those readers already acquainted with statistical mechanics or, alternatively, motivated to start using these ideas to think about biological problems, it is now time to move to chapter 3 (p. 77) and begin thinking about how the allostery concept helps us understand ion channels. Those that want to see the statistical mechanics protocol of Figure 2.1 in action are invited to read on.

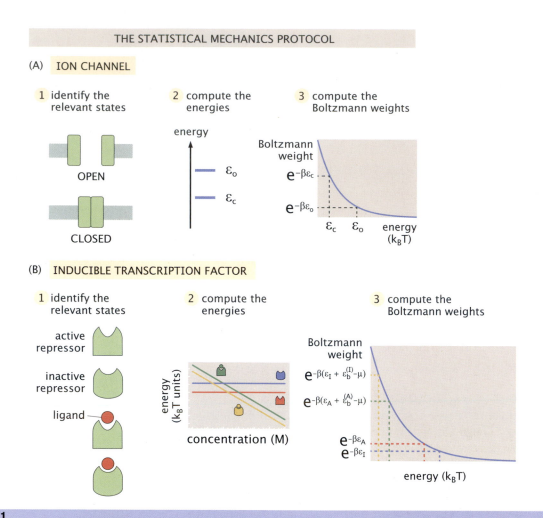

Figure 2.1

The statistical mechanical protocol. (A) A simple two-state model of an ion channel with energies ε_c and ε_o for the closed and open states, respectively. For equilibrium problems, the statistical mechanics protocol asks us to identify the relevant states (closed and open), compute their energies, and then compute their Boltzmann weights by exponentiating their energies. (B) An inducible transcription factor with one ligand binding site has four distinct states of activity/occupancy. The parameter μ is the chemical potential and will be discussed in detail later in the chapter. The relative probabilities of the different states are given by their Boltzmann weights.

2.1.1 Microstates

Statistical mechanics is a tool for computing the probabilities of the states of complex systems. One of the impressive features of a theory like statistical mechanics is its broad reach. The objective remains the same regardless of context: what are the probabilities of the different microscopic states the system can adopt? But the situations range over a huge array of different phenomena, whether in addressing questions such as the positions and velocities of the huge number of molecules jiggling around in a gas, or the conformations of a DNA molecule adsorbed onto a mica substrate, or the distribution of transcription factors across genomic DNA. For example, in the case of the gas molecules in the room currently occupied by the reader, each of those molecules that

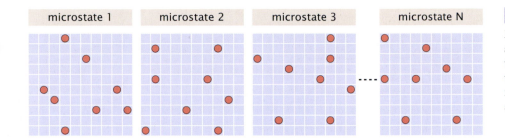

Figure 2.2

Microstates in lattice models of solution. Given L ligands distributed over Ω lattice sites, there are a number of distinct microstates. This figure shows some of the distinct microstates.

makes up that gas has a position (x_i, y_i, z_i) and a momentum (p_{xi}, p_{yi}, p_{zi}), where the subscript i tells us we are talking about the ith molecule. Each instantaneous snapshot of the positions and momenta of those molecules is known as a "microstate" of the system. Note that a given macroscopic state for our gas characterized by the parameters pressure, volume, and temperature corresponds to many distinct microstates.

Examples of other microstates abound. For example, if we think about the conformations of a protein subjected to pulling by an atomic force microscope, then every one of the possible macromolecular configurations for that molecule at each pulling force constitutes a microstate of the system corresponding to the macrostate labeled by the parameter L, the polymer length. An even more transparent molecular example is that of ion channels, which in the simplest incarnation can be thought of as admitting only two distinct microstates, as shown in Figure 2.1(A): closed and open! This list goes on, and the reader is invited to page through a molecular biology textbook and try to imagine the various microstates that correspond to familiar molecular processes such as transcription factor binding, the dynamics of G protein–coupled receptor signaling, and the various configurations allowed to the molecules making up the cytoskeleton. Regardless of which particular case study we consider, ultimately, it is the ambition of statistical mechanics to compute the probabilities of these different microstates.

The concrete example of microstates that will be central to everything we will do in this book is shown in Figure 2.2. Here we consider the arrangements of the molecules in solution as our first chance to count the number of different microstates available to a system. Note that for typical molecular concentrations (i.e., nM to mM), the mean spacing between molecules is larger than molecular dimensions. For example, it is left as an exercise for the reader to demonstrate that at nanomolar concentrations, the mean spacing between those molecules is roughly 1 μm, whereas in the case of millimolar concentrations, the mean spacing between molecules is roughly 10 nm. Our reason for making these simple estimates is to lend credence to the realism of the schematics in Figure 2.2, where we see that most of the lattice sites that represent our solution are not occupied by ligands.

The idea captured in Figure 2.2 is based upon a convenient fiction in which we pretend that a solution can be divided into tiny (i.e., ≈ 1 nm^3, corresponding to molecular dimensions) boxes, each of which is a slot where a ligand can sit. We call this a *lattice model* of the solution. Though it might seem ridiculously oversimplified, in fact, it captures the single most important facet of molecules in solution, namely, that there are many distinct molecular configurations.

The reason that we favor the lattice model of solution as opposed to the more realistic picture in which we allow ligand molecules to occupy a continuum of different positions in space is that by dividing up the solution into Ω distinct lattice sites, occupied by L ligands, we can easily count the number of ways of arranging the ligands without resorting to the mathematics of integral calculus. To see this, note that we have Ω choices for where we put the first ligand, $\Omega - 1$ choices for where we put the second ligand, and so on. Hence, the total number of choices for where we put our ligands is

$$W(L; \Omega) = \Omega \times (\Omega - 1) \times \cdots (\Omega - L + 1). \tag{2.1}$$

We can rewrite this equation more elegantly as

$$W(L; \Omega) = \frac{\Omega!}{(\Omega - L)!}. \tag{2.2}$$

But there is one last nuance for us to consider. Since all our ligands are identical, and their order is irrelevant, we can exchange them for one another without prejudice. As a result, we have to divide by the factor $L!$ to eliminate this overcounting, resulting in

$$W(L; \Omega) = \frac{\Omega!}{L!(\Omega - L)!}. \tag{2.3}$$

Now that we have learned how to enumerate the microstates of our system of interest, let's examine how to compute their probabilities.

2.1.2   The Fundamental Law of Statistical Mechanics

Given that the goal in statistical mechanics is to compute the probabilities of the different microstates, the big question is, how do we discriminate among these states? That is, what parameters confer lower and higher probability to our different microstates? The amazing insight of Maxwell, Boltzmann, and Gibbs among others was that the discriminatory variable is the energy of the microstate of interest. What they realized is that different states can be labeled according to their energy. States with higher energy are less probable. We can go further by exploiting the fundamental law of statistical mechanics, which is that the relative probability of the ith microstate and the jth microstate is of the form

$$\frac{p_i}{p_j} = \frac{e^{-\beta E_i}}{e^{-\beta E_j}}, \tag{2.4}$$

where $\beta = 1/k_B T$, $k_B = 1.38 \times 10^{-23} \text{ J} \cdot \text{K}^{-1}$ is Boltzmann's constant, T is the temperature in kelvin, and E_i and E_j are the energies of the two states in question. We can rewrite each probability as

$$p_i = ce^{-\beta E_i}, \tag{2.5}$$

where c is a proportionality constant that can be determined through the realization that the probabilities must sum to unity,

$$\sum_i p_i = c \sum_i e^{-\beta E_i} = 1. \tag{2.6}$$

This implies in turn that

$$c = \frac{1}{\sum_i e^{-\beta E_i}} = \frac{1}{Z}, \tag{2.7}$$

where the quantity Z is called the *partition function*. Hence, we can write our probabilities using the fundamental law

$$p_i = \frac{e^{-\beta E_i}}{\sum_i e^{-\beta E_i}}. \tag{2.8}$$

As Feynman (2018) says in his excellent book on statistical mechanics: "The fundamental law (i.e., our eqn. 2.8) is the summit of statistical mechanics, and the entire subject is either the slide-down from the summit, as the principle is applied to various cases, or the climb-up to where the fundamental law is derived." Indeed, the entirety of this book will constitute a pleasant slide down from this summit, and the reader interested in making the climb to that summit can see some of the suggested readings.

Estimate: The Boltzmann Law for Levitation

As a first example which is instructive both in the context of how to use the Boltzmann distribution and in telling us how to think about the energy scale $k_B T$, we consider the probability of levitating, by commenting quantitatively on the probability that a mass m will levitate to a height h. Specifically, recall that the gravitational potential energy of a mass m near the surface of the earth is $PE = mgh$. If we recall the Boltzmann law in the form of equation 2.4 and take as our zero of potential energy the surface of the earth itself, we can compute the probability of an object such as a human or a bacterium levitating to a height h as

$$p(h) = p(0)e^{-mgh/k_B T}, \tag{2.9}$$

as shown in Figure 2.3. This equation jibes with our intuition about the drop-off of atmospheric pressure with height, a phenomenon seen easily on any sports watch which reports changes in height with 5 m resolution. Effectively, the altimeter on such a watch is nothing more than a Boltzmann meter, reporting on precisely the same quantity as computed in equation 2.9 but instead of using a mass of 100 kg as for a human, using 5×10^{-26} kg for the typical mass of molecules in the atmosphere. Ultimately, it is the dimensionless ratio $mgh/k_B T$ that tells us whether levitation is likely or not. For cases in which this dimensionless number is much greater than 1, the likelihood of levitation is negligible. By way of contrast, when this dimensionless number is much smaller than 1, levitation is inevitable.

ESTIMATE

The probability introduced in equation 2.8 is really all there is to it. Figure 2.1 shows how to use the statistical mechanics protocol for two examples that will see action in upcoming chapters. Figure 2.1(A) shows us how to assess the relative probabilities of the closed state and the open state of ion channels. Concretely, using the Boltzmann law as embodied in equation 2.4, we can compute

Boltzmann and the true laws of levitation! The dimensionless parameter mgh/k_BT determines the accessibility of different states of levitation. For a human to levitate to a height of 1 μm, $\exp(-mgh/k_BT)$ is astronomically small, while for a bacterium to levitate to that same height is not remarkable at all. If we tune the strength of "gravity" by using a centrifuge, even a bacterium is no longer able to exploit thermal fluctuations and levitate.

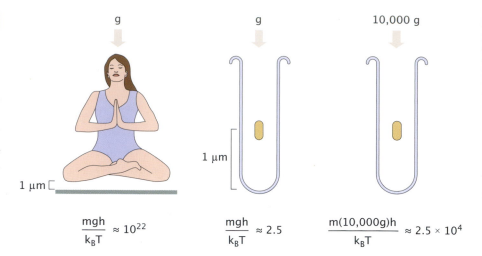

$$\frac{mgh}{k_BT} \approx 10^{22} \qquad \frac{mgh}{k_BT} \approx 2.5 \qquad \frac{m(10{,}000g)h}{k_BT} \approx 2.5 \times 10^4$$

the ratio of the probability the ion channel is open to the probability that the ion channel is closed as

$$\frac{p_{open}}{p_{closed}} = \frac{e^{-\beta\varepsilon_o}}{e^{-\beta\varepsilon_c}}, \tag{2.10}$$

where ε_c and ε_o are the energies of the closed and open states, respectively. Since we want to find an equation for p_{open} itself with no reference to p_{closed}, we can rewrite this equation as

$$p_{closed} = \frac{e^{-\beta\varepsilon_c}}{e^{-\beta\varepsilon_o}} p_{open}. \tag{2.11}$$

Since there are only two states, we have the very simple normalization condition that $p_{open} + p_{closed} = 1$, which we can rewrite as

$$p_{open} + \frac{e^{-\beta\varepsilon_c}}{e^{-\beta\varepsilon_o}} p_{open} = 1, \tag{2.12}$$

which can be simplified to the much nicer form

$$p_{open} = \frac{e^{-\beta\varepsilon_o}}{e^{-\beta\varepsilon_o} + e^{-\beta\varepsilon_c}}. \tag{2.13}$$

Note that the denominator is precisely what was asked of us in equation 2.8 and is a concrete example of the partition function.

Similarly, the same general reasoning can be applied in more complicated cases than this example of the ion channel. Figure 2.1(B) shows how a transcription factor can switch between an inactive and active conformation, with the free-energy balance between those two states tipped by the binding of an inducer molecule. We will examine this example in great detail in chapter 8 (p. 272).

2.1.3   The Dimensionless Numbers of Thermal Physics

One of the signature characteristics of the Boltzmann law as articulated in equation 2.4 is the central place given to the thermal energy $k_BT \approx 4 \times 10^{-21}$ J at

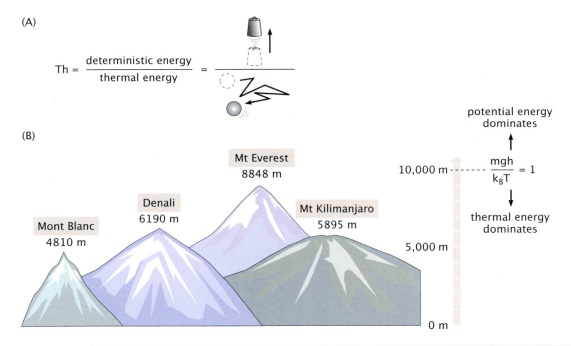

Figure 2.4

Dimensionless thermal numbers. (A) General concept of a thermal number which compares the deterministic energy scale of a problem (in this case gravitational potential energy of molecules of air) and the thermal energy scale. (B) Different regimes of the levitation number. For Le <1, thermal energy dominates, for Le ≈ 1, the two scales balance; and for Le > 1, gravitational potential energy dominates. For molecules such as O_2 and N_2, Le ≈ 1 occurs at roughly 10,000 m, slightly higher than Mt. Everest.

room temperature. Many biophysical processes, whether the binding of a ligand to a receptor or the melting of DNA in a PCR thermocycler, are the result of a competition between different physical effects. For example, in the case of a ligand binding to a receptor, the enthalpy gain when chemical contacts are made between the ligand and receptor comes at the cost of a reduction in the entropy of the ligands in solution, since the number of ligands jiggling around has been decreased by one.

To evaluate the strength of these kinds of competitions, we borrow one of the classic strategies of many branches of science and engineering, namely, the construction of dimensionless numbers such as the Reynolds number used in fluid mechanics. These dimensionless numbers, when equal to one, tell us that two competing effects have the same magnitude. For the purposes of statistical mechanics, we introduce what one might call the "thermal number" (Th) which compares some deterministic energy of interest with the energy of thermal motions, as shown in Figure 2.4(A). We will write the thermal number generically as Th and compute it as

$$\text{Th} = \frac{\text{deterministic energy}}{\text{thermal energy}}. \tag{2.14}$$

A concrete example of a deterministic energy illustrated in Figure 2.4(B) is the potential energy of an object such as a molecule as a function of its height.

Here, we can consider a dimensionless number which we will call the levitation number (Le) which compares gravitational potential energy with thermal energy.

But what exactly do we mean by thermal energy? One avenue for understanding the meaning of this idea is to reflect on the classic work of Robert Brown, who in 1828 described the persistent motions of microscopic particles in his paper entitled "A brief account of microscopical observations…, on the particles contained in the pollen of plants, and on the general existence of active molecules in organic and inorganic bodies." Here Brown reported on the constant and random jiggling of microscopic particles in solution that we now know are a manifestation of the fact that each molecule has an average kinetic energy

$$\frac{1}{2}mv^2 = \frac{3}{2}k_B T,$$ (2.15)

where T is the temperature in kelvin. Though the molecules in a solution have a distribution of speeds, we can estimate the typical speed of molecules in solution as

$$v = \sqrt{\frac{3k_B T}{m}}.$$ (2.16)

Estimate: Thermal Energy and Molecular Velocities
We can use the formula for molecular velocities in terms of temperature to obtain the molecular velocities as

$$v = \sqrt{\frac{4 \times 10^{-21} \text{ kg m}^2/\text{s}^2}{5 \times 10^{-26} \text{ kg}}} \approx \text{few} \times 10^2 \; \frac{\text{m}}{\text{s}},$$ (2.17)

where we introduce a convention of representing factors between 2 and 5 as "few," following Sanjoy Mahajan's excellent books.

To get a sense of this energy, it was noted that the energy stocked in the thermal motions of 1 L of gas in a room is enough to lift a bowling ball several meters high (see Lu, Mandal, and Jarzynski 2014). Let's see for ourselves. We can compute the energy stored in the thermal motions of all these molecules as

$$E = N_{molecules}E_{molecule} \approx N_{molecules} \times k_B T.$$ (2.18)

To make progress with this we need to know the number of molecules in a liter of air. We note that the density of air is 1/1000th that of water, meaning that the density of air is $\rho_{air} = 10^{-3}$ kg/L. The masses of the molecules making up the air are roughly 5×10^{-26} kg/molecule. Hence, the number of molecules is given by

$$\text{\# of molecules in 1 L of air} \approx \frac{10^{-3} \text{ kg/L}}{5 \times 10^{-26} \text{ kg/molecule}} = \frac{1}{5} \times 10^{23} \; \frac{\text{molecules}}{L}.$$ (2.19)

ESTIMATE

We thus find that the kinetic energy due to thermal motion of all these molecules in 1 L of gas is

$$E \approx N_{molecules} k_B T = \frac{1}{5} \times 10^{23} \text{ molecules} \times 4 \times 10^{-21} \frac{J}{molecule} \approx 100 \text{ J}. \tag{2.20}$$

Indeed, as was claimed, this is enough energy to lift a mass of 10 kg through a height of 1 m, as seen by equating this energy to the potential energy of that mass as

$$PE \approx (10 \text{ kg})(10 \text{ m/s}^2)(1 \text{ m}) = 100 \text{ J}. \tag{2.21}$$

Estimate: The Height of the Atmosphere

To get a sense of how the dimensionless ratio captured by the thermal number in equation 2.14 works, let's now revisit levitation. For the levitation problem we considered in Fig. 2.3, equation 2.14 now takes the concrete form of a number we will call the levitation number and define as

$$\text{Le} = \frac{\text{gravitational potential energy}}{\text{thermal energy}} = \frac{mgh}{k_B T}. \tag{2.22}$$

This dimensionless ratio is equal to one when the thermal energy and the gravitational potential energy are exactly balanced. For a molecule of air, we can rewrite this condition as

$$h = \frac{k_B T}{mg} \approx \frac{4 \times 10^{-21} \text{ J}}{5 \times 10^{-26} \text{ kg} \times 10 \text{ m/s}^2} \approx 10,000 \text{ m}, \tag{2.23}$$

a value entirely consistent with our intuition about the heights to which molecules in the atmosphere are still found.

ESTIMATE

Just as we balanced gravitational potential energy and thermal energy to construct the levitation number, we can consider balancing electrostatic potential energy with thermal energy in a number we will christen the Bjerrum number in honor of the Danish chemist who studied the balance between electrostatic forces and thermal energy. Bjerrum determined the length scale (l_B) known as the *Bjerrum length* at which these two energies are equal. Hence, we can define the Bjerrum number as

$$\text{Bj} = \frac{\text{electrostatic potential energy}}{\text{thermal energy}} = \frac{e^2}{4\pi \varepsilon_0 l_B D k_B T}, \tag{2.24}$$

where we have introduced the dielectric constant D, e is the charge on an electron, and ε_0 is the permittivity of vacuum. In this case, we can think of the Bjerrum length scale as telling us the distance opposite charges are allowed to wander away from each other by virtue of thermal energy. In the chapters to follow, part of our use of statistical mechanics will be qualitative and will have us repeatedly harkening back to these beautiful and telling dimensionless comparisons that tell us how the energy scale of some process of interest compares with the thermal energy scale set by the omnipresent $k_B T$.

2.1.4 Boltzmann and Probabilities

Once we have the probabilities dictated by Boltzmann in hand, as well as the intuitions that come from our understanding of the thermal energy $k_B T$, we can compute a whole range of interesting quantities. One of the first things we learn how to compute in statistical mechanics is the average energy of the system. Intuitively, to compute the average value of some quantity, we sum over all the possible outcomes of the system and weight each such outcome by its corresponding probability. When we exercise that logic for the energy, we find

$$\langle E \rangle = \sum_i E_i p(E_i), \tag{2.25}$$

where we recall that $p(E_i) = e^{-\beta E_i}/Z$, where Z is the partition function defined as $Z = \sum_i e^{-\beta E_i}$. In light of that definition we have

$$\langle E \rangle = \frac{\sum_i E_i e^{-\beta E_i}}{\sum_j e^{-\beta E_j}}. \tag{2.26}$$

Note that the numerator of that expression can be obtained as

$$\sum_i E_i e^{-\beta E_i} = -\frac{\partial}{\partial \beta} \sum_i e^{-\beta E_i} = -\frac{\partial}{\partial \beta} Z. \tag{2.27}$$

In light of these observations, we see that the average energy can thus be obtained by evaluating the logarithmic derivative of the partition function as

$$\langle E \rangle = -\frac{\partial}{\partial \beta} \ln Z. \tag{2.28}$$

In fact, for most of the examples in this book, rather than focusing on the average energy, we will be much more interested in questions such as the average number of ligands bound to our MWC molecule, a quantity known as the *occupancy*, or the probability that the MWC molecule is in the active state. We now turn to a brief introduction to the statistical mechanics of binding reactions which will serve as the foundation for most of the calculations in the remainder of the book by showing us how to derive these probabilities.

2.2 Case Study in Statistical Mechanics: Ligand–Receptor Binding

To be concrete in our use of the tools of statistical mechanics, we consider a binding reaction between a ligand and a receptor of the form

$$L + R \rightleftharpoons LR. \tag{2.29}$$

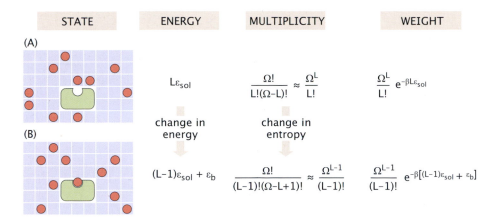

STATE	ENERGY	MULTIPLICITY	WEIGHT
(A)	$L\varepsilon_{sol}$	$\dfrac{\Omega!}{L!(\Omega-L)!} \approx \dfrac{\Omega^L}{L!}$	$\dfrac{\Omega^L}{L!}\, e^{-\beta L\varepsilon_{sol}}$
	change in energy	change in entropy	
(B)	$(L-1)\varepsilon_{sol} + \varepsilon_b$	$\dfrac{\Omega!}{(L-1)!(\Omega-L+1)!} \approx \dfrac{\Omega^{L-1}}{(L-1)!}$	$\dfrac{\Omega^{L-1}}{(L-1)!}\, e^{-\beta[(L-1)\varepsilon_{sol} + \varepsilon_b]}$

Figure 2.5

States-and-weights diagram for ligand-receptor binding. The cartoons show a lattice model of the solution for the case in which there are L ligands. In (A) the receptor is unoccupied. In (B) the receptor is occupied by a ligand, and the remaining $L-1$ ligands are free in solution. A given state has a weight dictated by its Boltzmann factor. The multiplicity refers to the number of different microstates that share that same Boltzmann factor (for example, all the states with no ligand bound to the receptor have the same Boltzmann factor).

In fact, operationally, almost everything we need to know for the many calculations we will do in the book is revealed in this most fundamental of examples. As seen in Figure 2.5, we can think of this ligand–receptor system as having only two classes of states, those in which the receptor is empty and those in which it is occupied.

2.2.1   Ligand Binding and the Lattice Model of Solutions

To examine the physics of Figure 2.5 using the fundamental law of statistical mechanics, we build on our analysis of Figure 2.2 to once again imagine there are L ligand molecules in the box with Ω lattice sites, as well as a single receptor with one binding site, as shown. For simplicity, we ignore any configurational degrees of freedom associated with the receptor itself, as well as rotation or vibrations of the ligand, though including these subtleties does not change the resulting insights. Our ambition is to compute the probability that a receptor is occupied by a ligand (p_{bound}) as a function of the number (or concentration) of ligands using the fundamental law of statistical mechanics embodied in the Boltzmann distribution. Note that every possible microstate within each of these two classes has exactly the same energy. That is, all the states for which the receptor is empty have the same energy, which we label $L\varepsilon_{sol}$, where ε_{sol} is the free energy of a single ligand in solution. Similarly, all those states in which the receptor is occupied have the same energy, given by $\varepsilon_b + (L-1)\varepsilon_{sol}$, where in this case one of the ligands is bound to the receptor with an energy ε_b, and the remaining $L-1$ ligands are free in solution. Further, recall that in equation 2.3 (p. 38) we already computed the number of microstates available to the ligands in solution.

To illustrate the logic of this calculation more clearly, Figure 2.5 shows the states available to this system, as well as their Boltzmann factors, multiplicities, and overall statistical weights. As noted earlier, the key point is that there are only two classes of states: (i) all those states for which there is no ligand bound to the receptor and (ii) all those states for which one of the ligands is bound to the receptor. What makes the bookkeeping simple is that although there are many realizations of each class of state, the Boltzmann factor is the same for all states of the same energy.

To compute the probability that a ligand is bound, we need to construct a ratio in which the numerator is the weight of all states in which one ligand is bound to the receptor, and the denominator is the sum over all states. This idea is represented graphically in Figure 2.5. What the first row of the figure shows is that there are a host of different states when the receptor is unoccupied. Specifically, the L ligands can be distributed among the Ω lattice sites in many different ways. The statistical weight of all these states is given by

$$\text{weight when receptor unoccupied} = \underbrace{\sum_{solution} e^{-\beta L \varepsilon_{sol}}}_{\text{free ligands}}, \qquad (2.30)$$

where we have introduced ε_{sol} as the energy for a ligand in solution. Since the Boltzmann factor is the same for each of these states, calculating this sum amounts to finding the number of arrangements of the L ligands among the Ω lattice sites, which we already did in equation 2.3 (p. 38), yielding

$$\sum_{solution} e^{-\beta L \varepsilon_{sol}} = \frac{\Omega!}{L!(\Omega - L)!} e^{-\beta L \varepsilon_{sol}}. \qquad (2.31)$$

There are also a host of different states in which the receptor is occupied: first, there are L different ligands that can bind to the receptor; second, the $L - 1$ ligands that remain behind in solution can be distributed among the Ω lattice sites in many different ways. In particular, we have

$$\text{weight when receptor occupied} = \underbrace{e^{-\beta \varepsilon_b}}_{\text{bound ligand}} \times \underbrace{\sum_{solution} e^{-\beta(L-1)\varepsilon_{sol}}}_{\text{free ligands}}, \qquad (2.32)$$

where we have introduced ε_b as the binding energy for the ligand and receptor. The summation $\sum_{solution}$ is an instruction to sum over all the ways of arranging the $L - 1$ ligands on the Ω lattice sites in solution with each of those states assigned the weight $e^{-\beta(L-1)\varepsilon_{sol}}$. Since the Boltzmann factor is the same for each of these states, calculating this sum amounts to finding the number of arrangements of the $L - 1$ ligands among the Ω lattice sites, which yields

$$\sum_{solution} e^{-\beta(L-1)\varepsilon_{sol}} = \frac{\Omega!}{(L-1)!(\Omega - (L-1))!} e^{-\beta(L-1)\varepsilon_{sol}}. \qquad (2.33)$$

The partition function is the sum over *all* possible arrangements of the system (both those with the receptor occupied and not) and is given by

$$Z(L, \Omega) = \underbrace{\sum_{solution} e^{-\beta L \varepsilon_{sol}}}_{\text{none bound}} + \underbrace{e^{-\beta \varepsilon_b} \sum_{solution} e^{-\beta(L-1)\varepsilon_{sol}}}_{\text{one ligand bound}}. \qquad (2.34)$$

In light of our earlier results, the partition function for our ligand–receptor problem can be written as

$$Z(L, \Omega) = e^{-\beta L \varepsilon_{sol}} \frac{\Omega!}{L!(\Omega - L)!} + e^{-\beta \varepsilon_b} e^{-\beta(L-1)\varepsilon_{sol}} \frac{\Omega!}{(L-1)!(\Omega - (L-1))!}.$$

(2.35)

We now simplify this result by using the approximation that

$$\frac{\Omega!}{(\Omega - L)!} \approx \Omega^L,$$

(2.36)

which is justified as long as $\Omega \gg L$. To see why this is a good approximation consider the case when $\Omega = 10^6$ and $L = 2$, resulting in

$$\frac{10^6!}{(10^6 - 2)!} = 10^6 \times (10^6 - 1) \approx (10^6)^2.$$

(2.37)

The error made by effecting this approximation can be seen by multiplying out all the terms in parentheses in equation 2.37, resulting in

$$\frac{10^6!}{(10^6 - 2)!} = (10^6)^2 - 10^6,$$

(2.38)

since the second term is insignificant compared with the first. An alternative way to develop intuition for this approximation is to think about seating 10 people in a 60,000-seat stadium. Effectively, each spectator has 60,000 choices of where to sit, so the total number of ways of seating the 10 spectators is approximately $60,000^{10}$.

With these results in hand, we can now write p_{bound} as

$$p_{bound} = \frac{e^{-\beta \varepsilon_b} \dfrac{\Omega^{L-1}}{(L-1)!} e^{-\beta(L-1)\varepsilon_{sol}}}{\dfrac{\Omega^L}{L!} e^{-\beta L \varepsilon_{sol}} + e^{-\beta \varepsilon_b} \dfrac{\Omega^{L-1}}{(L-1)!} e^{-\beta(L-1)\varepsilon_{sol}}}.$$

(2.39)

This result can be simplified by multiplying the top and bottom by $(L!/\Omega^L)e^{\beta L \varepsilon_{sol}}$, resulting in

$$p_{bound} = \frac{(\frac{L}{\Omega})e^{-\beta \Delta \varepsilon}}{1 + (\frac{L}{\Omega})e^{-\beta \Delta \varepsilon}},$$

(2.40)

where we have defined $\Delta \varepsilon = \varepsilon_b - \varepsilon_{sol}$. The overall volume of our system as shown in Figure 2.5 is $V_{box} = \Omega v$, where we have introduced the parameter v as the volume of the individual lattice sites. In light of these definitions, we can write the concentration as $c = L/V_{box}$. In addition, we also define $c_0 = \Omega/V_{box}$, a "reference" concentration (effectively, an arbitrary "standard state") corresponding to having all sites in the lattice occupied. To make a simple estimate of the parameters appearing in equation 2.40 we choose the size of the elementary boxes in our lattice model to be 1 nm³, which corresponds to $c_0 \approx 1$ M,

Figure 2.6

Langmuir binding curves. (A) The ligand–receptor problem described by the Langmuir binding curve. (B) Three different choices of K_d illustrate how the K_d sets the concentration at which the receptor will achieve an occupancy of 1/2.

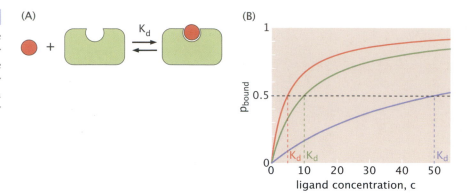

identical to the standard state used in many biochemistry textbooks. In light of our definition of c_0, we can rewrite p_{bound} as

$$p_{bound} = \frac{(\frac{c}{c_0})e^{-\beta\Delta\varepsilon}}{1 + (\frac{c}{c_0})e^{-\beta\Delta\varepsilon}}. \tag{2.41}$$

We can also write this result in the more traditional thermodynamic language of dissociation constants K_d as

$$p_{bound}(c) = \frac{\frac{c}{K_d}}{1 + \frac{c}{K_d}}, \tag{2.42}$$

where we have noted that $K_d = c_0 e^{\beta\Delta\varepsilon}$. This classic result goes under many different names depending upon the field (such as the Langmuir adsorption isotherm, or a Hill function with Hill coefficient $n = 1$). Regardless of name, this expression will be our point of departure for thinking about all binding problems, with its characteristic shape as shown in Figure 2.6. Note that a simple interpretation of the meaning of the K_d is that it is the concentration at which the binding is half-saturated.

Though we have derived this simple functional form in the context of ligand-receptor binding, it has much broader biological reach, as shown in Figure 2.7. One of the most important cases in which this same functional form shows up again and again is enzyme kinetics, where it goes under the title Michaelis-Menten kinetics, and will be the topic of section 6.2 (p. 205). But beyond its role in enzyme kinetics directly, this concept sees duty in analyzing the processes of the central dogma, the behavior of linear and rotary motors, and the rates of posttranslational modifications.

Though many problems of biological interest exhibit binding curves that are "sharper" than the Langmuir (or Michaelis-Menten) form (that is, they exhibit cooperativity), ultimately, even those curves are measured against the standard result derived here. In fact, the essence of many of the models we will describe throughout the book based upon the allostery concept will reveal effective cooperativity corresponding to more sensitive activity as a function of ligand concentration. Our treatment of binding thus far is cumbersome, with its requirement to keep track of the counting of the ligands in solution. With a little more investment in statistical mechanical infrastructure, we can simplify all our work in a way that will benefit us throughout the remainder of the book.

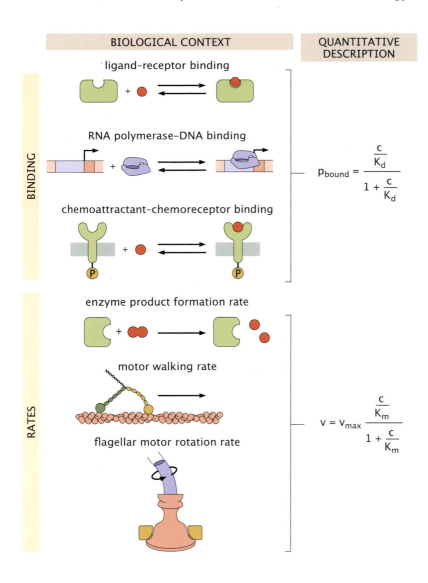

BIOLOGICAL CONTEXT   QUANTITATIVE DESCRIPTION

BINDING

ligand–receptor binding

RNA polymerase–DNA binding

$$p_{bound} = \frac{\dfrac{c}{K_d}}{1 + \dfrac{c}{K_d}}$$

chemoattractant–chemoreceptor binding

RATES

enzyme product formation rate

motor walking rate

$$v = v_{max}\,\frac{\dfrac{c}{K_m}}{1 + \dfrac{c}{K_m}}$$

flagellar motor rotation rate

Figure 2.7

The ubiquitous Langmuir binding curve or Michaelis-Menten activity curve (see sec. 6.2 (p. 205)). Examples of different biological processes in which this same functional form sees duty. Adapted from Wong et al. (2018).

2.3   Conceptual Tools of the Trade: Free Energy and Entropy

As we will see in examples in future chapters, such as the enzyme phosphofructokinase (chap. 6), or when we discuss combinatorial control by allosteric molecules (chap. 9), many of these situations involve multiple binding sites or more than one species of ligand. In these cases, the counting arguments carried out in the previous section become unwieldy. As a result, in this section, we introduce key ideas such as entropy, free energy, and the chemical potential that will permit us to tackle these more challenging problems more simply. After introducing the key conceptual ideas, we revisit the ligand-receptor problem of the previous section to see how these ideas play out concretely.

As already seen in our discussion of the dimensionless numbers of thermal physics (see sec. 2.1.3 on p. 40), many problems in statistical mechanics can be seen as the playing out of a competition between energetic and entropic contributions to the overall free energy. For example, in binding reactions, the K_d corresponds to that concentration at which the energy gain of binding a ligand to a receptor ($\Delta\varepsilon$) is equal to the entropy lost by removing one ligand from solution. This concentration of ligand corresponds to that choice of L for which the two terms in the denominator of equation 2.39 are equal. Notice also that this concentration corresponds to having half occupancy ($p_{bound} = 0.5$). At low concentrations, the entropic term is dominant, while at high enough concentrations, the energetic term dominates.

We can understand the outcome of a binding reaction as a competition between the conflicting demands of energy and entropy quantitatively by defining the free energy as

$$F = E - TS, \tag{2.43}$$

where E is the energy of the system, T is its temperature, and S is the entropy. (Note that we skirt the niceties of the distinction between Helmholtz and Gibbs free energies.) For our lattice model of solution, the free energy can be broken down into two parts. The energetic contribution to the free energy of the ligands in solution can be written in the simplest lattice model as

$$E(L) = L\varepsilon_{sol}, \tag{2.44}$$

where we are saying that the ligands are sufficiently dilute that they don't interact with one another and such that each time we put another ligand in solution, it costs an energy ε_{sol}. The entropic contribution to the free energy of the ligands in solution can be evaluated by appealing to the famed equation connecting number of microstates to the entropy (another brilliant insight from Boltzmann), namely,

$$S = k_B \ln W(L), \tag{2.45}$$

where $W(L)$ is the number of conformations available to the system (i.e., the number of microstates) when there are L ligands free in solution. But how do we compute the number of conformations available to our ligands in solution? This is precisely the counting exercise we performed earlier when first introducing the idealization of a lattice model of solution and culminating in equation 2.3 (p. 38).

To recap, if we imagine a toy model of a solution as containing Ω distinct boxes into which we may insert our L solute molecules, then the number of configurations in the unbound state is given by

$$W(L) = \frac{\Omega!}{L!(\Omega - L)!}. \tag{2.46}$$

As noted in equation 2.45, the entropy is related to the number of microstates W, and hence we are now poised to write the free energy as a function of the number of ligands in solution as

$$F(L) = L\varepsilon_{sol} - k_B T \ln \frac{\Omega!}{L!(\Omega - L)!}. \tag{2.47}$$

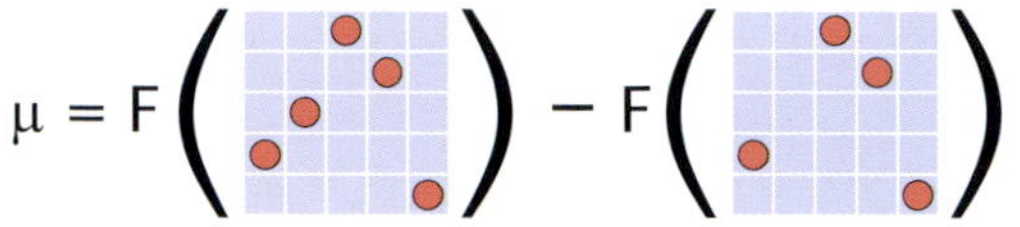

$$\mu = F\left(\begin{array}{c}\includegraphics\end{array}\right) - F\left(\begin{array}{c}\includegraphics\end{array}\right)$$

2.3.1   Resetting Our Zero of Energy Using the Chemical Potential

Figure 2.8

Chemical potential as a free-energy difference. The chemical potential μ is defined as the difference in free energy between the system when it has L molecules, defined as $F(L)$, and when it has $L-1$ molecules, with free energy $F(L-1)$.

In many ways, all this attention to the molecules in solution is a distraction from our main interest, which is the properties of the allosteric molecule itself. For example, we would like to know the probability that the allosteric molecule is active and how that probability depends upon the concentration of the ligands in solution. In the opening chapter we argued that one of our main objectives was to find ways to eliminate "uninteresting" degrees of freedom. The idea of the chemical potential that will be defined (see Fig. 2.8) and worked out in this section affords us precisely that opportunity.

The counting arguments of the previous section ultimately were used to assess the entropy of the ligands in solution and how much that entropy changes when one ligand leaves the solution to bind to the receptor. The chemical potential idea allows us to handle the free energy of the molecules in solution once and for all. To obtain the chemical potential, we compute the free-energy change when we lift a single ligand out of solution, as shown in Figure 2.8. For the little extra investment in adopting the chemical potential, we greatly simplify our description of the states and weights of the molecules that will interest us throughout the book.

As shown in Figure 2.8, the chemical potential is defined as the difference in free energy of the reservoir when it has L ligands and when it has $L-1$ ligands, namely,

$$\mu = F(L) - F(L-1). \tag{2.48}$$

When a ligand binds to a receptor, this quantity is precisely the change in free energy of the solution, which until now we have been computing explicitly using the counting arguments that allow us to figure out the number of configurations of the ligands in solution. We recall that the Helmholtz free energy for our system with L ligands was written as equation 2.47. This implies in turn that we have

$$\mu = L\varepsilon_{sol} - k_B T \ln \frac{\Omega^L}{L!} - (L-1)\varepsilon_{sol} + k_B T \ln \frac{\Omega^{L-1}}{(L-1)!}. \tag{2.49}$$

Using the rule of logarithms that $\ln A - \ln B = \ln(A/B)$, we can simplify our expression to the form

$$\mu = \varepsilon_{sol} + k_B T \ln \left(\frac{\Omega^{L-1}}{(L-1)!} \times \frac{L!}{\Omega^L}\right) = \varepsilon_{sol} + k_B T \ln \frac{L}{\Omega}, \tag{2.50}$$

which can be further simplified by realizing that the total volume of our box is given by $V_{box} = \Omega v$, where v is the volume of one of the little cells in our lattice model, seen in light blue in Figure 2.8. As a result, we recognize that we can rewrite $L/\Omega = (L/\Omega v)/(1/v) = c/c_0$, where we define $c_0 = 1/v$. Using these

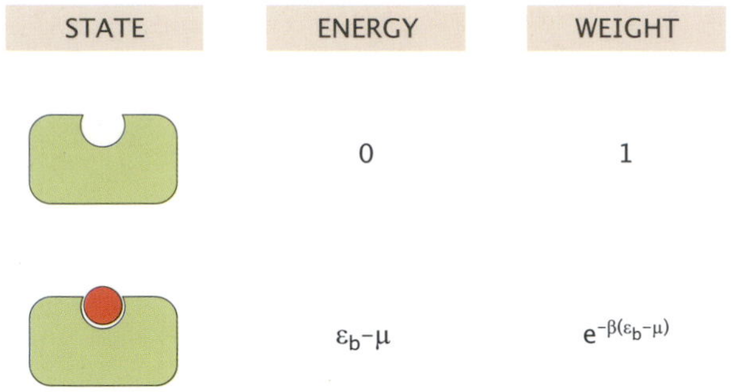

Figure 2.9

States and weights for receptor binding using the chemical potential.

STATE	ENERGY	WEIGHT
	0	1
	$\varepsilon_b - \mu$	$e^{-\beta(\varepsilon_b - \mu)}$

definitions, we can write the chemical potential simply as

$$\mu = \varepsilon_{sol} + k_B T \ln \frac{c}{c_0}. \tag{2.51}$$

The reason we have gone to all this trouble to reintroduce the free energy of the molecules in solution in terms of the chemical potential is to be able to do our states and weights much more simply. As shown in Figure 2.9, by using the chemical potential we have effectively shifted our definition of the zero of energy to correspond to the empty receptor. For the state in which the receptor is bound, we now have two contributions to the energy: the energy of binding ε_b and the free energy of removing a ligand from solution, μ. The probability of the bound state is given by

$$p_{bound} = \frac{e^{-\beta(\varepsilon_b - \mu)}}{1 + e^{-\beta(\varepsilon_b - \mu)}}. \tag{2.52}$$

Now, using the formula for the chemical potential given in equation 2.51, we can rewrite $p_{bound}(c)$ as

$$p_{bound}(c) = \frac{\frac{c}{c_0} e^{-\beta \Delta \varepsilon}}{1 + \frac{c}{c_0} e^{-\beta \Delta \varepsilon}}, \tag{2.53}$$

with $\Delta \varepsilon = \varepsilon_b - \varepsilon_{sol}$. By invoking the correspondence between thermodynamics and statistical mechanics, we have $K_d = c_0 e^{\beta \Delta \varepsilon}$, allowing us to write the binding curve in the classic Langmuir form already shown in equation 2.42.

Repeatedly throughout the book we will be interested in the average occupancy of some molecule such as the receptor considered here. Examples of more complex molecules include hemoglobin, which has four oxygen binding sites, or the nucleosome in which we could have even more than four binding sites for transcription factors. As a result, we need fluency in calculating these occupancies. For the case of the receptor considered in Figure 2.9, there are only two states of occupancy: $n = 0$ and $n = 1$. In light of this observation, the occupancy is defined formally as

$$\langle N_{bound} \rangle = \sum_{n=0}^{1} n p(n), \tag{2.54}$$

where we introduce the shorthand notation $p(n)$ to signify the probability of the state with n ligands. Given this definition, for this simplest case of a monovalent receptor, we can execute the sum explicitly by invoking equation 2.52, with the result

$$\langle N_{bound} \rangle = 0 \times \frac{1}{1 + e^{-\beta(\varepsilon_b - \mu)}} + 1 \times \frac{e^{-\beta(\varepsilon_b - \mu)}}{1 + e^{-\beta(\varepsilon_b - \mu)}}. \qquad (2.55)$$

In principle, that is all there is to it. However, once there is more than one binding site, both the combinatorics and the sums become cumbersome or even intractable.

As a way around this calculational challenge, we introduce one of the most powerful tricks from statistical mechanics, namely, the evaluation of averages by taking derivatives of the partition function. This idea made a first appearance in our equation 2.28 (p. 44), where we saw that we can obtain the average energy as a derivative with respect to the inverse temperature. Here, we pursue a similar strategy by noting that the average occupancy can be obtained as a derivative of a generalized partition function (known as the grand partition function) given by

$$\mathcal{Z} = \sum_i e^{-\beta(E_i - N_i \mu)}, \qquad (2.56)$$

with E_i as the energy of the state i and N_i as the number of particles of the state i. Note now that if we take the derivative with respect to μ, we have

$$\frac{1}{\beta} \frac{\partial \mathcal{Z}}{\partial \mu} = \sum_i N_i e^{-\beta(E_i - N_i \mu)}, \qquad (2.57)$$

implying in turn that

$$\langle N_{bound} \rangle = \frac{1}{\beta \mathcal{Z}} \frac{\partial \mathcal{Z}}{\partial \mu}, \qquad (2.58)$$

which for our simple ligand-receptor problem yields

$$\langle N_{bound} \rangle = \frac{e^{-\beta(\varepsilon_b - \mu)}}{1 + e^{-\beta(\varepsilon_b - \mu)}}. \qquad (2.59)$$

We will return to this same strategy repeatedly throughout the book because it takes painful and tedious calculations and turns them into routine manipulations, where instead of having to enumerate every state of occupancy we can evaluate a derivative of the grand partition function to find key properties of our allosteric molecules of interest.

We could continue and replay precisely the same logic used in describing conventional ligand-receptor binding in the context of MWC molecules, as shown in Figure 2.10. The beauty of the chemical potential idea is that it allows us to make no further mention of the physics of the molecules in solution, making it possible to banish any reference to ε_{sol} and the many annoying multiplicity factors.

These occupancy problems can be reexamined in light of our thinking on dimensionless parameters as expressing the outcome of different competing factors, with that competition especially clear in the language of the chemical potential. In this case, we compare the competing influences involved in ligand-receptor binding. Specifically, we argue that in equation 2.59, when $p_{bound} = 1/2$ corresponds to that concentration in which the energy gain upon binding *exactly* offsets the entropy loss incurred because one ligand is stolen

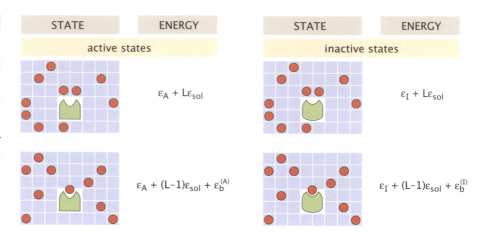

Figure 2.10

States and energies of the simplest MWC molecule. The left column shows the active states of the MWC molecule, and the right column shows the inactive states. Here we account for the energies of the ligand in solution and acknowledge the difference in binding energy of a ligand depending upon whether it is in the active ($\varepsilon_b^{(A)}$) or the inactive ($\varepsilon_b^{(I)}$) state.

from the solution, corresponding to $\varepsilon_b = \mu$. We can write this condition in dimensionless form using what I will call the Langmuir number as

$$\text{La} = \frac{\text{change in energy of ligand upon binding}}{\text{change in entropy of reservoir on losing ligand}} = -\frac{\Delta \varepsilon_b}{k_B T \ln \frac{c}{c_0}}, \quad (2.60)$$

where the negative sign comes from the fact that in binding, $\varepsilon_b = \varepsilon_b - \varepsilon_{sol} < 0$. A Langmuir number greater than 1 implies that binding dominates the entropy term. Convevsely, a Langmuir number less than 1 implies that the entropy term dominates.

2.4   The MWC Concept in Statistical Mechanical Language

Statistical mechanics tells us that the statistical weight of a given state in equilibrium is $\exp(-E_{state}/k_B T)$. Throughout the book, we will repeatedly return to the same basic graphical and calculational protocol, namely, enumerating the microstates of different energies in states-and-weights diagrams such as introduced in Figure 2.5, and then using those statistical weights to compute the probabilities of interest. As seen in Figure 2.10, the states of the simplest MWC molecule are identified both by their activity and the presence or absence of a bound ligand. For the inactive state with no ligand bound, we assign an energy ε_I, while for the active state with no ligand bound we assign an energy ε_A. Further, the binding energy of a ligand to the MWC molecule is *different* depending upon whether that molecule is in the inactive or the active state. That the binding energy in the active and inactive states is *different* is an absolutely central facet that defines the character of the MWC model.

To see how these ideas play out in the context of the MWC molecule, Figure 2.11 shows us the much simpler form of the states and weights that emerges when we shift our zero of energy to correspond to L ligands in solution (i.e., no ligands bound to the receptor). In this case, now when we pluck a molecule out of solution to bind to our MWC molecule, we incur an effective

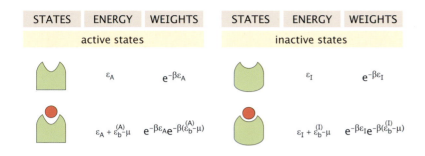

Figure 2.11

States and weights of the simplest MWC molecule using the chemical potential concept. The left column shows the active states of the MWC molecule, and the right column shows the inactive states of the MWC molecule. Those states with a bound ligand incur a cost μ to take the ligand out of the reservoir that is partially compensated for by the binding energy ε_b.

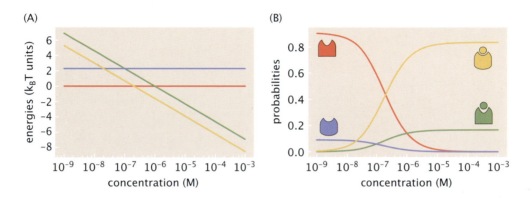

Figure 2.12

Energies and probabilities of different states as a function of concentration of ligand. (A) Free energies of the four states of the simplest MWC molecule as a function of ligand concentration. (B) Probabilities of the four states as a function of ligand concentration. The crossing points in these graphs correspond to those concentrations at which the energies of two states cross. Parameters used in the plots are $\varepsilon_I = 2.3\ k_B T$, $\varepsilon_A = 0$, $K_I = 2 \times 10^{-8}$ M, and $K_A = 10^{-6}$ M.

energy cost μ which we include in the energy definition in the Boltzmann formula, resulting in an equivalent but more streamlined treatment of the states and weights.

The MWC model works precisely because it acknowledges how the energies of the different states change as a function of the concentration of ligands. Figure 2.11 shows the energy of the four states of our simplest MWC molecule. Using the definition of chemical potential given in equation 2.51, we can write the energies of the MWC molecule when bound to a ligand as

$$\text{energy when active and bound} = \varepsilon_A + \Delta\varepsilon_b^{(A)} - k_B T \ln \frac{c}{c_0}, \qquad (2.61)$$

and

$$\text{energy when inactive and bound} = \varepsilon_I + \Delta\varepsilon_b^{(I)} - k_B T \ln \frac{c}{c_0}, \qquad (2.62)$$

where we defined $\Delta\varepsilon_b^{(A)} = \varepsilon_b^{(A)} - \varepsilon_{sol}$ and $\Delta\varepsilon_b^{(I)} = \varepsilon_b^{(I)} - \varepsilon_{sol}$. Figure 2.12(A) shows how these free energies vary as a function of ligand concentration. The attractive feature of this diagram is that it shows how as we increase the ligand concentration, the free energy of the unbound states is beaten out eventually by the free energies of the bound states.

As the Boltzmann distribution tells us, once we know the energies of the different states we can also compute their probabilities. Using the states and weights of Figure 2.11, we can compute the probabilities of each of the different states. By invoking our result that $\mu = \varepsilon_{sol} + k_B T \ln c/c_0$, the probability of the active state with no ligand bound is given by

$$p_{A,0}(c) = \frac{e^{-\beta \varepsilon_A}}{e^{-\beta \varepsilon_A}(1 + \frac{c}{c_0}e^{-\beta \Delta \varepsilon_b^{(A)}}) + e^{-\beta \varepsilon_I}(1 + \frac{c}{c_0}e^{-\beta \Delta \varepsilon_b^{(I)}})}, \qquad (2.63)$$

and similarly for the probability of the inactive state with no ligand bound,

$$p_{I,0}(c) = \frac{e^{-\beta \varepsilon_I}}{e^{-\beta \varepsilon_A}(1 + \frac{c}{c_0}e^{-\beta \Delta \varepsilon_b^{(A)}}) + e^{-\beta \varepsilon_I}(1 + \frac{c}{c_0}e^{-\beta \Delta \varepsilon_b^{(I)}})}. \qquad (2.64)$$

The probabilities of the active state and inactive state with ligand bound are similarly found by recourse to Figure 2.11, yielding

$$p_{A,b}(c) = \frac{e^{-\beta \varepsilon_A} \frac{c}{c_0} e^{-\beta \Delta \varepsilon_b^{(A)}}}{e^{-\beta \varepsilon_A}(1 + \frac{c}{c_0}e^{-\beta \Delta \varepsilon_b^{(A)}}) + e^{-\beta \varepsilon_I}(1 + \frac{c}{c_0}e^{-\beta \Delta \varepsilon_b^{(I)}})} \qquad (2.65)$$

and

$$p_{I,b}(c) = \frac{e^{-\beta \varepsilon_I} \frac{c}{c_0} e^{-\beta \Delta \varepsilon_b^{(I)}}}{e^{-\beta \varepsilon_A}(1 + \frac{c}{c_0}e^{-\beta \Delta \varepsilon_b^{(A)}}) + e^{-\beta \varepsilon_I}(1 + \frac{c}{c_0}e^{-\beta \Delta \varepsilon_b^{(I)}})}. \qquad (2.66)$$

In light of the various statistical weights provided in Figure 2.11, we are now poised to derive the central equation that embodies the MWC concept, namely, the probability of being in the active state, given by

$$p_{active}(c) = p_{A,0}(c) + p_{A,b}(c) = \frac{e^{-\beta \varepsilon_A}(1 + \frac{c}{c_0}e^{-\beta \Delta \varepsilon_b^{(A)}})}{e^{-\beta \varepsilon_A}(1 + \frac{c}{c_0}e^{-\beta \Delta \varepsilon_b^{(A)}}) + e^{-\beta \varepsilon_I}(1 + \frac{c}{c_0}e^{-\beta \Delta \varepsilon_b^{(I)}})}. \qquad (2.67)$$

Specifically, what we have done is to note that the states shown in the left side of Figure 2.11 are active. Hence, to find the probability of the receptor being in the active state, we simply take the sum of the statistical weights for these two active states and divide by the sum over all statistical weights. This follows the usual prescription in any probability calculation, namely, to compute the ratio of the weight of the states of interest to the weight of all states. Note that, as expected, if the binding energies for the inactive and active states are the same, the factor $(1 + \frac{c}{c_0}e^{-\beta \Delta \varepsilon_b})$ cancels, and we see that p_{active} is independent of the ligand concentration.

As shown in Figure 2.13, we can reconcile the perspectives of statistical mechanics and thermodynamics of binding by invoking the language of dissociation constants, as we did when discussing simple binding in equation 2.42

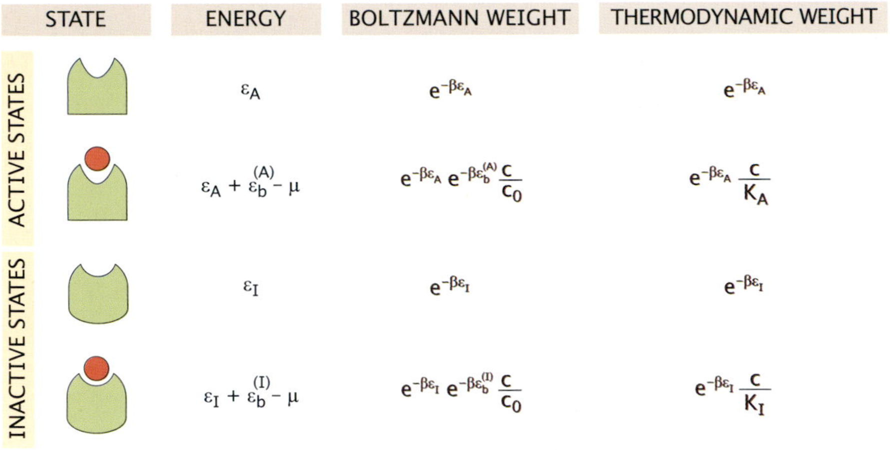

STATE		ENERGY	BOLTZMANN WEIGHT	THERMODYNAMIC WEIGHT
ACTIVE STATES		ε_A	$e^{-\beta\varepsilon_A}$	$e^{-\beta\varepsilon_A}$
		$\varepsilon_A + \varepsilon_b^{(A)} - \mu$	$e^{-\beta\varepsilon_A} e^{-\beta\varepsilon_b^{(A)}} \dfrac{c}{c_0}$	$e^{-\beta\varepsilon_A} \dfrac{c}{K_A}$
INACTIVE STATES		ε_I	$e^{-\beta\varepsilon_I}$	$e^{-\beta\varepsilon_I}$
		$\varepsilon_I + \varepsilon_b^{(I)} - \mu$	$e^{-\beta\varepsilon_I} e^{-\beta\varepsilon_b^{(I)}} \dfrac{c}{c_0}$	$e^{-\beta\varepsilon_I} \dfrac{c}{K_I}$

Figure 2.13

States and weights for the simplest MWC molecule. Comparison of the states and weights in statistical mechanics language and thermodynamic language.

(p. 48). To be precise, we can rewrite the expression for the activity of the MWC molecule using the language of K_ds as

$$p_{active} = \frac{e^{-\beta\varepsilon_A}(1 + \frac{c}{K_A})}{e^{-\beta\varepsilon_A}(1 + \frac{c}{K_A}) + e^{-\beta\varepsilon_I}(1 + \frac{c}{K_I})}. \tag{2.68}$$

Figure 2.12(B) shows the probability of this activity as a function of the ligand concentration, a quantity often measured in experiments. One strategy for proceeding once we have such activity curves is to use them to fit some measured activity curve to find the microscopic parameters of the MWC model. However, we attach only secondary importance to the question of the specific values of the model parameters and argue that it is *general* insights that are most important. For example, in coming chapters, we will illustrate how our statistical mechanical models will allow us to understand a given allosteric input-output function, giving insights into properties such as the leakiness, the dynamic range, the concentration at which the input-output function achieves its half-maximal response (EC_{50}), and the sensitivity. By focusing on these phenotypic properties, we gain general insights into both physiological and evolutionary adaptation of the organisms that harbor allosteric pathways.

2.5 Cooperativity and Allostery

One of the signature phenotypic characteristics of activity curves like those shown in Figure 1.21 (p. 21) is their sensitivity. *Sensitivity* refers to the relationship between an increment in output for a given increment of concentration. Perhaps the most useful way to think of this is as the fold-change in output activity as a function of the fold-change in input concentration, namely,

$$\text{sensitivity} = \frac{\frac{\Delta p}{p}}{\frac{\Delta c}{c}}. \tag{2.69}$$

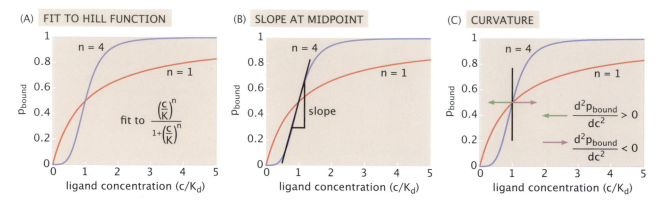

Figure 2.14

Three variations on the theme of cooperativity. (A) Fit to a Hill function. (B) Evaluated from the slope at the midpoint between the minimum and maximum values of the activity curve. (C) Determined from the curvature of the binding (or activity) curve.

This can be rewritten in the language of derivatives as

$$\text{sensitivity} = \frac{c}{p}\frac{dp}{dc},\tag{2.70}$$

which can also be represented mathematically as

$$\text{sensitivity} = \frac{d\ln p_{active}}{d\ln c}.\tag{2.71}$$

Operationally, this means that if we plot activity versus concentration on a log-log plot, then the slope of that activity curve is a measure of the sensitivity. Often, we invoke the slope of this plot at its midpoint between minimal and maximal activity to describe the "cooperativity" of the response.

There are many ways of talking about cooperativity, each of which has its own strengths and weaknesses but all of which when taken together help us better understand the nature of input-output response curves. Several different views of cooperativity are highlighted in Figure 2.14. Figure 2.14(A) shows one way of describing cooperativity, namely, by fitting the sigmoidal activity curve to a so-called Hill function

$$p_{active} = \frac{\left(\frac{c}{K_d}\right)^n}{1+\left(\frac{c}{K_d}\right)^n},\tag{2.72}$$

where n is the Hill coefficient. A Hill coefficient of $n > 1$ signals a cooperative response, while $n = 1$ corresponds to the Langmuir binding curve seen earlier in the chapter and known for its noncooperative response. Note that unlike in the case of the MWC model, the parameter n seen in the Hill function has no simple interpretation related to the number of binding sites available for ligands.

A second method of characterizing cooperativity in activity curves is by evaluating the slope of the activity curve on a log-log plot at the midway point between its minimal and maximal responses, as shown in Figure 2.14(B). Note that if we apply this method to a Hill function itself, that slope determines the

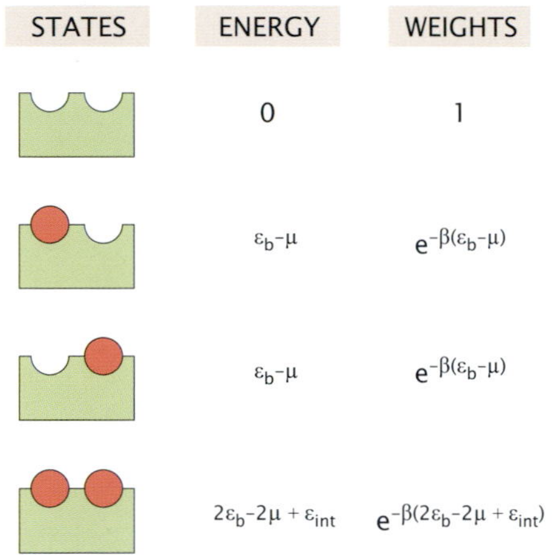

STATES	ENERGY	WEIGHTS
	0	1
	$\varepsilon_b - \mu$	$e^{-\beta(\varepsilon_b - \mu)}$
	$\varepsilon_b - \mu$	$e^{-\beta(\varepsilon_b - \mu)}$
	$2\varepsilon_b - 2\mu + \varepsilon_{int}$	$e^{-\beta(2\varepsilon_b - 2\mu + \varepsilon_{int})}$

Figure 2.15

States and weights for a nonallosteric two-site receptor. There are a total of four different states, corresponding to the two choices of occupancy for each of the two sites. The doubly occupied state includes an interaction energy ε_{int} to account for cooperativity.

Hill coefficient. The final method we entertain here for characterizing the existence of cooperativity utilizes the change in sign of the curvature with increasing concentration, as shown in Figure 2.14(C) and will be demonstrated in the next few sections. Throughout the book, we will pass back and forth between these different perspectives.

2.5.1 Cooperativity and Hill Functions

For many readers, their exposure to the notion of cooperativity will have been in the context of Hill functions. Hill functions are one of the most ubiquitous ideas for capturing the features of input-output curves in biology. Often, the first instinct of those who have some sort of data featuring a concentration titration is to "fit" that data to some functional form that captures the sensitivity in a simple way. Specifically, these data fits are of the form given in equation 2.72.

To see how cooperative effects arise in traditional binding problems (i.e., no allostery), consider a receptor with two binding sites, such as shown in Figure 2.15. In principle, there are four allowed states of the system: unoccupied, singly occupied on the left site, singly occupied on the right site, and doubly occupied. If we carry out the full statistical mechanical treatment of this problem using the methods worked out throughout the chapter, the partition function is given by

$$Z = 1 + 2e^{-\beta(\varepsilon_b - \mu)} + e^{-\beta(2\varepsilon_b - 2\mu + \varepsilon_{int})}, \qquad (2.73)$$

where ε_{int} is an "interaction energy" that either penalizes ($\varepsilon_{int} > 0$) or stabilizes ($\varepsilon_{int} < 0$) the doubly occupied state. We can rewrite this in thermodynamic language as

$$Z = 1 + 2\frac{c}{K_d} + \left(\frac{c}{K_d}\right)^2 \omega, \qquad (2.74)$$

where now we introduce the cooperativity factor $\omega = e^{-\beta \varepsilon_{int}}$. With the partition function in hand, we can now write the probabilities of each state as

$$p_0 = \frac{1}{1 + 2e^{-\beta(\varepsilon_b - \mu)} + e^{-\beta(2\varepsilon_b - 2\mu + \varepsilon_{int})}} = \frac{1}{1 + 2\frac{c}{K_d} + (\frac{c}{K_d})^2 \omega} \qquad (2.75)$$

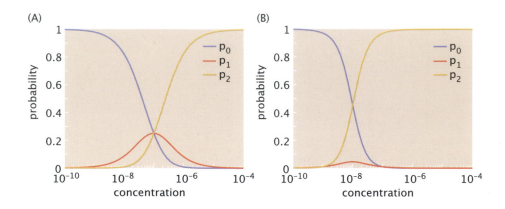

Figure 2.16

Probabilities of the different states of the nonallosteric two-state receptor. (A) Probabilities of the different states of occupancy for the case where there is no cooperativity ($\varepsilon_{int} = 0$, $\omega = 1$) between the binding sites. Note that since there are two states of single occupancy, we have $p_0 + 2p_1 + p_2 = 1$. (B) Probabilities of the different states of occupancy of the receptor for the case where the cooperativity is characterized by $\varepsilon_{int} = -4.6\ k_B T$ and $\omega \approx 100$. In the presence of cooperativity, the probability of the singly bound intermediate is suppressed, suggesting the approximation of ignoring that state altogether, as in a Hill function.

for the unoccupied state,

$$p_1 = \frac{e^{-\beta(\varepsilon_b - \mu)}}{1 + 2e^{-\beta(\varepsilon_b - \mu)} + e^{-\beta(2\varepsilon_b - 2\mu + \varepsilon_{int})}} = \frac{\frac{c}{K_d}}{1 + 2\frac{c}{K_d} + (\frac{c}{K_d})^2 \omega} \qquad (2.76)$$

for the probability of either one of the two singly occupied states, and

$$p_2 = \frac{e^{-\beta(2\varepsilon_b - 2\mu + \varepsilon_{int})}}{1 + 2e^{-\beta(\varepsilon_b - \mu)} + e^{-\beta(2\varepsilon_b - 2\mu + \varepsilon_{int})}} = \frac{(\frac{c}{K_d})^2 \omega}{1 + 2\frac{c}{K_d} + (\frac{c}{K_d})^2 \omega} \qquad (2.77)$$

for the doubly occupied state.

It is interesting to examine the probabilities of these different states as a function of the concentration of ligand. Figure 2.16 shows the probabilities of these different states for two examples, one in which there is no cooperativity ($\omega = 1$) and one in which there is a strong cooperative binding. Note that in the second case the probability of the singly bound state (p_1) is very low for all concentrations, providing a rationale for neglecting that state altogether and resulting in a binding curve like that provided by a Hill function.

Further reflection on Figure 2.16 in the absence of the cooperative interaction energy terms reveals that all three curves are equal at the same point, satisfying the condition

$$p_0(c_*) = p_1(c_*) = p_2(c_*). \qquad (2.78)$$

Using the fact that all three probabilities share the partition function (eqn. 2.74) as their denominator, we see we can rewrite this condition as

$$1 = \frac{c_*}{K_d} = (\frac{c_*}{K_d})^2, \qquad (2.79)$$

implying that $c_* = K_d$. As a result, we have the condition

$$p_0 = p_1 = p_2 = \frac{1}{1 + 2\frac{c_*}{K_d} + (\frac{c_*}{K_d})^2} = \frac{1}{4}, \qquad (2.80)$$

indicating that the probabilities of all four states are the same.

For the cooperative case, we note that there is a shift in the concentration at which $p_0 = p_2$. The concentration at which these two probability curves cross is dictated by the condition

$$\frac{1}{Z} = \frac{(\frac{c_*}{K_d})^2 \omega}{Z}, \qquad (2.81)$$

which can be solved for $c_* = K_d/\sqrt{\omega}$, providing a precise statement of how the cooperativity parameter shifts the midpoint concentration.

Our reason for discussing Hill functions is that many of the MWC molecules that will form the substance of the case studies considered throughout the book have been subjected to careful experiments that result in titration curves showing how the occupancy or activity varies as a function of ligand concentration. These curves, in turn, are often fit to Hill functions. One of the most interesting examples of this kind of thinking concerns the cyclic GMP–gated ion channels that form a key part of the signal transduction cascade in vision. One of the key outcomes of such fits is a Hill coefficient, which in the case of those channels is roughly $n \approx 3$. The problem begins when attempts are then made to interpret such coefficients in terms of the mechanistic underpinnings of the molecular function, with hypotheses such as that three ligands bind the channel in order to drive the gating.

The problem with such phenomenological fits is that they are nearly completely bereft of the ability to provide intuition and predictive power. Worse yet, in my view, they can even be misleading, since they lead to attempts to interpret the significance of the Hill parameter, for example, in terms of the binding sites. As we will argue throughout the book, the spirit of the statistical mechanical dissection of many of the problems we will consider here is to try to identify the set of dominant microscopic states (i.e., active and inactive, empty, and ligand bound) and to compute the statistical weights of the entirety of these states.

2.5.2   Cooperativity in the MWC Model

Thus far, we have primarily considered the MWC model in the context of allosteric molecules with only a single binding site, as described in section 2.4 (p. 54). For some purists, this case does not even merit consideration as an official MWC model. Indeed, for the MWC model to reveal its own version of a sigmoidal activity curve, we need to consider more than a single binding site for the ligands, as shown in Figure 2.17. By summing the relevant statistical weights, we find

$$p_{active}(c) = \frac{e^{-\beta \varepsilon_A}(1 + \frac{c}{K_A})^2}{e^{-\beta \varepsilon_A}(1 + \frac{c}{K_A})^2 + e^{-\beta \varepsilon_I}(1 + \frac{c}{K_I})^2}. \qquad (2.82)$$

To make our analysis of cooperativity in this context more transparent, we adopt a simplifying notation in which we define $x = c/K_A$, $L = e^{-\beta(\varepsilon_I - \varepsilon_A)}$, and

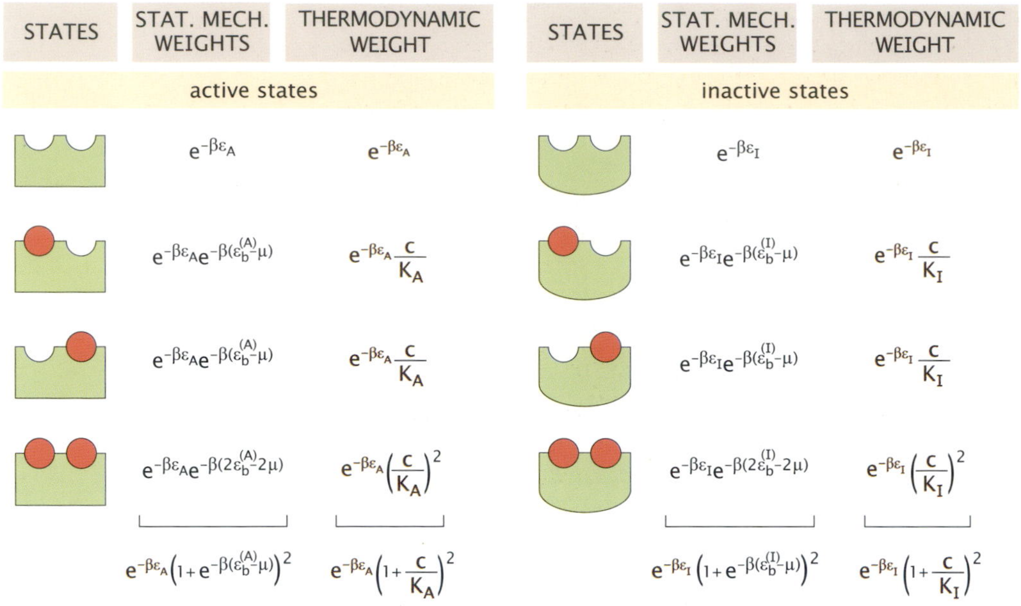

STATES	STAT. MECH. WEIGHTS	THERMODYNAMIC WEIGHT	STATES	STAT. MECH. WEIGHTS	THERMODYNAMIC WEIGHT
active states			inactive states		
	$e^{-\beta\varepsilon_A}$	$e^{-\beta\varepsilon_A}$		$e^{-\beta\varepsilon_I}$	$e^{-\beta\varepsilon_I}$
	$e^{-\beta\varepsilon_A}e^{-\beta(\varepsilon_b^{(A)}-\mu)}$	$e^{-\beta\varepsilon_A}\dfrac{c}{K_A}$		$e^{-\beta\varepsilon_I}e^{-\beta(\varepsilon_b^{(I)}-\mu)}$	$e^{-\beta\varepsilon_I}\dfrac{c}{K_I}$
	$e^{-\beta\varepsilon_A}e^{-\beta(\varepsilon_b^{(A)}-\mu)}$	$e^{-\beta\varepsilon_A}\dfrac{c}{K_A}$		$e^{-\beta\varepsilon_I}e^{-\beta(\varepsilon_b^{(I)}-\mu)}$	$e^{-\beta\varepsilon_I}\dfrac{c}{K_I}$
	$e^{-\beta\varepsilon_A}e^{-\beta(2\varepsilon_b^{(A)}-2\mu)}$	$e^{-\beta\varepsilon_A}\left(\dfrac{c}{K_A}\right)^2$		$e^{-\beta\varepsilon_I}e^{-\beta(2\varepsilon_b^{(I)}-2\mu)}$	$e^{-\beta\varepsilon_I}\left(\dfrac{c}{K_I}\right)^2$
	$e^{-\beta\varepsilon_A}\left(1+e^{-\beta(\varepsilon_b^{(A)}-\mu)}\right)^2$	$e^{-\beta\varepsilon_A}\left(1+\dfrac{c}{K_A}\right)^2$		$e^{-\beta\varepsilon_I}\left(1+e^{-\beta(\varepsilon_b^{(I)}-\mu)}\right)^2$	$e^{-\beta\varepsilon_I}\left(1+\dfrac{c}{K_I}\right)^2$

Figure 2.17

States and weights for a two-site receptor that can exist in both an active and an inactive conformation. The left column shows the four states of occupancy for the active conformation, and the right column shows the four states of occupancy for the inactive conformation.

$f = K_A/K_I$. In light of these definitions, we can rewrite the activity curve for an MWC molecule with two binding sites as

$$p_{active}(x) = \frac{(1+x)^2}{(1+x)^2 + L(1+fx)^2}. \tag{2.83}$$

Note that L, f, and x are all positive. For the more general case with n binding sites, this becomes

$$p_{active}(x) = \frac{(1+x)^n}{(1+x)^n + L(1+fx)^n}. \tag{2.84}$$

We see that as $x \to 0$, we determine the leakiness (i.e., the activity in the absence of ligand) as

$$\lim_{x\to 0} p_{active}(x) = \frac{1}{1+L}. \tag{2.85}$$

Similarly, in the large concentration limit $x \to \infty$, we can write the saturation as

$$\lim_{x\to\infty} p_{active}(x) = \frac{1}{1+f^n L}. \tag{2.86}$$

Both these phenotypic properties will take a place of prominence throughout the book.

In order to assess the qualitative features of the activity curve related to cooperativity, it is useful to examine the properties of the MWC function by evaluating its first and second derivatives. Recall from Figure 2.14(C) (p. 58) that a change in the sign of the second derivative signals the presence of cooperativity, and we adopt the strategy of using that feature as our window into the cooperativity of MWC molecules. We begin with the case of $n = 1$ ligand binding sites, with the result that

$$\frac{dp_{active}(x)}{dx} = \frac{L+1}{(L(1+fx)+(1+x))^2} > 0, \tag{2.87}$$

and

$$\frac{d^2 p_{active}(x)}{dx^2} = \frac{-2(L+1)(Lf+1)}{(L(1+fx)+(1+x))^3} < 0. \tag{2.88}$$

Because the curvature as measured by the second derivative is less than zero, the MWC activity curve plotted on a linear scale for the case $n = 1$ does not have the characteristic sigmoidal shape of a cooperative activity curve. In the MWC context, the way that such sigmoidal properties arise is by virtue of there being more than a single binding site for the regulatory ligand.

The algebra quickly becomes messy and opaque if we consider the case of general n, so to be concrete and make the point about cooperativity, we consider the case in equation 2.83 with $n = 2$. In this case, the second derivative of the activity curve yields the still messy form

$$\frac{d^2 p_{active}}{dx^2} = \frac{2(f-1)L \left(\begin{array}{c} f^3 L x^2 (2x+3) + 3f^2 L x(x+2) \\ +f \left(3L + (2x-1)(x+1)^2 \right) - L + 3(x+1)^2 \end{array} \right)}{\left(L(fx+1)^2 + (x+1)^2 \right)^3}. \tag{2.89}$$

We can develop intuition for this result by focusing only on the numerator, since the denominator is strictly positive, and we are interested only in the sign of $d^2 p_{active}/dx^2$. If we consider reasonable parameter values such as $f = K_A/K_I = 10^{-2}$ and $L = 10$, then we find that as a function of x, $d^2 p_{active}/dx^2$ is positive for small x and becomes negative for large x, indicating precisely the kind of sign change characteristic of cooperative (sigmoidal) behavior seen in Figure 2.14(C). As described in the paper by Mello and Tu (2005) (see Further Reading), a simple general expression can be obtained for the effective Hill coefficient in terms of the number of binding sites n and the ratio K_A/K_I. The reader is encouraged to examine their treatment of the problem.

2.6   Internal Degrees of Freedom and Ensemble Allostery

This chapter offers a series of brief vignettes on the uses of statistical physics that will be important for our investigation of the allostery phenomenon in biology. For the most part, these vignettes have been of a practical nature, instructing us in how to invoke the tools of statistical mechanics to learn about probabilities and averages of interest. Here, we tackle a deep topic of much broader significance that strikes right at the heart of modern thinking on allostery specifically but on biology more broadly, namely, how do we decide the "right" level of description of a given biological problem.

In many ways, the long history of physics is in part a story of a quest for simplicity in the form of finding the "natural variables" of a given problem. For example, in the early days of mechanics, it took a prodigious effort to figure out how to handle extended bodies. Newton, in his so-called Superb Theorem, showed that a spherical mass such as Earth behaved gravitationally as though all its mass were concentrated at the center. This result rigorously justified the apparently absurd idea that when one takes all the little chunks of Earth, each separately acting gravitationally on Newton's falling apple, all those little chunks are equivalent to a single *effective* mass all concentrated at Earth's center. Amazing! Next, in the hands of scientists like Euler, the natural variables for the mechanics of solids and fluids was captured in beautiful kinematic concepts such as the strain tensor or the rate of strain tensor. In this case, instead of treating the full molecular complexity of the motions of every molecule in the fluid or solid, we replace such explicitness with macroscopic quantities such as the strain or strain rate, which integrate out the molecular configurations.

These same ideas saw action well beyond mechanics. The rise of thermodynamics and the theory of heat also demanded that scientists confront the question of which degrees of freedom are important. The beginning of the resolution of these struggles is evident in the title of one of the important papers of Rudolf Clausius, namely, "The Nature of the Motion which we call Heat." As the kinetic theory of heat came to the fore, it was possible to once again integrate out the molecular degrees of freedom and in so doing, under certain approximations such as that the gas be sufficiently dilute, show that key macroscopic relations such as the ideal gas law ($pV = nRT$) followed, with quantities like the temperature standing in for the underlying kinetic energy of the constituent molecules. We took this historical aside for a very important reason, namely, as a reminder to the reader that eliminating degrees of freedom is often not the result of naivete but, rather, the exact opposite, the result of extreme sophistication. And that now brings us to allostery.

Earlier in the chapter (see sec. 2.3.1, p. 51), we introduced the idea of the chemical potential as a way to "integrate out" the degrees of freedom of the ligands when they are in solution. Our first encounter with the ligand-receptor problem in Figure 2.5 (p. 45) had us treating the statistics of the ligands in solution explicitly, leading to a host of ugly multiplicity factors. By way of contrast, when we tackled the exact same problem in the language of chemical potentials in Figure 2.9 (p. 52), all reference to the ligands in solution had been "integrated out" through the parameter μ, which captures the entropy change of the reservoir when a ligand is taken from the solution and bound to the receptor.

Another way of thinking of the consequence of integrating out the degrees of freedom of the ligands in solution is that we have replaced the energy of the distinct microstates with the free energy of these microstates, where the free energy includes the entropic contribution of the ligands in solution. Now, when we use the Boltzmann weights, the "energies" that appear in the exponentials are really free energies, reflecting the critical point that one person's Boltzmann energy is another person's free energy. The remainder of this section will reveal how this same basic idea is repeated over and over again in statistical physics, and it will help us rigorously analyze recent developments in the study of allostery that sometimes go under the title "ensemble allostery."

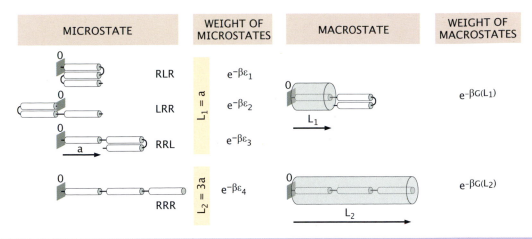

Figure 2.18

Integrating out the internal states of a polymer. If we think of a polymer being stretched to a particular macrostate of a length L, there are many distinct microscopic states of that polymer corresponding to that length. We can represent the Boltzmann weight either by considering the individual microstates separately or by grouping them according to the length of the corresponding macrostate and using the free energy to capture the hidden entropy associated with the many microstates.

To be concrete, we begin with the important example of the stretching of macromolecules such as DNA in the presence of an applied force, such as is imposed by an optical tweezers device. The idea of these experiments is that a piece of DNA is attached to a bead which is in turn "trapped" by laser light in a microscope. As a result, it is then possible to measure the force-extension relation for this DNA in much the same way that for a spring we determine the relation between the applied force and the stretch of the spring, canonized in Hooke's law. In the context of such experiments, the "state" of the DNA polymer is reported in terms of a single macroscopic variable, namely, its length, as shown in the right side of Figure 2.18, and yet this macrostate comprises an infinite number of microscopic configurations consistent with that macrostate. As shown in the left column of the figure, there are three different microscopic ways that we can produce a chain with macroscopic length $L = a$.

Amazingly, statistical mechanics teaches us how to write the Boltzmann distributions for those macrostates without ever having to mention the "hidden variables" (i.e., the many microscopic configurations making up a given macrostate) explicitly. To see that, note that the partial partition function for the macrostate of length L is given by

$$Z(L) = \sum_{\text{state's length } L} e^{-\beta \varepsilon_i} = e^{-\beta G(L)}. \tag{2.90}$$

For the discrete random-walk model shown in Figure 2.18, given a polymer with N segments, we know precisely how to enumerate all the states of a given macroscopic length using the binomial coefficients. The partial partition function implies, in turn, that

$$G(L) = -k_B T \ln Z(L), \tag{2.91}$$

Figure 2.19

Ensemble allostery. Both the active and inactive states have many distinct vibrational states represented here by springs stretched to different lengths. Statistical mechanics shows us how to bunch all such states together into the effective energies G_{active} and $G_{inactive}$.

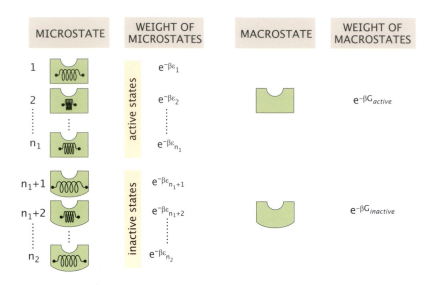

meaning we can now write a rigorous statistical mechanical description of the chain, referencing only the variable L, without ever mentioning the internal degrees of freedom. But what does all this have to do with allosteric molecules?

In recent years, with the advent of new methods in structural biology such as those based on nuclear magnetic resonance, it has become possible to peer into the hidden degrees of freedom of allosteric molecules, which exist in a state of incessant motion. This vision of allostery has given rise to the notion of "ensemble allostery," with the argument being that there are many hidden microscopic states corresponding to a given macrostate of activity. This is precisely the point made in our representation of the polymer in Figure 2.18, where we see many internal (and hidden) microscopic states corresponding to each macroscopic state of length L. Though we will discuss this in great detail in chapter 11 (p. 347), here we note the elements of the argument and why it is so critically connected to the rigorous idea of integrating out degrees of freedom.

In Figure 2.19 we show how many different distinct vibrational states of an allosteric molecule correspond to the same macrostate, which we label either active or inactive. As will be carried out rigorously in chapter 11, we can write the partition function for the active states as

$$Z_A = \sum_{n=1}^{n_1} e^{-\beta\varepsilon_n} = e^{-\beta\varepsilon_1} + e^{-\beta\varepsilon_2} + \cdots e^{-\beta\varepsilon_{n_1}} = e^{-\beta G_{active}}, \qquad (2.92)$$

where the index n labels the internal states (such as the different quantum modes of vibration) of the allosteric molecule. This sum implies in turn that

$$G_{active} = -k_B T \ln \sum_{n=1}^{n_1} e^{-\beta\varepsilon_n}, \qquad (2.93)$$

demonstrating that our use of the language "integrating out" degrees of freedom represents what really is going on. Note again that we have *rigorously*

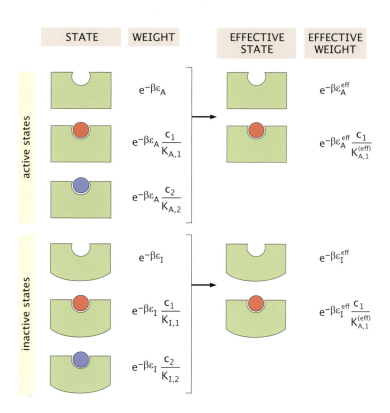

| | STATE | WEIGHT | EFFECTIVE STATE | EFFECTIVE WEIGHT |

Figure 2.20

Hidden variables and effective parameters for competitor binding. The binding of a competitor molecule can be absorbed *exactly* into an effective $K^{(eff)}$ for the binding of the substrate of interest (shown in red) and a corresponding effective offset energy, as shown in equations 2.96–2.99.

mapped the ensemble model onto a corresponding *effective* two-state model, as shown in Figure 2.19. My reason for putting this argument front and center early in the book is that it is mission critical to understand that integrating out degrees of freedom can be done in a way that is not wishful thinking or some horrible approximation but, rather, rigorously and precisely, in a way that captures hidden internal degrees of freedom that might be accessed experimentally.

We now go further in our reflections on the rigorous construction of effective theories and the concomitant emergence of effective parameters within them. Thus far, we have treated the binding site for our substrate of interest on some allosteric molecule as though it were reserved for only one species. However, as seen in Figure 2.20, there can be a competitor molecule that seeks the attention of our substrate binding site. The full states and weights for that problem are shown in the left column of that figure. In light of those states and weights, the activity of the molecule can be written as

$$
p_{active}(c_1, c_2) = \frac{e^{-\beta \varepsilon_A} \left(1 + \frac{c_1}{K_{A,1}} + \frac{c_2}{K_{A,2}}\right)^n}{e^{-\beta \varepsilon_A} \left(1 + \frac{c_1}{K_{A,1}} + \frac{c_2}{K_{A,2}}\right)^n + e^{-\beta \varepsilon_I} \left(1 + \frac{c_1}{K_{I,1}} + \frac{c_2}{K_{I,2}}\right)^n} \quad (2.94)
$$

where c_1 is the concentration of species 1, and c_2 is the concentration of species 2—the competitor, and n is the number of binding sites per allosteric molecule. This equation was introduced as equation 1.4 (p. 31).

It is a matter of a few steps of algebra to see that this equation can be rewritten as

$$p_{active}(c_1; c_2) = \frac{e^{-\beta \varepsilon_A^{(eff)}} \left(1 + \frac{c_1}{K_{A,1}^{(eff)}}\right)^n}{e^{-\beta \varepsilon_A^{(eff)}} \left(1 + \frac{c_1}{K_{A,1}^{(eff)}}\right)^n + e^{-\beta \varepsilon_I^{(eff)}} \left(1 + \frac{c_1}{K_{I,1}^{(eff)}}\right)^n}. \tag{2.95}$$

In writing the equation in this way we have introduced a collection of new parameters

$$\varepsilon_A^{(eff)}(c_2) = \varepsilon_A - k_B T \ln(1 + \frac{c_2}{K_{A,2}}) \tag{2.96}$$

and

$$\varepsilon_I^{(eff)}(c_2) = \varepsilon_I - k_B T \ln(1 + \frac{c_2}{K_{I,2}}) \tag{2.97}$$

for the effective energies of the active and inactive states, respectively, and

$$K_{A,1}^{(eff)} = K_{A,1} \ln(1 + \frac{c_2}{K_{A,2}}) \tag{2.98}$$

and

$$K_{I,1}^{(eff)} = K_{I,1} \ln(1 + \frac{c_2}{K_{I,2}}) \tag{2.99}$$

for the effective K_ds for the active and inactive states, respectively. This result shows that if some competitor was present at a fixed concentration in all our experiments, we could successfully "fit the model" to the data with a single set of effective parameters. This mapping is exact and would allow us to completely ignore the presence of the competitor species. My view is that ignoring this other species is not a weakness showing the model is "wrong" but is instead a strength showing us how to construct a simpler effective theory.

A second category of effective theory arises in the case of some regulatory ligand, as shown in Figure 2.21. Here we see that in addition to the main substrate there is a second effector molecule. Given these states and weights, we can write the activity of the molecule as a function of concentration of both substrate and effector as

$$p_{active}(c_1, c_2) = \frac{e^{-\beta \varepsilon_A}(1 + \frac{c_1}{K_{A,1}^{(eff)}})^n (1 + \frac{c_2}{K_{A,2}^{(eff)}})^n}{e^{-\beta \varepsilon_A}(1 + \frac{c_1}{K_{A,1}^{(eff)}})^n (1 + \frac{c_2}{K_{A,2}^{(eff)}})^n + e^{-\beta \varepsilon_I}(1 + \frac{c_1}{K_{I,1}^{(eff)}})^n (1 + \frac{c_2}{K_{I,2}^{(eff)}})^n}. \tag{2.100}$$

We can formally rewrite this equation in the form

$$p_{active}(c_1; c_2) = \frac{e^{-\beta \varepsilon_A^{(eff)}}(1 + \frac{c_1}{K_{A,1}^{(eff)}})^n}{e^{-\beta \varepsilon_A^{(eff)}}(1 + \frac{c_1}{K_{A,1}^{(eff)}})^n + e^{-\beta \varepsilon_I^{(eff)}}(1 + \frac{c_1}{K_{I,1}^{(eff)}})^n}. \tag{2.101}$$

The cost of making this simplification is that we now have to adopt effective parameters of the form

$$\varepsilon_A^{(eff)}(c_2) = \varepsilon_A - k_B T \ln(1 + \frac{c_2}{K_{A,2}}) \tag{2.102}$$

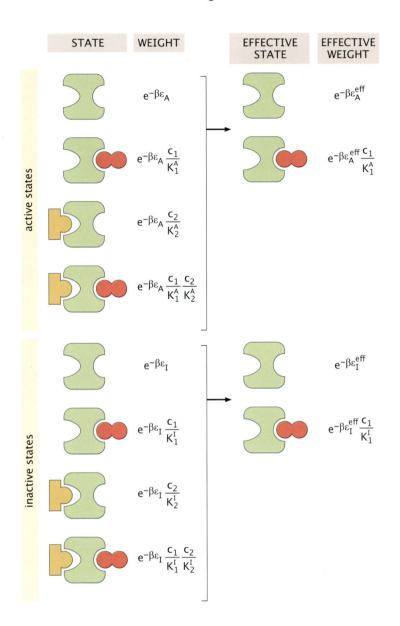

STATE	WEIGHT	EFFECTIVE STATE	EFFECTIVE WEIGHT

active states

$e^{-\beta\varepsilon_A}$

$e^{-\beta\varepsilon_A}\dfrac{c_1}{K_1^A}$

$e^{-\beta\varepsilon_A}\dfrac{c_2}{K_2^A}$

$e^{-\beta\varepsilon_A}\dfrac{c_1}{K_1^A}\dfrac{c_2}{K_2^A}$

$e^{-\beta\varepsilon_A^{eff}}$

$e^{-\beta\varepsilon_A^{eff}}\dfrac{c_1}{K_1^A}$

inactive states

$e^{-\beta\varepsilon_I}$

$e^{-\beta\varepsilon_I}\dfrac{c_1}{K_1^I}$

$e^{-\beta\varepsilon_I}\dfrac{c_2}{K_2^I}$

$e^{-\beta\varepsilon_I}\dfrac{c_1}{K_1^I}\dfrac{c_2}{K_2^I}$

$e^{-\beta\varepsilon_I^{eff}}$

$e^{-\beta\varepsilon_I^{eff}}\dfrac{c_1}{K_1^I}$

Figure 2.21

Hidden variables and effective parameters. The binding of a regulatory ligand (yellow) can be accounted for without approximations into an effective energy for the active and inactive states.

and

$$\varepsilon_I^{(eff)}(c_2) = \varepsilon_I - k_B T \ln\left(1 + \frac{c_2}{K_{I,2}}\right). \tag{2.103}$$

As with the example of Figure 2.20, this mapping is exact and, should we wish, would allow us to completely ignore the effector molecules by burying their effect into the effective parameters of the model that acknowledges only the substrate molecules.

Ultimately, this is one of the most important sections of the whole book and carries a much bigger message than the specific applications to the MWC model of allostery. The key point that is largely missing from current biological thinking is the importance of effective theories and their associated effective parameters. It is routine to hear arguments about the supposed shortcomings

Figure 2.22

Separation of time scales in the mechanics of materials. On short time scales of seconds, minutes, or even days, a lead pipe is under mechanical equilibrium, with the sum of all forces equal to zero, $\sum_i \mathbf{F}_i = 0$, a stricture due to Newton. But over decades, the pipe deforms owing to mass transport, through a process of creep. There is a separation of time scales that permits equilibrium thinking on "short" time scales. Reprinted with permission from by Frost, H. J. and M. F. Ashby (1982), *Deformation-Mechanism Maps*, p. 143, fig. 19.1, Oxford: Pergamon Press.

of theoretical approaches that either ignore some presumed important factor or, alternatively, are not "mechanistic." The view I present here is that in some cases these arguments are deeply misguided. In fact, they miss out on the joy and wonder resulting from rigorous simplifications that can be made with full intent and that result in theoretical constructs celebrating what is ignored rather than bemoaning it.

2.7 Beyond Equilibrium

Thus far, this chapter has focused on that part of the allosterician's toolkit that focuses strictly on equilibrium. That is, we have built up a series of tools that determine the equilibrium probabilities when some system reaches its terminal privileged state. But as noted by Julien Tailleur in a lecture at the College de France, saying "nonequilibrium physics" is like saying "non-elephant biology." Tailleur was channeling, in turn, a comment from the great mathematician Stanislaw Ulam, who noted in a different context, "Using a term like nonlinear science is like referring to the bulk of zoology as the study of non-elephant animals." Almost everything in the physical world is out of equilibrium, so at first cut it seems strange indeed that so much of our thinking about molecular processes in biological systems can fruitfully use equilibrium ideas. Though this book will not be able to settle these issues by establishing the entire suite of theoretical tools that help us understand equilibrium (i.e., the terminal privileged states), kinetic descriptions of the approach to equilibrium, nonequilibrium steady states, and full-blown nonequilibrium phenomena, we nevertheless attempt to acknowledge the important role of time evolution in complex systems and to see how the investment of energy in processes such as ATP hydrolysis allows these systems to do things completely forbidden to equilibrium systems.

One of the justifications for the approach of using equilibrium ideas is a separation of time scales, such as that indicated in Figure 2.22. This figure shows old lead pipes which over time have sagged through a diffusive process known as material creep. The point is that even though at any instant the lead pipe is in apparent mechanical equilibrium, with the sum of all forces equal to zero, over long time scales the pipe not only supports the flow of water within it but itself flows! In light of this separation of scales, the unreasonable effectiveness of equilibrium thinking becomes less mysterious, since often on some time scale of interest we can think of a process as achieving a quasi-equilibrium.

Though I confess that much of the approach in this book will focus on the use of tools from equilibrium statistical physics for the investigation of allostery,

ligand–receptor complex

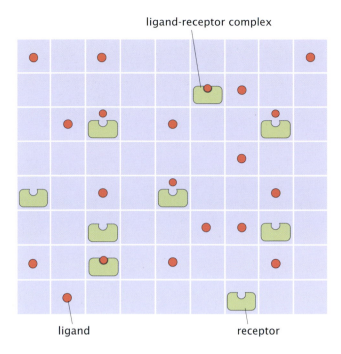

ligand   receptor

Figure 2.23

Lattice model for the kinetics of the simplest ligand–receptor interaction. The system is divided into Ω distinct boxes. There are L ligands and R receptors in solution, and the binding reaction between them to form the complex LR can occur when a ligand and receptor are found at the same lattice site. When a ligand and receptor are found in the same box, the probability that they will react to form LR in time Δt is given by $k'_{on}\Delta t$.

we will also return repeatedly to kinetic approaches to help us understand the transient behavior of our allosteric molecules of interest. To foreshadow how such thinking will go, we consider a ligand and receptor whose concentrations are [L] and [R], which are either free in solution or bound in the LR complex, whose concentration is [LR]. To model the reaction

$$L + R \rightleftarrows LR \qquad (2.104)$$

we consider a lattice model as shown in Figure 2.23 in which there are a total of Ω distinct "lattice" sites, L ligands, and R receptors. We treat the solution as an ideal solution of ligand and receptor, where the probability of occupancy of an elementary box by ligand is L/Ω, and the probability of occupancy of a box by receptor is R/Ω. Furthermore we assume that for L and R to form a complex they must be in the same box. Given these assumptions, in time Δt the change in the number of ligand–receptor complexes due to binding of ligand to receptor can be written as

$$\Delta N_{LR} = \underbrace{\left(-(k_{off}\Delta t)N_{LR}\right)}_{\text{unbinding term}} + \left(\underbrace{\Omega}_{\text{no. of boxes}} \times \underbrace{\frac{N_L}{\Omega}\frac{N_R}{\Omega}}_{\text{box occupancy prob.}} \times (k'_{on}\Delta t) \right),$$

$$(2.105)$$

where the first term is the decay process, in which complexes decay into free ligands and receptors, while the second term represents the rate at which LR is produced and is obtained by working out the rate per elementary box times the total number of such boxes; k'_{on} is the rate at which ligand–receptor pairs that are colocalized in the same elementary box transform to a ligand–receptor complex.

Kinetics of the simplest MWC molecule. Transitions between the four allowed states of the simplest MWC molecule. The two states on top are "active," and the two states on the bottom are "inactive". Binding of the ligand alters the equilibrium between the active and inactive states.

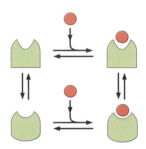

This result can be recast in the more familiar differential form by dividing the left- and right-hand sides by Δt and then by the volume Ωv, where v is the volume of each elementary box in our lattice model. In the limit of small Δt, we have

$$\frac{d}{dt}\left(\frac{N_{LR}}{\Omega v}\right) = -k_{off}\left(\frac{N_{LR}}{\Omega v}\right) + \frac{\Omega}{\Omega v}\frac{N_L}{\Omega v}\frac{N_R}{\Omega v}v^2 k'_{on}, \qquad (2.106)$$

where in the second term on the right we have divided and multiplied by v twice to convert to concentration variables such as $[L] = N_L/\Omega v$, $[R] = N_R/\Omega v$, and $[LR] = N_{LR}/\Omega v$. These manipulations transform equation 2.105 into

$$\frac{d[LR]}{dt} = -k_{off}[LR] + k_{on}[L][R], \qquad (2.107)$$

where the bimolecular on-rate is related to the lattice model rate constant by $k_{on} = v k'_{on}$ and has units of $M^{-1}\,s^{-1}$. This heuristic derivation shows how to link simple lattice models with macroscopic rate equations defined in terms of concentration variables.

The rate-equation formalism described here provides a useful opportunity to review what we already know about the equilibria of reactions. In particular, we can use our dynamical equations to obtain the equilibrium conditions in which the concentrations are no longer changing over time. It is important to note that individual molecules are still undergoing the reactions, but the net number of transformations in each direction is equal, so the overall concentration does not change. We can express this idea mathematically as $d[LR]/dt = 0$, resulting in

$$-k_{off}[LR]_{eq} + k_{on}[L]_{eq}[R]_{eq} = 0. \qquad (2.108)$$

This equation provides a relation between the equilibrium concentrations ($[L]_{eq}$, $[R]_{eq}$, $[LR]_{eq}$) and the relevant rate constants, namely,

$$K_d = \frac{[L]_{eq}[R]_{eq}}{[LR]_{eq}} = \frac{k_{off}}{k_{on}}. \qquad (2.109)$$

More generally, when considering the dynamics of MWC molecules, we will have to consider a richer set of kinetic processes, as shown schematically in Figure 2.24. We will have our first detailed analysis of this problem in the next chapter, where we will examine the kinetics of ion channel opening and closing not only from an equilibrium perspective but also from a kinetic one.

2.8   Summary

As noted in the preface, models are "not meant to be descriptions, pathetic descriptions, of nature; they are designed to be accurate descriptions of our pathetic thinking about nature." This chapter acknowledged that fundamentally, this is a book about calculations, and so, the present chapter laid the critical groundwork by providing the mathematical grammar for everything that will follow. Our central tools are those of statistical physics, and the central practical lesson of this chapter is how to do calculations in statistical physics. My most important message at a deeper philosophical level is that theoretical physics has taught us how to be "self conscious" in our theory construction in the sense that we are clear on what we have ignored, and we examine the precise implications of a particular version of our pathetic thinking. Often, ignoring degrees of freedom is not the consequence of deplorable naivete (biological or otherwise) but, rather, is a measure of extreme sophistication, as illustrated by the Gibbs chemical potential or the Einstein model of vibrational entropy, examples in which the pathetic thinking has stood the test of time. In the pages that follow, we will use the ideas developed in this chapter over and over again. By the time that chapter 5 rolls around, we will have repeated our same basic calculational strategy so many times that sometimes I will no longer do the calculations but instead will describe only what such calculations might look like. With that said, we now launch into a series of exciting case studies, beginning with how ligands gate ion channels in contexts ranging from muscle contraction to vision.

2.9   Further Reading

Bialek, W. S. (2013) *Biophysics: Searching for Principles*. Princeton: NJ: Princeton University Press. This excellent book goes deep on nearly everything and has a very nice description of allostery in the context of chemotaxis.

Dill, K. A., and S. Bromberg (2011) *Molecular Driving Forces*. New York: Garland Science, Taylor & Francis Group. This great book on statistical physics perfects the fine art of using lattice models; it is also worth reading on whatever they are offering. Though I have heard students complain about its simplicity, from my perspective this means they have missed its deep sophistication. Simplicity like this is not just hastily penned.

Graham, I., and T. Duke (2005) "The logical repertoire of ligand-binding proteins." *Phys. Biol.* 2:159–165. This excellent paper shows the generality of statistical mechanical approaches to allostery.

Gunawardena, J. (2014) "Time-scale separation: Michaelis and Menten's old idea, still bearing fruit." *FEBS J.* 281:473–488. This excellent article brings unity to many of the topics to be described throughout the book by offering a graph-theoretic perspective on the dynamics of the macromolecules of the cell. I deeply regret not being able to bring this approach to bear more explicitly on the story told here but could not muster the personal competence that was needed to do so. The reader is urged to consult a number of Gunawardena's excellent papers to see the unity of thinking brought on the heels of the graph-theoretic perspective.

Martins, B. M. C., and P. S. Swain (2011) "Trade-offs and constraints in allosteric sensing." *PLoS Comput. Biol.* 7(11): e1002261. This deep article explores the phenotypic properties inherent in the MWC molecule in a way that goes beyond and complements what is done in this chapter.

Marzen, S., H. G. Garcia, and R. Phillips (2013) "Statistical mechanics of Monod-Wyman-Changeux (MWC) models." *J. Mol. Biol.* 425:1433–1460. This article provides a mathematically richer treatment of the topics covered in this chapter.

Mello, B. A., and Y. Tu (2005) "An allosteric model for heterogeneous receptor complexes: Understanding bacterial chemotaxis responses to multiple stimuli." *Proc. Natl. Acad. Sci.* 102:17354–17359. The Supplemental Information of this paper has an excellent treatment giving intuitive (and rigorous) discussions of the phenotypic parameters relevant to MWC models, with special reference to the effective Hill coefficient described in the chapter.

Phillips, R., J. Kondev, J. Theriot, and H. Garcia (2013) *Physical Biology of the Cell*. New York: Garland Science, Taylor & Francis Group. Though it is embarrassing to reference another book I participated in writing, much of the notation, the figures, and philosophy of what is done here is borrowed from there.

2.10 REFERENCES

Feynman, R. P. (2018) *Statistical Mechanics: A Set of Lectures*. Boca Raton, FL: CRC Press, Taylor & Francis Group.

Frost H. J., and M. F. Ashby (1982) *Deformation-Mechanism Maps*. Oxford: Pergamon Press.

Lu, Z., D. Mandal, and C. Jarzynski (2014) "Engineering Maxwell's demon." *Phy. Today* 67(8):60.

Wong, F., A. Dutta, D. Chowdhury, and J. Gunawardena (2018) "Structural conditions on complex networks for the Michaelis-Menten input-output response." *Proc. Natl. Acad. Sci.* 115:9738–9743.

PART II

THE LONG REACH
OF ALLOSTERY

SIGNALING AT THE CELL MEMBRANE: ION CHANNELS

3

> Shut up and calculate.
>
> —David Mermin

3.1 How Cells Talk to the World

There are few videos more famous in the world of cell biology than the classic sequence from Rogers showing a neutrophil hunting down a bacterium (see Figure 1.4, p. 7). I especially like this video because it makes me think of so many different parts of biology (and physics). As a source for reflection, it takes us back all the way to the nineteenth century and the work of Elie Metchnikoff, who formulated precise ideas about the nature of immunity on the basis of cells that patrol the organism looking for foreign invaders. This classic video also points the way forward for research generations to come by exhibiting the delicate choreography relating cell signaling and motility. Our key message from this video is that there is some form of molecular communication from the environment to the neutrophil itself.

For the purposes of this chapter, the key point we make with the example of Figure 1.4 is that the cell membrane of the neutrophil is decorated with membrane proteins that take stock of the environment around the cell. As previously highlighted in Figure 1.2 (p. 4), the nature of membrane receptors associated with cell signaling runs the gamut from ligand-gated ion channels (the primary subject of this chapter) to membrane receptors such as G protein–coupled receptors that interact with molecules on the cytosolic side of the cell membrane, giving rise to signaling cascades. This chapter is foundational in the sense that all the tools we develop here will be repeated in other molecular contexts throughout the book. As already argued in the previous chapter, within the confines of the MWC model there is one equation to rule them all with different examples, all based on the same underlying statistical mechanical and kinetic principles, and it is the ambition of this chapter to show in detail how those principles can be implemented.

Figure 3.1

Nerves and muscle contraction. (A) A view of the contact between a nerve and muscle, highlighting a neuromuscular junction. (B) Schematic of neurotransmitter release and the gating of ion channels by the binding of ligands. (A) adapted by permission from Springer Nature via Copyright Clearance Center: Desaki, J. and Y. Uehara (1981), "The overall morphology of neuromuscular junctions as revealed by scanning electron microscopy," *J. Neurocytol.*

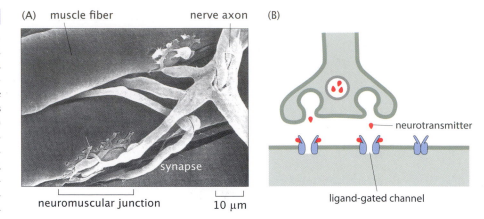

3.2 Biological Processes and Ion Channels

Concentration gradients across membranes are one of the most powerful ways in which living organisms store both energy and information. For example, there is more than a 20-fold difference in the concentration of K^+ ions across many cell membranes. By themselves, the lipid bilayers that make up these membranes are highly impermeable to the passage of ions. However, as part of the scheme for exploiting such gradients either as a source of energy or to communicate information (such as in the passage of an action potential), membranes can transiently change their permeability. These transient changes in membrane permeability are mediated by a class of membrane proteins known as ion channels, which open and close in response to driving forces such as voltage, membrane tension, or the concentration of ligands.

Indeed, ion channels serve as one of the most natural examples for thinking of proteins from the two-state allosteric perspective. Though a great variety of subtleties arise in describing channels, the simplest picture is that a given channel can exist in one of two states, either closed, in which there is no ion permeability, or open, in which the flow of certain ions is now permitted. More nuanced descriptions of such channels acknowledge important features such as the presence of more than two states or the existence of transient inactivated states, but these topics are beyond the scope of this introduction.

Perhaps the most familiar example of ion channels and their gating is found in our own human motility. Every time we use our muscles to move, our motion is the consequence of huge numbers of channels transiently opening, and releasing ions that eventually result in excitation at our neuromuscular junctions, as indicated schematically in Figure 3.1. As seen in the figure, our muscles are "wired up" through connections to the axons of nerves that travel over length scales of tens of centimeters. When an action potential reaches the neuromuscular junction at the terminal end of such an axon, it results in the release of neurotransmitters across the synapse between the nerve and muscle cells, as shown in Figure 3.1(B). These neurotransmitters, in turn, diffuse across the synapse, where they then bind onto ligand-gated ion channels, causing them to transiently open and resulting in a change in membrane potential across the muscle fiber.

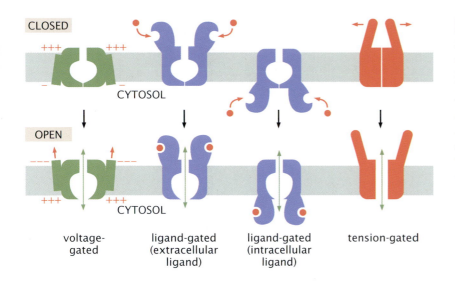

CLOSED

CYTOSOL

OPEN

CYTOSOL

voltage-gated   ligand-gated (extracellular ligand)   ligand-gated (intracellular ligand)   tension-gated

Figure 3.2

Driving forces for ion-channel gating. Ion channels are gated by a variety of different mechanisms including transmembrane voltage, ligands, and membrane tension. From a statistical mechanics perspective, these differences in mechanism are all included in the way that ε_{closed} and ε_{open} depend upon the driving force Fd, with a different force F and "displacement" d for each mechanism.

Ion channels are important in circumstances much more wide and varied than in the context of the neuromuscular junction just described. Indeed, these channels are a ubiquitous part of the molecular machinery of cells of all types, relevant in contexts ranging from the survival of bacteria under osmotic shock to the motility of *Paramecium*. As seen in Figure 3.2, channels are induced to switch back and forth between the closed and open states by a variety of different driving forces. For example, as seen in the left panel of the figure, changes in the membrane potential are responsible for gating a wide variety of channel types. As shown in the right side of the figure, channels can also be gated by mechanical tension in the cell membrane.

As shown in the middle panel of Figure 3.2, ligand-gated ion channels change their open probability depending on the concentration of ligands. To measure these changes in open probability, beautiful single-molecule patch-clamp experiments were developed that make it possible to measure the current passing through the channels as a function of time. Figure 3.3 gives a schematic representation of the patch-clamp technique. As seen in the figure, ion channels are present in the membrane of interest (often an oocyte from a frog in which the relevant channels have been expressed), and a patch of membrane is excised through suction into a pipette. Electrodes are used to establish the potential difference that drives the ions and to measure the current passing through the channel as a function of the driving force that gates those channels. The channel spontaneously switches between the closed and open states, as revealed by the jumps in current across the membrane. Our goal here is to see how the tools of statistical mechanics in conjunction with our conceptual understanding of two-state allostery can help us explain such data.

To compute the relative probabilities of the closed and open states, we invoke the Boltzmann distribution, which tells us that we assign probabilities in accordance with the energies of the competing states. From a statistical mechanics perspective, our simple starting point is the idea that the open state has an energy ε_{open}, and the closed state has a corresponding energy ε_{closed}. According to the protocol of statistical mechanics introduced in Figure 2.1

Figure 3.3

The patch-clamp technique used to make single-molecule measurements of ion-channel currents. Current trace shown in the lower right for the nicotinic acetyl-choline receptor is adapted from Labarca et al. (1995).

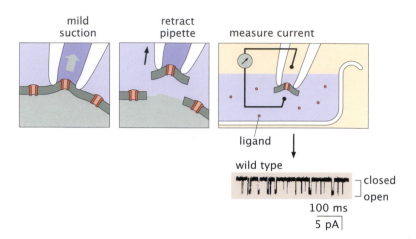

(p. 36), these definitions mean that we can write the probability of the open state as

$$p_{open} = \frac{e^{-\beta\varepsilon_{open}}}{e^{-\beta\varepsilon_{open}} + e^{-\beta\varepsilon_{closed}}}, \tag{3.1}$$

where the denominator represents the partition function $Z = \sum_{states} e^{-\beta E_{states}}$, where in this case there are only two states to consider, namely, closed and open, giving rise to the simple form for the partition function shown in the denominator.

The preceding expression is too abstract, since it says nothing about how the external world can signal the channel that it needs to open up. The answer to this puzzle is the realization that the energy of the open or closed state, for example, can be controlled by external stimuli such as changes in the membrane potential, ligand concentration, or membrane tension, as indicated schematically in Figure 3.2. We can rewrite our expression for the open probability as

$$p_{open} = \frac{1}{1 + e^{-\beta\Delta\varepsilon}}, \tag{3.2}$$

where we have defined the energy difference $\Delta\varepsilon = \varepsilon_{closed} - \varepsilon_{open}$. We can capture each of the different driving forces shown in Figure 3.2 using the general form

$$\Delta\varepsilon = \varepsilon_0 + Fd, \tag{3.3}$$

where ε_0 is the free-energy difference between the open and closed states in the absence of any driving force F, and Fd is meant to conjure up thoughts of (force) $\times$ (distance)= work. Note that, in general, $\varepsilon_0 < 0$, favoring the closed state, and $Fd > 0$, favoring the open state. Given this abstract dependence of the energy difference between the closed and open states on the driving force F, we can now rewrite our equation for the open probability as

$$p_{open} = \frac{1}{1 + e^{-\beta(\varepsilon_0 + Fd)}}. \tag{3.4}$$

Note that with increasing F, we reach a sufficiently large F such that $\varepsilon_0 + Fd = 0$, and $p_{open} = 1/2$. For larger driving forces, the open state has a higher probability than the closed state.

How do we write *Fd* concretely for real-world driving forces? For the case of mechanosensitive channels, for example, the driving force for gating is tension τ in the surrounding membrane, and the energy associated with the driving force can be written as

$$Fd = \tau \, \Delta A, \qquad (3.5)$$

where ΔA is the change in area upon channel gating. The quantity $Fd = \tau \, \Delta A$ reports on the tension × area work characteristic of stretched membranes. Similarly, in the presence of a transmembrane potential, the energy change can be written as

$$Fd = Q \Delta V, \qquad (3.6)$$

where Q is the charge, and ΔV is the change in voltage across the membrane. Our foray into MWC models focuses us instead on the case in which the driving force behind channel gating is the binding of a ligand to the channel. In the next few sections, we provide a detailed exploration of how ligands can provide an analogous driving force for ion-channel gating to the cases of tension and voltage just described.

3.3 Ligand-Gated Channels

A channel with open and closed states whose equilibrium depends on whether a small molecule (the "ligand") has bound to it is called a *ligand-gated ion channel*. These channels have binding sites for their regulatory ligands and often are more likely to be open in the ligand-bound state than when in the ligand-free state. As a result, the ligand serves to drive the population toward the open state. Not only are these channels a superb topic of statistical mechanical study, but they are also central players in many important parts of biology, as was introduced with the neuromuscular junction in Figure 3.1 and will be described in more detail now. Note that although we will use the general term "ligand" for the molecules that shift the free-energy balance between closed and open states, words such as "effector" and "agonist" are used in the literature, as well. The term *effector* usually refers to some small regulatory ligand that can either favor or disfavor the active state relative to the inactive state. The term *agonist* is more specific, since it refers to the case of ligands that activate some allosteric molecule by binding to a regulatory site within that molecule. For simplicity, I will stick to the generic term ligand, and in each case, the context will make it clear where that molecule binds on some allosteric molecule of interest and how it alters the relative probabilities of the inactive and active states.

Perhaps the most notable examples of ligand-gated ion channels show up in the context of neuroscience. Several important examples are shown in Figure 3.4. One of these channels is the so-called nicotinic acetylcholine receptor, a channel with a pentameric structure and with two subunits (often denoted as α) that have the acetylcholine binding sites. These channels are key players in the cellular communication that takes place between neurons. In particular, they are found on the postsynaptic side of the neuromuscular junction. As hinted at in Figure 3.1, the command to move a muscle is the result of a nerve impulse traveling down a neuron to the neuromuscular junction. There, the

Figure 3.4

Key examples of ligand-gated ion channels. (A) Nicotinic acetylcholine receptor, revealing its heteropentameric structure with two binding sites for acetylcholine. (B) cGMP-gated ion channel. These channels have four cGMP binding sites.

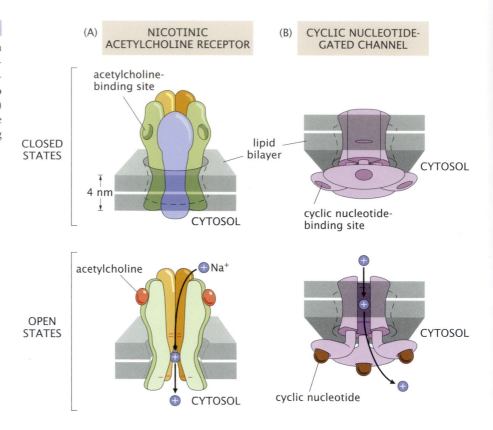

(A) NICOTINIC ACETYLCHOLINE RECEPTOR

(B) CYCLIC NUCLEOTIDE-GATED CHANNEL

arrival of the action potential has the consequence of releasing neurotransmitters across the synaptic cleft, where they bind with the nicotinic acetylcholine receptors on the extracellular membrane that then transiently open, resulting in a change in membrane potential in the muscle itself. The nicotinic acetylcholine receptors are a representative example of a much broader range of channels that include γ-aminobutyric acid (GABA) and glycine receptors. Generally, when open, they are permeable to both sodium and potassium ions.

The kinds of currents measured in patch-clamp experiments on these channels were shown in Figure 3.3. The idea is that for each concentration of acetylcholine, the current is measured using these electrophysiological methods. By measuring the fraction of time that the channel is in the open state, one can then obtain the open probability itself. Specifically, for each concentration of acetylcholine, we have a different fraction of time that the channels are in the open state, with the open probability defined as

$$p_{open}(c) = \frac{\sum t_{open}^{(i)}(c)}{t_{tot}}, \qquad (3.7)$$

where $t_{open}^{(i)}$ is the duration of the ith open interval, and t_{tot} is the total time of the measurement. Figure 3.5 gives an example of the normalized current as a function of the concentration of acetylcholine, defined as

$$\text{normalized current} = \frac{I - I_{min}}{I_{max} - I_{min}}. \qquad (3.8)$$

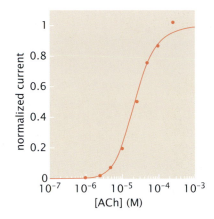

Figure 3.5

Peak ion-channel currents as a function of ligand concentration. Adapted from Labarca et al. (1995).

From a theoretical perspective, we will make contact with measurements of the normalized current by computing the normalized gating probability as a function of the acetylcholine concentration, which we will define as

$$\bar{p}_{open}(c) = \frac{p(c) - p_{min}}{p_{max} - p_{min}}. \qquad (3.9)$$

A second important example of a ligand-gated ion channel is the cyclic GMP–gated channels associated with photoreceptors (see Figure 3.4). These channels are found in the plasma membrane of the outer segment of both rod and cone cells, as shown in Figure 3.6. Signal transduction in our visual systems proceeds through a series of coupled reactions that begin when a photon is absorbed by a rhodopsin molecule, which is part of the richly decorated membrane discs in photoreceptors, as illustrated in Figure 3.6. The initial consequence of this photon absorption is a photoisomerization reaction that takes retinal from the 11-*cis* form to the unkinked all-trans form. This change in the shape of the retinal has a concomitant influence on the rhodopsin molecule, activating it so that it induces a signal cascade.

The relevant ligand for gating the cGMP-gated channel is guanosine 3′,5′-cyclic monophosphate (which we abbreviate as cGMP). These nucleotides are engaged in regulating a signaling cascade in the rod outer segment, and it is their concentration that dictates the electrical process we perceive as vision. As a result of the signal cascade, once the rhodopsin has suffered a photon-induced conformational change, the concentration of cGMP drops, shifting the equilibrium of these channels toward the closed state and thus inhibiting the current influx that characterizes the behavior of these channels in the dark. The net effect of all these processes stimulated by the absorption of a single photon is that the membrane potential of the cell changes by roughly 1 mV for a period of roughly 200 ms.

The question we want to sort out now is how to think about these processes from the standpoint of statistical mechanics. In particular, this is our first opportunity to take a detailed look at all the molecular states that can be adopted by an MWC molecule (in this case a ligand-gated ion channel) and to compute how the relative probabilities of these states change as a function of ligand concentration.

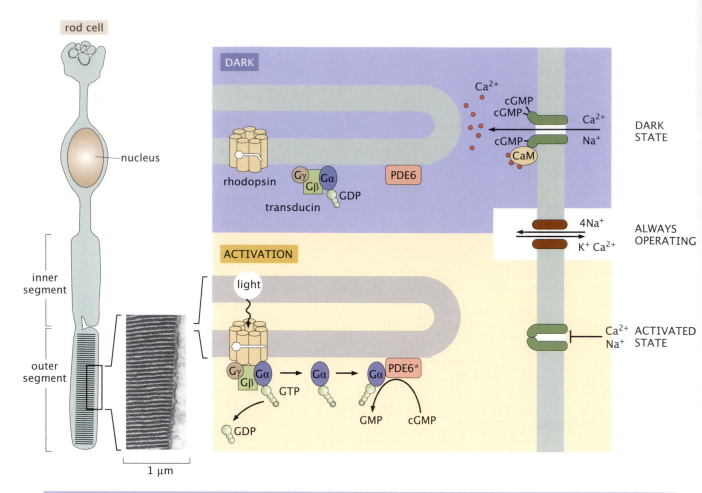

Figure 3.6

Photoreceptor outer segment and ion-channel gating. The schematic of the rod cell shows the outer segment. The electron microscopy image shows the stacks of closely spaced membranes that house the light-detection and signaling apparatus shown on the right. The cartoon on the right shows how the presence or absence of light alters the molecular state of the system.

3.4 Statistical Mechanics of the MWC Channel

Taking our inspiration from the nicotinic acetylcholine receptor, we attempt to build both a diagrammatic and mathematical caricature of some of the key features of its gating transition. As we will do repeatedly throughout the book, we merge the diagrammatic and mathematical pictures in the form of states-and-weights diagrams like that shown in Figure 3.7. These diagrams enumerate all the different microscopic states that the system of interest can take and also assign a statistical weight to each such state.

The first point to notice is that the "bare" energy (i.e., in the absence of the ligand) of the channel is different for the closed and open states. We assign energies ε_{closed} and ε_{open} to these states, with $\varepsilon_{closed} < \varepsilon_{open}$, implying that in the absence of ligand, the closed state is favored. In particular, in the absence of ligand (concentration $c = 0$), we can translate this intuition into its

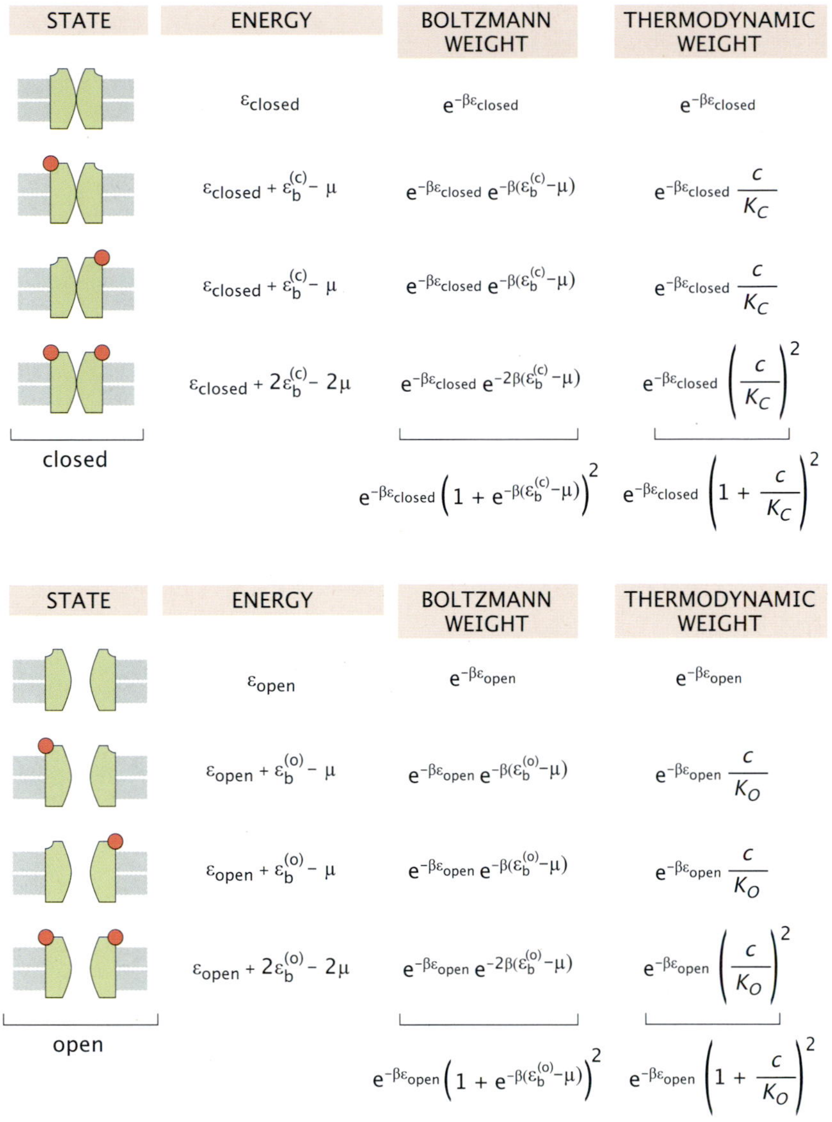

STATE	ENERGY	BOLTZMANN WEIGHT	THERMODYNAMIC WEIGHT
	ε_{closed}	$e^{-\beta\varepsilon_{closed}}$	$e^{-\beta\varepsilon_{closed}}$
	$\varepsilon_{closed} + \varepsilon_b^{(c)} - \mu$	$e^{-\beta\varepsilon_{closed}}\, e^{-\beta(\varepsilon_b^{(c)}-\mu)}$	$e^{-\beta\varepsilon_{closed}}\, \dfrac{c}{K_C}$
	$\varepsilon_{closed} + \varepsilon_b^{(c)} - \mu$	$e^{-\beta\varepsilon_{closed}}\, e^{-\beta(\varepsilon_b^{(c)}-\mu)}$	$e^{-\beta\varepsilon_{closed}}\, \dfrac{c}{K_C}$
	$\varepsilon_{closed} + 2\varepsilon_b^{(c)} - 2\mu$	$e^{-\beta\varepsilon_{closed}}\, e^{-2\beta(\varepsilon_b^{(c)}-\mu)}$	$e^{-\beta\varepsilon_{closed}}\left(\dfrac{c}{K_C}\right)^2$
closed		$e^{-\beta\varepsilon_{closed}}\left(1 + e^{-\beta(\varepsilon_b^{(c)}-\mu)}\right)^2$	$e^{-\beta\varepsilon_{closed}}\left(1 + \dfrac{c}{K_C}\right)^2$

STATE	ENERGY	BOLTZMANN WEIGHT	THERMODYNAMIC WEIGHT
	ε_{open}	$e^{-\beta\varepsilon_{open}}$	$e^{-\beta\varepsilon_{open}}$
	$\varepsilon_{open} + \varepsilon_b^{(o)} - \mu$	$e^{-\beta\varepsilon_{open}}\, e^{-\beta(\varepsilon_b^{(o)}-\mu)}$	$e^{-\beta\varepsilon_{open}}\, \dfrac{c}{K_O}$
	$\varepsilon_{open} + \varepsilon_b^{(o)} - \mu$	$e^{-\beta\varepsilon_{open}}\, e^{-\beta(\varepsilon_b^{(o)}-\mu)}$	$e^{-\beta\varepsilon_{open}}\, \dfrac{c}{K_O}$
	$\varepsilon_{open} + 2\varepsilon_b^{(o)} - 2\mu$	$e^{-\beta\varepsilon_{open}}\, e^{-2\beta(\varepsilon_b^{(o)}-\mu)}$	$e^{-\beta\varepsilon_{open}}\left(\dfrac{c}{K_O}\right)^2$
open		$e^{-\beta\varepsilon_{open}}\left(1 + e^{-\beta(\varepsilon_b^{(o)}-\mu)}\right)^2$	$e^{-\beta\varepsilon_{open}}\left(1 + \dfrac{c}{K_O}\right)^2$

Figure 3.7

States and weights for an MWC model of a ligand-gated ion channel. This is a model of an ion channel with two binding sites for the ligand, such as in the nicotinic acetylcholine receptor. The top panel shows the various states of occupancy for the closed state, while the bottom panel shows the various states of occupancy for the open state.

mathematical incarnation as

$$p_{open}(c=0) = \frac{e^{-\beta\varepsilon_{open}}}{e^{-\beta\varepsilon_{open}} + e^{-\beta\varepsilon_{closed}}}. \qquad (3.10)$$

Effectively, this expression provides a measure of the "leakiness" of the channel, namely, how much current will flow through that channel even in the absence of the ligands that gate it. Figure 3.8 shows an example of the current traces from nicotinic acetylcholine receptor in the absence of acetylcholine, demonstrating spontaneous transitioning between the closed and open states, indicative of leakiness. The time scales for these spontaneous openings are shown in Figure 3.8. Our main question now is, how is this open probability amended by the presence of ligands.

The MWC framework suggests that ligands can bind the channel in both the closed and open states, as indicated by the microscopic states depicted in

Figure 3.8

Leakiness of the nicotinic acetylcholine receptor. Current traces showing the spontaneous opening of the channels even in the absence of the driving force due to binding of acetylcholine. Adapted from Jackson (1986).

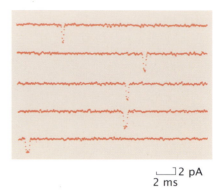

2 pA
2 ms

Figure 3.7. Further, since we draw our inspiration from the case of the nicotinic acetylcholine receptor, we posit two distinct binding sites, which implies a total of eight states. As Figure 3.7 shows, both the closed and open states can be ligand free, singly occupied on each of the two binding sites, and doubly occupied. Note that statistical mechanics tells us how to reckon the probabilities of each one of these eight states by assigning a statistical weight to the ith state of the form $weight_i = e^{-\beta(\varepsilon_i - n_i \mu)}$, where ε_i is the energy of the ith state, n_i is the number of bound ligands in that state (i.e., drawn from the particle reservoir), and μ is the chemical potential of those ligands, which we recall from the previous chapter is nothing more than jargon for the free-energy cost to take a ligand from solution and attach it to a receptor. One of the strokes of genius in the MWC framework is that there are different binding energies for the closed and open states, given by $\varepsilon_b^{(c)}$ and $\varepsilon_b^{(o)}$, respectively. In light of all these definitions and using the states and weights of Figure 3.7, we can then write the open probability itself as

$$p_{open} = \frac{e^{-\beta \varepsilon_{open}}(1 + e^{-\beta(\varepsilon_b^{(o)} - \mu)})^2}{e^{-\beta \varepsilon_{open}}(1 + e^{-\beta(\varepsilon_b^{(o)} - \mu)})^2 + e^{-\beta \varepsilon_{closed}}(1 + e^{-\beta(\varepsilon_b^{(c)} - \mu)})^2}. \tag{3.11}$$

It is sometimes useful to adopt a more macroscopic perspective in which the binding reactions are described in terms of dissociation constants K_d rather than the microscopic binding energies $\varepsilon_b^{(o)}$ and $\varepsilon_b^{(c)}$, as we discussed in the previous chapter. To effect this change of perspective given equation 3.11, we recall that the chemical potential can be written in the form $\mu = \mu_0 + k_B T \ln c/c_0$. Substituting this into equation 3.11 and recalling that we can define K_d through $K_d = c_0 e^{\beta \Delta \varepsilon}$, where we have defined $\Delta \varepsilon = \varepsilon_b - \mu_0$, we arrive at the equivalent expression for the open probability,

$$p_{open}(c) = \frac{e^{-\beta \varepsilon_{open}}(1 + \frac{c}{K_O})^2}{e^{-\beta \varepsilon_{open}}(1 + \frac{c}{K_O})^2 + e^{-\beta \varepsilon_{closed}}(1 + \frac{c}{K_C})^2}. \tag{3.12}$$

This expression has a familiar look, as will the expressions that arise in each of our case studies, and can also be tailored to have the appearance of the original MWC model by noting that we can also introduce

$$L = e^{-\beta(\varepsilon_{closed} - \varepsilon_{open})}, \tag{3.13}$$

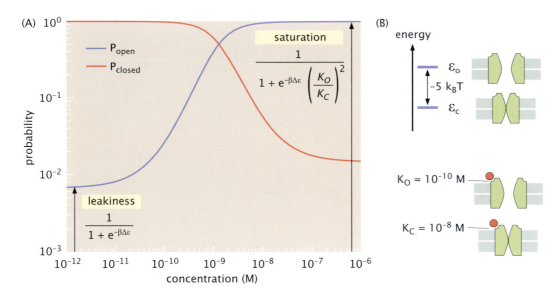

Figure 3.9

Ion-channel probabilities and parameters. (A) Probability of open and closed states of the ligand-gated ion channel as a function of ligand concentration. Leakiness is exhibited by the open probability in the limit of low ligand concentration. Saturation is exhibited by the open probability in the limit of high concentration. (B) Parameters used in the figure are $K_O = 10^{-10}$ M, $K_C = 10^{-8}$ M, and $\Delta\varepsilon = -5k_B T$, resulting in a leakiness of 0.007 and a saturation of 0.985.

where in the MWC model this parameter is used to measure the relative probabilities of the two states in the absence of ligand. In this case, the open probability is given by

$$p_{open}(c) = \frac{(1 + \frac{c}{K_O})^2}{(1 + \frac{c}{K_O})^2 + L(1 + \frac{c}{K_C})^2}. \tag{3.14}$$

The open probability as a function of ligand concentration is plotted in Figure 3.9, which shows some of the detailed features of this model, such as the fact that even in the absence of ligand, there is a finite probability of being in the open state, and similarly, that even in saturating conditions, the channel is not guaranteed to be open.

As noted earlier, there are many important phenotypic properties of allosteric molecules and their input-output functions that can be seen in Figure 3.9. We can think of the leakiness as the probability that the channels will be open even in the absence of ligand, and already implicitly expressed in equation 3.10 but rewritten more transparently as

$$\text{leakiness} = p_{min} = p_{open}(c = 0) = \frac{1}{1 + e^{-\beta\Delta\varepsilon}}, \tag{3.15}$$

where $\Delta\varepsilon = \varepsilon_{closed} - \varepsilon_{open}$. The dynamic range concerns itself with the opposite limit, namely, when the concentration is so large that the bound states dominate the unbound states. In this case, we have

$$\text{dynamic range} = \lim_{c \to \infty} p_{open}(c) - p_{open}(c = 0). \tag{3.16}$$

Figure 3.10

Ion-channel currents as a function of ligand concentration. Data for the wild-type channel is shown, as well as for four different mutants containing $n = 1, 2, 3$, and 4 point mutations to the channel. The parameters used in the graph are $K_C = 60 \times 10^{-6}$ M, $K_O = 0.1 \times 10^{-9}$ M, and $\Delta\varepsilon = -23.7, -19.2, -14.6, -8.5, -4.0$ ($k_B T$ units) for $n = 0, 1, 2, 3, 4$, respectively. Adapted from Labarca et al. (1995).

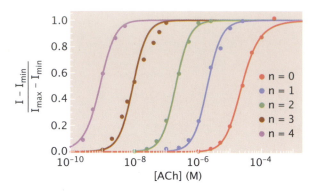

The first of these terms, which we will call the *saturation*, can be found by examining the open probability in the presence of saturating levels of acetylcholine, namely,

$$\text{saturation} = p_{max} \equiv p_{\text{open}}(c \to \infty) = \frac{1}{1 + e^{-\beta\Delta\varepsilon}\left(\frac{K_O}{K_C}\right)^2}. \qquad (3.17)$$

Using our equation for the open probability (eqn. 3.14), we find the dynamic range as

$$\text{dynamic range} = \frac{1}{1 + e^{-\beta\Delta\varepsilon}\left(\frac{K_O}{K_C}\right)^2} - \frac{1}{1 + e^{-\beta\Delta\varepsilon}}. \qquad (3.18)$$

Both the leakiness and saturation are revealed in graphical form in Figure 3.9.

To give a feel for how this model works in the context of actual ion-channel data, Figure 3.10 shows the normalized current passing through both wild-type and mutant nicotinic acetylcholine receptors as a function of the acetylcholine concentration. The mutants considered correspond to replacing leucine with serine at a position in the M2 transmembrane helices of the pentameric nicotinic acetylcholine receptor. This particular mutation was chosen because it corresponds to a conserved residue in nicotinic acetylcholine receptors, glycine receptors, GABA$_A$ receptors, serotonin receptors, and glutamate receptors, and hence it was of interest to explore its role in controlling channel function. Specifically, the mutant receptors leading to the data considered here arise from imposing different numbers of mutations (shown as $n = 1, n = 2, n = 3$, and $n = 4$ in the figure) that are distant from the acetylcholine binding region and that are thought to vary only the free-energy difference between the closed and open states. As seen in the plot, there is a stereotyped response with some regime of concentrations over which the normalized current goes from zero to 1, with different numbers of mutations successively shifting the concentration to gate the channels to lower concentrations. However, plotting the data in the scaled way it is plotted in Figure 3.10 can be quite misleading, since it masks the possible leakiness (i.e., the channel can be open without ligand, as shown in Figure 3.8) and lack of saturation (i.e., the channel can be closed even with "saturating" concentrations of ligand), both of which are exhibited in Figure 3.9.

To make contact with the experimental data, we have to find a way to use the MWC model to plot a quantity like the normalized current. To that end, we

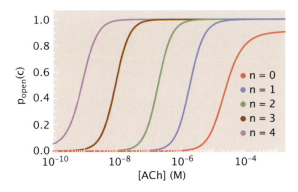

Figure 3.11

MWC model for ion-channel currents as a function of ligand concentration. Normalized open probability (p_{open}) for a wild-type channel and four mutants as a function of acetylcholine concentration as determined by the MWC model.

exploit the definition of the normalized open probability given in equation 3.9 (p. 83). Note that in this case we have defined p_{min}, which gives the minimum probability that the channel is open for the MWC model, and p_{max}, which provides the corresponding maximum probability that the channel is open. Using these definitions, we now have a function that goes from zero to 1 in the same way as the normalized current and can be written as

$$\bar{p}_{open}(c) = \frac{\frac{(1+\frac{c}{K_O})^2}{(1+\frac{c}{K_O})^2+e^{-\beta\Delta\varepsilon}(1+\frac{c}{K_C})^2} - \frac{1}{1+e^{-\beta\Delta\varepsilon}}}{\frac{1}{1+e^{-\beta\Delta\varepsilon}(\frac{K_O}{K_C})^2} - \frac{1}{1+e^{-\beta\Delta\varepsilon}}}. \tag{3.19}$$

Note that this equation depends upon three parameters: $\Delta\varepsilon$, the energy difference between the closed and open states in the absence of ligand; K_C, the dissociation constant for binding of acetylcholine to the channels when they are closed; and K_O, the dissociation constant for binding of acetylcholine to the channels when they are open. The curves in Figure 3.10 correspond to a one-parameter family of fits in which all five curves have the same values of K_C and K_O, and each curve has its own distinct value of $\Delta\varepsilon$.

As seen in Figure 3.10, the model framework described here provides what appears to be a satisfactory reflection on the experimental data for several mutants of the nicotinic acetylcholine receptor. Unfortunately, the way that the data are plotted in terms of normalized currents deprives us of the opportunity to look more deeply into the mechanistic properties of both the wild-type and mutant channels. Figure 3.11 shows the open probabilities themselves that correspond to the normalized currents shown in Figure 3.10. Note that the probabilities allow us to see how the leakiness (see Figure 3.12) and dynamic range vary among different mutants.

There are other key biophysical parameters that characterize the functional response of MWC molecules more generally, and ion channels in particular. Specifically, we now go further to consider the EC_{50} (i.e., the concentration at which the channels are at half saturation, as shown in Figure 3.13) and the slope of $\bar{p}_{open}$ at EC_{50}. The concentration $c = EC_{50}$ is defined as the concentration at which

$$p_{open}(EC_{50}) = \frac{p_{open}^{min} + p_{open}^{max}}{2}. \tag{3.20}$$

As seen in Figure 3.10, there is a systematic pattern of shifts in the EC_{50}s as additional mutations are added to the different channel subunits. The measured

Figure 3.12

Predicted leakiness of ion-channel mutants. The ion-channel mutants considered here reveal a progression of leakiness values with each additional mutation.

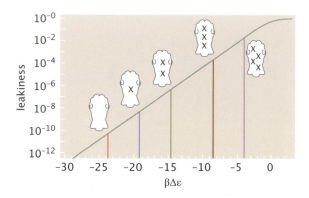

Figure 3.13

Schematic of the definition of EC_{50}. Two hypothetical dose-response curves for ion-channel gating show the probability of channel opening as a function of ligand concentration. EC_{50} is that concentration at which the open probability is midway between its lowest value p_{min} and its maximum value p_{max}.

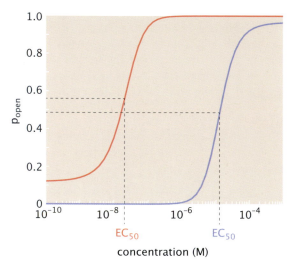

EC_{50} values for all the different mutants of the nicotinic acetylcholine receptor are shown in Figure 3.14. These data show that each mutation results in roughly a 10-fold shift in the EC_{50} value, the midpoint between the leakiness and the saturation. The quantitative nature of the MWC model allows us to make quantitative contact with the data and to predict how the model parameters change with mutations.

One approach to understanding the mechanistic origins of EC_{50} in the MWC model is to solve for it directly, since the condition given in equation 3.20 can be expressed as a quadratic equation in the unknown EC_{50}. However, the result is not particularly enlightening in its full glory. Instead, we consider the limit $K_C \to \infty$, which says the affinity of ligands for the closed state is negligible, and we calculate EC_{50} in this regime. In this limit, we have

$$p_{open}^{K_C \to \infty}(c) = \frac{\left(1 + \frac{c}{K_O}\right)^2}{\left(1 + \frac{c}{K_O}\right)^2 + e^{-\beta \Delta \varepsilon}}. \tag{3.21}$$

As a result, we have the condition for the concentration at which the midpoint probability occurs as

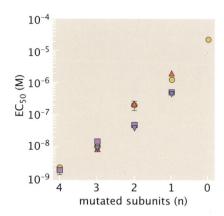

Figure 3.14

EC_{50} for different mutants of the nicotinic acetylcholine receptor. For each case (i.e., single mutants, double mutants, etc.) several different mutants are made in the pentameric channel. Standard error of the mean is shown in those cases where they are larger than the symbols. Adapted from Labarca et al. (1995).

$$\frac{1}{2}(p_{min} + p_{max}) = \frac{\left(1 + \frac{EC_{50}}{K_O}\right)^2}{\left(1 + \frac{EC_{50}}{K_O}\right)^2 + e^{-\beta\Delta\varepsilon}}. \tag{3.22}$$

This condition leads to a quadratic equation for EC_{50}. If we work in the limit that $p_{max} = 1$, largely consistent with the curves shown in Figure 3.11, and use equation 3.15 for p_{min}, the left side simplifies to the form

$$\frac{1}{2}(p_{min} + p_{max}) \approx \frac{1}{2}\left(\frac{1}{1 + e^{-\beta\Delta\varepsilon}} + 1\right) = \frac{1}{2}\left(\frac{2 + e^{-\beta\Delta\varepsilon}}{1 + e^{-\beta\Delta\varepsilon}}\right). \tag{3.23}$$

If we now define the new variable

$$x = \left(1 + \frac{EC_{50}}{K_O}\right), \tag{3.24}$$

then our condition for EC_{50} becomes a simple quadratic equation for the variable x of the form

$$\frac{1}{2}\left(\frac{2 + e^{-\beta\Delta\varepsilon}}{1 + e^{-\beta\Delta\varepsilon}}\right) = \frac{x^2}{x^2 + e^{-\beta\Delta\varepsilon}}. \tag{3.25}$$

We can solve this equation for EC_{50} in succession, first obtaining

$$x = \sqrt{2 + e^{-\beta\Delta\varepsilon}}, \tag{3.26}$$

and then solving for EC_{50} itself we find

$$EC_{50} = K_O\left(\sqrt{2 + e^{-\beta\Delta\varepsilon}} - 1\right). \tag{3.27}$$

This theoretical result is shown in Figure 3.15. This compact formula can be further simplified if we make the approximation that $e^{-\beta\Delta\varepsilon} \gg 1$, an approach well justified for all the channels considered except the 4-subunit mutant. After making this approximation, we can further simplify our result to the form

$$EC_{50} \approx K_O e^{-\beta\Delta\varepsilon/2}. \tag{3.28}$$

Figure 3.15

EC_{50} of nAChR receptor mutants. The data points correspond to the 0-, 1-, 2-, 3-, and 4-mutated subunit receptors, while the curve shows the results of the MWC model predictions.

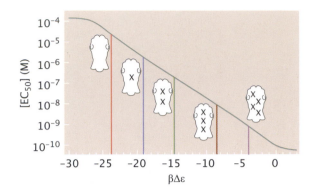

Formally, this approximate result amounts to taking the first term from a Taylor series about $e^{\beta\Delta\varepsilon} \approx 0$. For reference, for all five curves shown in Figure 3.10 and the corresponding EC_{50}s shown in Figure 3.14, this approximate EC_{50} is at most a factor of 2 off from the exact EC_{50} computed from the best fit.

The fits of the parameter $\Delta\varepsilon$ from Figure 3.10 reveal that the values associated with the 0-, 1-, 2-, 3-, and 4-mutated subunit receptors increased linearly by roughly 4.6 $k_B T$ (corresponding to roughly a 10-fold change in the EC_{50} values with each additional Leu9Ser mutation). This succession of $\Delta\varepsilon$ values implies that the EC_{50} should decrease exponentially for each additional mutated subunit, as seen in Figure 3.15, consistent with what was shown experimentally (see Figure 3.14).

One other important phenotypic characteristic of an MWC molecule is the sensitivity of its input-output response. We highlighted a number of different ways of characterizing the sensitivity/cooperativity of molecular switches in Figure 2.14 (p. 58). The effective Hill coefficient h equals twice the log-log slope of the normalized current evaluated at $c = EC_{50}$, namely,

$$h = 2\frac{d}{d\log c}\log\left(\frac{p_{open}(c) - p_{open}^{min}}{p_{open}^{max} - p_{open}^{min}}\right)_{c=[EC_{50}]}. \tag{3.29}$$

In the context of the MWC model itself, the resulting expressions are cumbersome and resist immediate intuitive interpretation. Alternatively, if we recall the sigmoidal properties of Hill functions, we can develop some intuition for this definition. The sensitivity exhibited by a Hill function is based on a phenomenological description of the concentration dependence of the channel gating behavior as

$$p_{open}(c) = \frac{\left(\frac{c}{K}\right)^n}{1 + \left(\frac{c}{K}\right)^n}, \tag{3.30}$$

a fitting form often used to capture the results of experimental measurements on the dose-response curves of ion channels. In the case of a Hill function, we note that

$$h = 2\frac{d\log p}{d\log c} = 2\frac{c}{p}\frac{dp}{dc}. \tag{3.31}$$

If we use equation 3.30 as our description of the open probability, then we find

$$\frac{dp}{dc} = \frac{n\left(\frac{c}{K}\right)^{n-1}\frac{1}{K}}{\left(1+\left(\frac{c}{K}\right)^n\right)^2},$$ (3.32)

which implies in turn that

$$2\frac{c}{p}\frac{dp}{dc} = \frac{2n}{1+\left(\frac{c}{K}\right)^n}.$$ (3.33)

If we evaluate this expression at the Hill function midpoint $c = K$, we find that $h = n$, consistent with our intuition about the meaning of the Hill coefficient as a measure of the sensitivity of the response. As noted previously, the EC_{50} determines how the normalized current shifts left and right, while the effective Hill coefficient corresponds to the slope at EC_{50}. Together, these two properties determine the approximate window of ligand concentrations for which the normalized current transitions from 0 to 1. The same definition given in equation 3.29 that we have now seen culminates in an intuitive result for the Hill coefficient when our activity function is described by a Hill function also reports on the sensitivity of more general input-output curves, such as those offered by the MWC model. In the limit $1 \ll e^{-\beta\Delta\varepsilon} \ll \left(\frac{K_C}{K_O}\right)^n$ (n is the number of ligand binding sites on the channel), which we will show is relevant for both the nicotinic acetylcholine receptor and cGMP-gated ion channels considered in this chapter, the effective Hill coefficient is given by $h \approx n$, a compact and very intuitive result.

What we have seen thus far in the chapter is the way that the MWC model provides us with a clear picture of the biophysical parameters that determine the physiological and evolutionary response of a given molecular switch (in this case an ion channel switching between closed and open states). Within the MWC model considered here, those key biophysical properties of the channel described earlier can be approximated to leading order as

$$\text{leakiness} \approx e^{\beta\varepsilon},$$ (3.34)

which describes the current even in the absence of ligand and shown in Figure 3.8, and

$$\text{dynamic range} \approx 1,$$ (3.35)

which tells us that in the large concentration limit these channels will always be open. We have already seen that the EC_{50} can be succinctly captured as

$$EC_{50} \approx K_O e^{-\beta\epsilon/n},$$ (3.36)

and, finally, the effective Hill coefficient takes the especially pleasing and simple form

$$h \approx n.$$ (3.37)

What these various calculations have shown is the power of the MWC framework to provide intuitive explanations of key quantities such as the leakiness, dynamic range, EC_{50}, and the effective Hill coefficient. In addition, they give us a single framework for thinking about a whole suite of different mutants without each new mutant becoming a completely new adventure in parameter fitting. An even more compelling vision of the unity of this perspective is provided by the method of data collapse, which we consider now.

3.5   Data Collapse, Natural Variables, and the Bohr Effect

Many of the problems we will examine throughout this book feature families of activity curves or families of binding curves. Specifically, what I mean by this is that in the spirit of tunable parameters (i.e., knobs), experiments on a given system admit of a series of measurements in which some parameter is changed resulting in a "shift" to the binding (or activity) curves. The most naive example of this is shown in Figure 3.16 for ligand–receptor binding. Figure 3.16(B) shows a family of curves for simple binding described by the equation

$$p_{bound}(c) = \frac{\frac{c}{K_d}}{1 + \frac{c}{K_d}}, \tag{3.38}$$

where the family of curves differ in that each one is characterized by a different K_d. In this case, the key point is the realization that the "natural" variable for this problem is the dimensionless concentration given by $x = c/K_d$ that collapses the curves. In anthropomorphic terms, the receptor "doesn't care" about the concentration that we pipette into our tube during some experiment. The receptor "cares" about the value of the concentration relative to the K_d. If we now replot the data using

$$p_{bound}(x) = \frac{x}{1 + x}, \tag{3.39}$$

then each binding curve falls on one universal curve, as seen in Figure 3.16(C). This may seem like we are stating the obvious, but to my way of thinking a more generous view is that by plotting concentration in units of K_d we are acknowledging that each different receptor has its own distinction between low and high concentrations.

This idea of finding the natural variables of a problem has a long tradition in physics and chemistry. A less trivial example than the preceding one was revealed in the context of liquids and gases. Specifically, detailed experimental work by thermodynamicists revealed the equations of state for each working substance such as oxygen or methane. By equation of state we mean a curve that shows how the phase boundary relating liquid and gas depends upon the density. Figure 3.17 shows a classic example from Guggenheim (1945) of the so-called law of corresponding states. The idea of this empirical result is that the equation of state relating macroscopic variables such as the temperature and density is the same for many different substances when they are plotted in dimensionless variables where these quantities are scaled by their values at the critical point. The beauty of the law of corresponding states illustrated in Figure 3.17 is that it shows us how to define new temperature and density variables for any substance such that all equations of state fall onto one master curve. Similar plots can be made for other kinds of materials such as magnetic materials and superconductors.

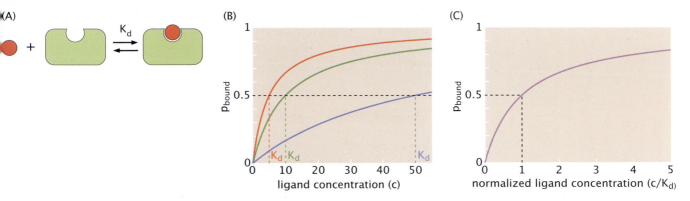

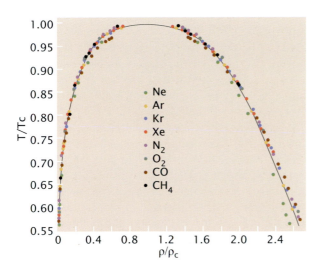

Figure 3.16

Data collapse for simple binding. (A) The ligand–receptor binding equilibrium. (B) Simple binding curves for a series of mutant receptors differing only in their binding affinity for their ligand. (C) The same curves now plotted with respect to the natural unit $x = c/K_d$.

Figure 3.17

Law of corresponding states. Phase boundary between the liquid and gas phases for a number of different atomic and molecular species is shown in the reduced temperature and density plane. Adapted from Guggenheim (1945).

   The key point in these examples is that at the outset one has the impression that each substance has its own peculiarities, so there doesn't seem to be a unifying framework for viewing the data. And yet, once the natural variables of the problem are identified, we see that different substances have essentially the same equations of state. Are there similar ideas that come into play in the context of allostery? We explore that question in the context of our ion-channel data now.

3.5.1   Data Collapse and the Ion-Channel Bohr Effect

The question of data collapse for MWC molecules is even more nuanced than that revealed by the law of corresponding states. To get a sense of what "data collapse" means in the context of simple binding curves, consider the discussion culminating in Figure 3.16. For MWC molecules, the corresponding data collapse is at once more subtle and revealing. We begin by recalling that the probability an MWC ion channel with n ligand binding sites is open is given by

$$p_{open} = \frac{e^{-\beta \varepsilon_{open}}(1 + \frac{c}{K_O})^n}{e^{-\beta \varepsilon_{open}}(1 + \frac{c}{K_O})^n + e^{-\beta \varepsilon_{closed}}(1 + \frac{c}{K_C})^n}. \tag{3.40}$$

This can be simplified further and put into a functional form convenient for exploring the data collapse by dividing top and bottom by the numerator, resulting in

$$p_{open} = \frac{1}{1 + e^{-\beta \Delta \varepsilon} \frac{\left(1 + \frac{c}{K_C}\right)^n}{\left(1 + \frac{c}{K_O}\right)^n}}, \tag{3.41}$$

where once again we define $\Delta \varepsilon = \varepsilon_{closed} - \varepsilon_{open}$. A convenient and completely equivalent general form that emphasizes the notion of data collapse arises if we rewrite p_{open} as

$$p_{open} = \frac{1}{1 + e^{-\beta F_{Bohr}}} \tag{3.42}$$

where we define the Bohr parameter

$$F_{Bohr} = \Delta \varepsilon - nk_B T \ln \frac{\left(1 + \frac{c}{K_C}\right)}{\left(1 + \frac{c}{K_O}\right)}. \tag{3.43}$$

The reason for this nomenclature is inspired by work from Mirny using the MWC model to examine the wrapping of the genomic DNA in eukaryotes in macromolecular assemblies known as nucleosomes (see chap. 10, p. 316). In the context of such highly compacted DNA, the binding of proteins to nucleosomal DNA as a function of the concentration of these proteins yields a binding curve. Mirny notes that just as the binding of oxygen to hemoglobin depends upon the pH or the concentration of carbon dioxide in what is known as the Bohr effect (see Figure 7.2, p. 233), the binding of proteins to nucleosomal DNA will depend upon how tightly the DNA is wrapped around the histones making up the nucleosome. The adhesion between the DNA and the histones can depend upon posttranslational modifications to the histones that are *analogous* to the way that the binding between oxygen and hemoglobin can be tuned by the carbon dioxide concentration seen in the Bohr effect.

Let's explore this analogy and the data collapse it implies more deeply. Just as earlier we argued that binding curves were most naturally expressed in terms of the dimensionless concentration c/K_d, we now argue that the natural variable for our MWC activity curves is the Bohr parameter. Figure 3.18 shows how the mapping from bare concentration to the effective concentration-dependent free energy allows us to collapse all data onto one master curve. The left panel of the figure shows that when all five curves have the same open probability, they have the same free-energy difference between open and closed states, as shown by the free-energy axis in the middle. For example, consider the situation when $\bar{p}_{open} = 0.2$, corresponding to the lowest horizontal dotted line in the figure. Since all five curves have the same probability of being open (i.e., $\bar{p}_{open} = 0.2$), by equation 3.42, this means they all have the same value of F_{Bohr}.

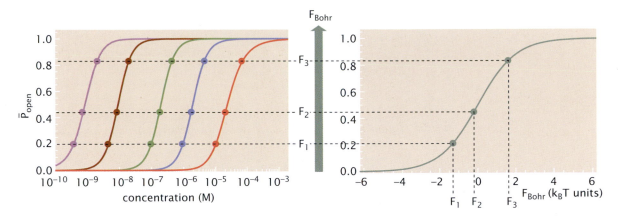

Figure 3.18

Families of activity curves and their corresponding data collapse. The curves on the left are a family of channel-open probabilities for different mutants. A given open probability corresponds to an energy given by equation 3.43 and shown by the vertical arrow. The curve on the right is a plot of the open probability according to equation 3.42 in terms of the free-energy difference between the closed and open states and given by the Bohr parameter.

Hence, if instead of plotting open probability with respect to pipetted concentration we plot it with respect to the free-energy difference measured by the Bohr parameter, all curves collapse onto a single master curve. We will consider this same idea repeatedly throughout the book, constantly trying to see the difference between "pipettor's variables" and the "natural variables" of the system itself. Indeed, ultimately, the most important single mission statement of this book is the need to devise tools with broad reach in biology that teach us how to develop effective theories of biological systems that intentionally suppress reference to the entirety of the degrees of freedom of those systems.

The measurements of Figure 3.10 showed us the normalized currents for both wild-type and mutant versions of the nicotinic acetylcholine receptor. When plotted as in Figure 3.10, each dose-response curve when seen by itself shows how as the concentration of ligand (in this case acetylcholine) is varied, the ion-channel open probability is altered; however, as shown, the data give no sense of how the different mutations are actually altering the channel properties. The argument we are making here is that by appealing to the underlying MWC model, we can ask questions such as whether the mutations alter only the free-energy difference between the open and closed states in the absence of ligand or, rather, whether they actually change the dissociation constants for the two states.

To see how hypotheses about the different mutants can be tested on the basis of the statistical mechanical model of the opening probability, we use the equation for the open probability in a form that reveals the generic features of the MWC molecule and is given in equation 3.41. Using this equation, how do we replot the data of Figure 3.10 in a way that reveals the possible data collapse? For each data point from Figure 3.10 we compute its Bohr parameter by using equation 3.43 for the relevant concentration c and use the parameter $\Delta\varepsilon$ relevant to the mutant of interest. This amounts to implementing the hypothesis that the mutations have an isolated effect, altering only the energy difference

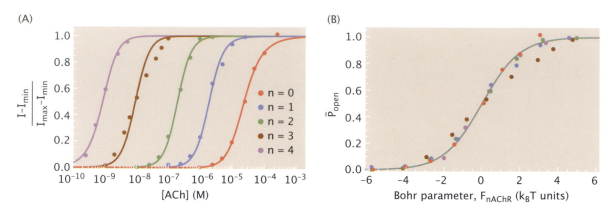

Figure 3.19

Data collapse for ion-channel currents as a function of ligand concentration. By computing the Bohr parameter associated with every data point in Figure 3.5, we can now replot the data as shown here with the interesting consequence that all the data appear to fall on one master curve.

between the closed and open states and leaving the K_ds untouched. As seen in Figure 3.19, following this procedure for this particular dataset results in a very convincing data collapse, all based on the single parameter $\Delta\varepsilon$, though for one of the curves ($n = 3$) in the figure, there seems to be some systematic differences between the theory and the experimental data.

3.6   Rate Equation Description of Channel Gating

Thus far, our picture of channel gating has been built in terms of equilibrium thinking and culminated in equation 3.4, which uses the tools of statistical mechanics to compute the probability of the open state as a function of the energy difference between the closed and open states. Figure 3.2 (p. 79) shows us how the difference in energy between the closed and open states can be tuned by driving forces such as voltage, ligand concentration, or membrane tension. We now consider a complementary view of channel gating (and this complementarity between equilibrium statistical mechanics and rate equations will be exploited again and again throughout the book) in which we examine the rates of transitions between the different states. In his Nobel Lecture, in the context of different ways of formulating quantum mechanics, Richard Feynman (1965) noted that although these different formulations might yield the same answers when considering traditional problems such as the hydrogen atom or the harmonic oscillator, they were "psychologically inequivalent" when launching into the unknown. One reason it is so important to develop this complementary kinetic approach to the description of allosteric molecules is that it will be psychologically inequivalent for thinking about conformational changes and their connection to protein activity. For example, in the context of ion channels, the kinetic view will permit us to handle a number of interesting additional features such as the distributions of times spent in the different conformational states and will give us a more versatile language for discussing the various channel-opening substates to be discussed at the end of the chapter.

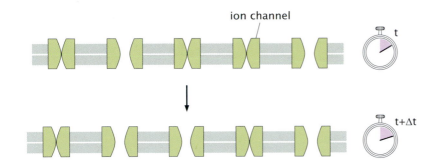

ion channel

t

t+Δt

Figure 3.20

Channel-gating kinetics. A patch of membrane with a collection of ion channels, some of which are open, some of which are closed. At every time step, channels can either switch their state or stay in the same state.

To exploit the rate-equation paradigm for the problem of ion channel dynamics, we begin with the simplest case by appealing to Figure 3.20. The idea is that we have a patch of membrane occupied by a total of N distinct two-state channels. Once we have this simple two-state ion channel in hand, we will then turn to the richer dynamics of the MWC ion channel in which the two states can themselves be subject to different states of ligand occupancy. For now, our goal is to write a rate equation that characterizes the time evolution of the probability of being in the open state, and we do this by computing $p_{open}(t) = N_O(t)/N$, where $N_O(t)$ is the number of open channels at time t, which will effectively determine that probability. We adopt a strategy in which time is discretized into steps of duration Δt, and at each time step the channel can undergo a transition from its current state, or it can remain in the same state. The "reaction" of interest is of the form

$$O \underset{k_+}{\overset{k_-}{\rightleftharpoons}} C, \tag{3.44}$$

where O signifies the open state, C signifies the closed state, and k_+ and k_- are the rate constants that determine the probability of a change of state during a given time step. In particular, if the channel is closed, the probability it will make a transition to the open state in the time step Δt is $k_+ \Delta t$. Similarly, if the channel is open, the probability it will make a transition to the closed state in the time step Δt is $k_- \Delta t$. The change in the number of open channels in a given time step can be written as

$$\Delta N_O = \underbrace{-k_- \Delta t N_O}_{O \rightarrow C} + \underbrace{k_+ \Delta t N_C}_{C \rightarrow O}. \tag{3.45}$$

If we divide both sides by Δt, we find the rate of change in the number of open channels as

$$\frac{\Delta N_O}{\Delta t} = -k_- N_O + k_+ N_C, \tag{3.46}$$

which we can further simplify by dividing this equation by N and using $p_O = N_O/N$ and $p_C = N_C/N$. If we examine the limit as $\Delta t \rightarrow 0$, we obtain the differential equation

$$\frac{dp_O}{dt} = -k_- p_O + k_+ p_C, \tag{3.47}$$

our first example of a chemical master equation, a dynamical description of the probabilities over time. If we now exploit the fact that $p_O + p_C = 1$, which amounts to the statement that the channels are either open or closed (i.e., it is a two-state system) we may rewrite this as

$$\frac{dp_O}{dt} = -k_- p_O + k_+(1 - p_O). \tag{3.48}$$

This equation is more transparent if written in the form

$$\frac{dp_O}{dt} = -(k_- + k_+)p_O + k_+. \tag{3.49}$$

As we have already seen in the previous section on data collapse, we will often strive to find the "natural variables" for a given problem. For the dynamical equation described here, one route to finding the natural variables of the problem is to rewrite our description in dimensionless variables. For example, though the time elapsed can be measured in seconds, these absolute units are less interesting than comparing the elapsed time to the time scales implied by the rate constants k_- and k_+ which have units of 1/time. To see this, we begin by dividing both sides of equation 3.49 by $k_- + k_+$ and defining the dimensionless time $\bar{t} = (k_- + k_+)t$. In light of this definition of a dimensionless time, we can rewrite equation 3.49 as

$$\frac{dp_O}{d\bar{t}} = -p_O + \bar{k}, \tag{3.50}$$

where we have defined a second dimensionless variable $\bar{k}$ as

$$\bar{k} = \frac{k_+}{k_- + k_+}. \tag{3.51}$$

If we now define $\bar{p} = p_O - \bar{k}$ (or $p_O = \bar{p} + \bar{k}$), then our differential equation adopts the simple form

$$\frac{d\bar{p}}{d\bar{t}} = -\bar{p}, \tag{3.52}$$

which can be solved immediately by separation of variables. Specifically, we can write the solution to this equation as $\bar{p}(\bar{t}) = \bar{p}_0 e^{-\bar{t}}$ directly by inspection. Now, we can rewrite the equation in the form that includes all quantities with their original dimensions as

$$p_O(t) = \frac{k_+}{k_- + k_+} + \bar{p}_0 e^{-(k_- + k_+)t}. \tag{3.53}$$

This result is the general solution for the open probability of the ion channels as a function of time. To go further, we have to use the initial conditions to determine the constant $\bar{p}_0$, which will be different for different starting conditions. As an example, if we consider the initial condition that $p_O(0) = 1$ (that is, all the channels are open at $t = 0$), we can determine the constant $\bar{p}_0$, resulting in

$$p_O(t) = \frac{k_+}{k_- + k_+} + \frac{k_-}{k_- + k_-} e^{-(k_- + k_+)t}. \tag{3.54}$$

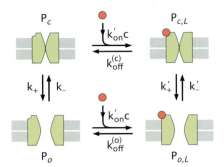

Figure 3.21

Kinetic scheme for the simplest MWC ion channel. For an MWC ion channel with only one ligand binding site, there are four states that the channel can transition between: closed (c), closed with ligand bound (c,L), open (o), and open with ligand bound (o,L). Each transition is characterized by its own rate constant.

Regardless of the initial conditions, the system decays exponentially to its steady-state value. Note that in the long-time limit as $t \to \infty$, the second term in equation 3.54 goes to zero, and hence we are left with

$$p_O(t \to \infty) = \frac{k_+}{k_- + k_+} = \frac{1}{1 + \frac{k_-}{k_+}}. \tag{3.55}$$

Given this preliminary exercise in exploring the probabilities of two-state ion channels, we now turn to the specific case of interest here, namely, the properties of allosteric, ligand-gated ion channels.

Our starting point for generalizing the kinetic scheme just described is to acknowledge the presence of more states than simply the closed and open ones. As shown in Figure 3.21, if we consider the simplest of MWC ion channels with only one binding site for the ligand, we will have four states, corresponding to the channel being either empty or occupied by a single ligand in both the closed and open states. As seen in the figure, the transitions between these states are governed by the rate constants that tell us the probability of such transitions during each time step Δt.

Given these four states, the goal of our kinetic analysis is to find the probabilities of each of them over time, as shown in Figure 3.22. To describe these states, we use the notation introduced in Figure 3.21. In the absence of ligands, we have the closed and open states, represented by $p_c(t)$ and $p_o(t)$, respectively. We will represent the closed and open states bound to ligand using the notation $p_{c,L}(t)$ and $p_{o,L}(t)$, where the additional subscript L refers to the fact that a ligand is bound to the channel. Each one of these four probabilities has its own dynamical equation. In particular, using the kinetic scheme shown in Figure 3.21, we can write the four governing equations for the four states. Note that each state has a total of four arrows connecting it to other states. Each such arrow corresponds to a flux either into or out of the particular state of interest. The rate of transition from states without ligand to those with ligand is of the form $k'_{on}c$, since these rates are proportional to the concentration of ligand c. Using the kinetic diagram as a guide to these fluxes and adopting the shorthand notation that $k_{on} = k'_{on}c$, we can write our dynamical equations as

$$\frac{dp_c(t)}{dt} = -k_{on}p_c + k_{off}^{(c)}p_{c,L} - k_+p_c + k_-p_o \tag{3.56}$$

and

$$\frac{dp_o(t)}{dt} = -k_{on}p_o + k_{off}^{(o)}p_{o,L} + k_+p_c - k_-p_o \tag{3.57}$$

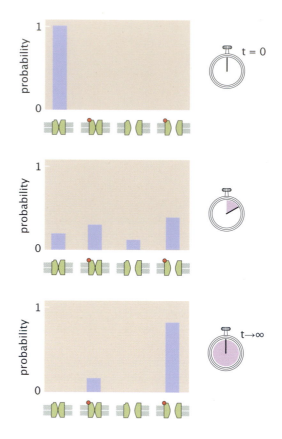

Figure 3.22

Time evolution of the probability distribution. In the initial condition, shown at the top, all channels are in the closed, unbound state. The transient condition (middle) shows the distribution of probabilities some finite time after addition of ligand. The steady-state distribution is shown in the bottom and is obtained in the long-time limit for saturating concentration of ligand.

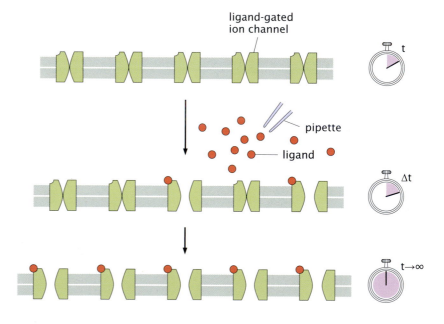

Figure 3.23

Channel-gating kinetics. A patch of membrane with a collection of ion channels, some of which are open, some of which are closed. At every time step, channels can either switch their state or stay in the same state. After addition of ligand, the distribution changes. At fixed ligand concentration at very long times the distribution reaches a steady state.

for the dynamics of the ligand-free closed and open states, respectively. Similarly, the probabilities of the ligand-bound states are given by

$$\frac{dp_{c,L}(t)}{dt} = k_{on}p_c - k_{off}^{(c)}p_{c,L} - k_+'p_{c,L} + k_-'p_{o,L} \tag{3.58}$$

and

$$\frac{dp_{o,L}(t)}{dt} = k_{on}p_o - k_{off}^{(o)}p_{o,L} + k_+'p_{c,L} - k_-'p_{o,L} \tag{3.59}$$

A critical feature of the model is that the off rates for the ligand-bound state may be different in the closed and open states. This is the kinetic analogue of the equilibrium picture introduced earlier, in which we argued that the binding energy is different in the closed and open states. Another key feature of these kinetic equations as written is that we have adopted the notation k_{on} for the on rate for ligand binding. In writing the on rate in this way we have hidden two important features of the problem. First, obviously, this on rate depends upon the concentration of ligand and really can be thought of as $k_{on} = k_{on}'c$, where c is the concentration of ligand. The second hidden feature of the way we have written the kinetics is that we have taken the on rate to be indifferent to whether the channel is in the closed or open state, effectively making the assumption of a diffusion-limited on rate.

To explore the implications of the dynamical equations we have written for the MWC ion channel, we have to navigate to some reasonable part of the dauntingly large parameter space of our seven different rate constants (and note, this is for an oversimplified kinetic scheme with only one binding site, as shown in Figure 3.21, p. 101). To be concrete, each of the arrows in Figure 3.21 has associated with it a particular rate parameter we need to know to solve our dynamical equations. We begin by attacking the leakiness of the MWC molecule, namely, the fraction of time the channel is open even in the absence of ligand. Our earlier states-and-weights approach can now be tailored to a channel with a single ligand binding site, as shown in Figure 3.24. Using statistical mechanics allows us to determine the ratio of the probabilities of the ligand-free open and closed states as

$$\frac{p_o^{eq}}{p_c^{eq}} = \frac{e^{-\beta\varepsilon_o}}{e^{-\beta\varepsilon_c}}, \tag{3.60}$$

where we introduce the notation p_o^{eq} to distinguish these as the equilibrium probabilities. We can write this same statement in the language of the principle of detailed balance, which describes the balance of fluxes between all pairs of states. The key idea is that the probability of being in a given state times the rate of transitioning from that state to some other state is equal to the probability of being in that other state times the transition rate back to the original state. Figure 3.25(A) implements the detailed balance condition by noting that

$$p_o k_- = p_c k_+. \tag{3.61}$$

Together, these two equations then imply a constraint on our rates of the form

$$\frac{k_+}{k_-} = e^{-\beta(\varepsilon_o - \varepsilon_c)}, \tag{3.62}$$

STATE
ENERGY
BOLTZMANN WEIGHT
THERMODYNAMIC WEIGHT
CLOSED
OPEN
ε_{closed}
$e^{-\beta\varepsilon_{closed}}$
$e^{-\beta\varepsilon_{closed}}$
$\varepsilon_{closed} + \varepsilon_b^{(c)} - \mu$
$e^{-\beta\varepsilon_{closed}} e^{-\beta(\varepsilon_b^{(c)} - \mu)}$
$e^{-\beta\varepsilon_{closed}} \frac{c}{K_C}$
ε_{open}
$e^{-\beta\varepsilon_{open}}$
$e^{-\beta\varepsilon_{open}}$
$\varepsilon_{open} + \varepsilon_b^{(o)} - \mu$
$e^{-\beta\varepsilon_{open}} e^{-\beta(\varepsilon_b^{(o)} - \mu)}$
$e^{-\beta\varepsilon_{open}} \frac{c}{K_O}$

Figure 3.24

States and weights for a hypothetical ion channel with a single ligand binding site.

(A)
k_+
k_-
$w_c = e^{-\beta\varepsilon_{closed}}$
$w_o = e^{-\beta\varepsilon_{open}}$
(B)
k'_+
k'_-
$w_{c,L} = e^{-\beta\varepsilon_{closed}} e^{-\beta(\varepsilon_b^{(c)} - \mu)}$
$w_{o,L} = e^{-\beta\varepsilon_{open}} e^{-\beta(\varepsilon_b^{(o)} - \mu)}$
(C)
k_{on}
$k_{off}^{(c)}$
$w_c = e^{-\beta\varepsilon_{closed}}$
$w_{c,L} = e^{-\beta\varepsilon_{closed}} e^{-\beta(\varepsilon_b^{(c)} - \mu)}$
(D)
k_{on}
$k_{off}^{(o)}$
$w_o = e^{-\beta\varepsilon_{open}}$
$w_{o,L} = e^{-\beta\varepsilon_{open}} e^{-\beta(\varepsilon_b^{(o)} - \mu)}$

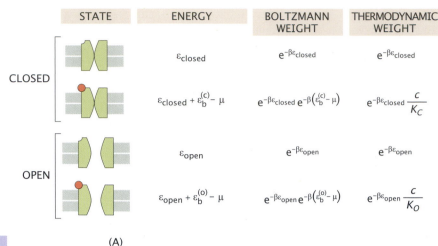

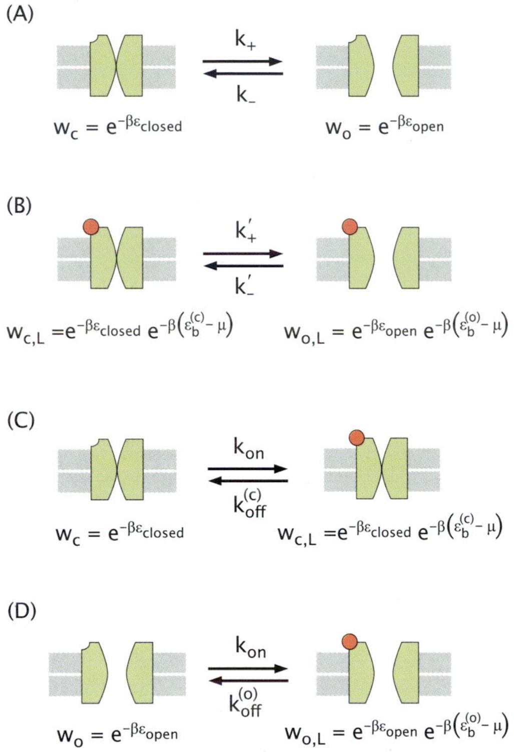

Figure 3.25

Balancing fluxes to determine kinetic parameters. For each of the pair of nodes in the kinetic diagram shown in Figure 3.21, we impose the condition that the statistical weight (w_i for node i) associated with that node times the rate constant for the arrow exiting that node has to be equal to the statistical weight of the partner node times the rate constant associated with the arrow exiting that node.

which reexpresses the leakiness, but now in the language of rate constants. As an order of magnitude guess, we imagine that the lifetime of the open state in the absence of ligand is of the order 10–100 μs. Figure 3.8 (p. 86) shows that sometimes the open states can last even longer. Our guess of 10–100 μs translates into a rate $k_- \approx 1/t_{open} \approx 10^5 - 10^4\ \mathrm{s}^{-1}$. We can thus determine k_+ on the basis of the constraint equation given in equation 3.62, obtaining the condition

$$k_+ = k_- e^{-\beta(\varepsilon_o - \varepsilon_c)} = \frac{1}{100} k_- \approx 10^2\ \mathrm{s}^{-1}, \tag{3.63}$$

where we have taken $\varepsilon_o - \varepsilon_c = -4.6\ k_B T$, resulting in the factor 1/100 from the rule of thumb that $e^{2.3} \approx 10$.

For the on rates for ligand binding corresponding to the right-pointing arrows shown in Figures 3.25(C) and (D), we resort to the simplest argument, which is to assume these rates are provided by the diffusion-limited on rate of $k'_{on} = 4\pi Da \approx 10^9 \text{ M}^{-1}\text{s}^{-1}$. As can be seen from Figure 3.5, the typical concentration for eliciting half maximal response of the ion channels is roughly 10^{-5} M, implying that the actual on rate has a value of the order of

$$k_{on} = k'_{on}c \approx 10^9 \text{ M}^{-1}\text{s}^{-1} \times 10^{-5}\text{M} \approx 10^4 \text{ s}^{-1}. \qquad (3.64)$$

To determine the rate constants $k^{(o)}_{off}$ and $k^{(c)}_{off}$ we appeal once again to Figures 3.25(C) and (D). From our knowledge of channels such as the nicotinic acetylcholine receptor, we take the dissociation constants for the open and closed states of the channel as $K_O \approx 10^{-10}$ M, and $K_C \approx 10^{-4}$ M. We can use these results in conjunction with the detailed balance condition to write

$$k^{(c)}_{off} = k'_{on} \times K_C \approx 10^9 \text{ M}^{-1}\text{s}^{-1} \times 10^{-4} \text{ M} \approx 10^5 \text{ s}^{-1}, \qquad (3.65)$$

and similarly,

$$k^{(o)}_{off} = k'_{on} \times K_O \approx 10^9 \text{ M}^{-1}\text{s}^{-1} \times 10^{-10} \text{ M} \approx 10^{-1} \text{ s}^{-1}. \qquad (3.66)$$

We can use a similar argument to that used to find k_+/k_- in equation 3.62 to establish a ratio of the rates of transition between the closed and open state for the case in which there is a ligand bound, as shown in Figure 3.25(B). In particular, we appeal to the states and weights in the statistical mechanics model that allows us to write this ratio following Figure 3.25(B) as

$$\frac{k'_+}{k'_-} = e^{-\beta(\varepsilon_o - \varepsilon_c)} \frac{e^{-\beta(e^{(o)}_b - \mu)}}{e^{-\beta(e^{(c)}_b - \mu)}}. \qquad (3.67)$$

Using the definition of the chemical potential and the K_d we can rewrite this expression in the simpler form

$$\frac{k'_+}{k'_-} = \frac{k_+}{k_-} \frac{K_C}{K_O}. \qquad (3.68)$$

We can use our knowledge of channels such as the nicotinic acetylcholine receptor, which has two ligand binding sites rather than the single binding site of the simple model considered here, to express the ratio of the K_ds as

$$\frac{K_C}{K_O} \approx \frac{10^{-4}}{10^{-10}} \approx 10^6. \qquad (3.69)$$

Note that now, in light of all these results, we can rewrite our ratios of off rates in the ligand-bound state as

$$\frac{k'_+}{k'_-} = \frac{k_+}{k_-} \frac{k^{(c)}_{off}}{k'_{on}} \frac{k'_{on}}{k^{(o)}_{off}}, \qquad (3.70)$$

Table 3.1 Model rate constants used in examining the dynamics of the single-site MWC ion channel.

Rate constant	s^{-1}
k_{on}	10^4
$k_{off}^{(o)}$	10^{-1}
$k_{off}^{(c)}$	10^5
k_+	10^2
k_-	10^4
k'_+	10^6
k'_-	10^2

which we can rewrite as

$$k'_+ k_- k'_{on} k_{off}^{(o)} = k'_- k_+ k_{off}^{(c)} k_{on}, \tag{3.71}$$

a result sometimes known as the *cycle condition*. The entirety of the rates we have estimated are summarized in Table 3.1. Note, however, that these rates are for a one-ligand ion-channel model and are intended to show the style of thinking rather than as relevant parameters for any particular real-world channel.

As we will see at the end of the chapter, kinetic studies have been central to delving more deeply into the physics of allosteric transitions. In particular, kinetic studies ask more of our understanding by demanding that we respond to lifetime distributions. We can take a number of different approaches to solving the dynamical equations we have written for the probabilities of the different states exhibited in Figure 3.21. Perhaps the most straightforward option is simply to resort to integrating these equations numerically. In this case, the solutions evolve over time, as shown in Figure 3.26. One of the most important features of these solutions is their long-time behavior. Specifically, what these solutions tell us in the long-time limit is the equilibrium solutions that we have already worked out earlier in the chapter. A second approach we can take with our kinetic constants in hand is to use stochastic simulations, such as the Gillespie algorithm, that allow us to explore specific realizations of the dynamical trajectories of our systems. At the end of this chapter we will consider a class of kinetic experiments in which ligands are tethered to the channels, making a more systematic analysis of the allosteric transitions possible. Further, throughout the book, we will exploit the psychological inequivalence of the kinetic perspective as a way to hold up our study of allostery to a different kind of intellectual light.

3.7 Cyclic Nucleotide–Gated Channels

As seen in Figure 3.4 (p. 82), in addition to channels such as the nicotinic acetylcholine receptor already described, the cyclic GMP–gated channels form another class of physiologically very important receptors. As previously seen in

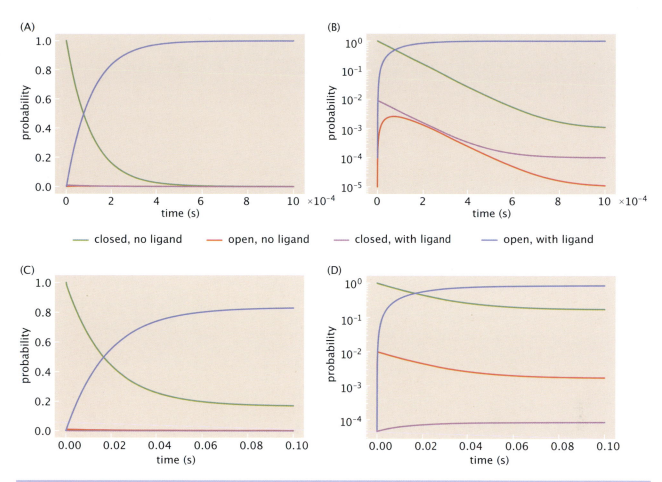

Figure 3.26

Probabilities as a function of time for the simplest MWC ion channel using the parameters in Table 3.1. (A) Probability of different states of the channel over time. The initial condition is $p_c(0) = 1$, and the concentration of ligand is 10^{-5} M. (B) Probabilities of different states as a function of time plotted on a log scale. (C) Probability of different states of the channel over time with the same parameters as in (A) and a concentration of ligand of 5×10^{-8} M. (D) Probabilities of different states shown in (C) as a function of time plotted on a log scale.

the context of photoreceptors in Figure 3.6 (p. 84), these channels are gated in response to the binding of nucleotides. Analogous channels are also important in the context of other senses such as olfaction and gustation, and in hormone response, though the ligands can be cAMP rather than cGMP.

Interestingly, precisely the model described earlier for nicotinic acetylcholine receptor can be tailored to the cGMP channels, but this time with four binding sites per channel instead of two, as shown in Figure 3.27. As seen in the figure, a great deal of molecular degeneracy results because are four distinct states corresponding to single ligand occupancy, six distinct states of double occupancy, and so on.

Using the states and weights depicted in Figure 3.27, we obtain the open probability

$$p_{open} = \frac{(1 + \frac{c}{K_O})^4}{(1 + \frac{c}{K_O})^4 + e^{-\beta \varepsilon}(1 + \frac{c}{K_C})^4}. \tag{3.72}$$

Figure 3.27

States and weights for a cyclic nucleotide–gated ion channel. The left column shows the various states of ligand occupancy for the closed channel, and the right column shows the various states of ligand occupancy for the open channel.

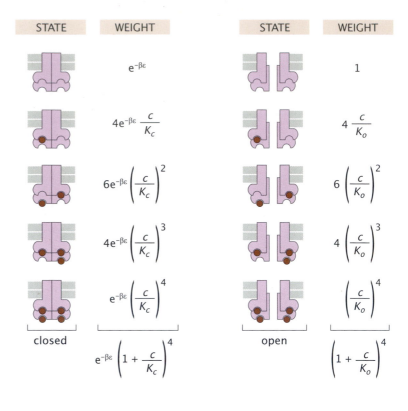

STATE	WEIGHT	STATE	WEIGHT
	$e^{-\beta\varepsilon}$		1
	$4e^{-\beta\varepsilon}\dfrac{c}{K_c}$		$4\dfrac{c}{K_o}$
	$6e^{-\beta\varepsilon}\left(\dfrac{c}{K_c}\right)^2$		$6\left(\dfrac{c}{K_o}\right)^2$
	$4e^{-\beta\varepsilon}\left(\dfrac{c}{K_c}\right)^3$		$4\left(\dfrac{c}{K_o}\right)^3$
	$e^{-\beta\varepsilon}\left(\dfrac{c}{K_c}\right)^4$		$\left(\dfrac{c}{K_o}\right)^4$

closed $e^{-\beta\varepsilon}\left(1+\dfrac{c}{K_c}\right)^4$ open $\left(1+\dfrac{c}{K_o}\right)^4$

Figure 3.28

Open probability of the bovine retinal CNG channel. The probability of the channel being opened is measured as a function of the concentration of cGMP. The fit corresponds to an MWC model using equation 3.72 for the case of four binding sites per channel. No error bars were reported in the original data. Adapted from Goulding, Tibbs, and Siegelbaurn (1994).

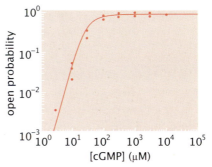

Note that the key difference between this result and that advanced earlier in the context of the nicotinic acetylcholine receptor is the exponent of 4 seen here rather than the 2 seen earlier, reflecting the difference in the number of binding sites. A comparison between this model and experimental measurements of ligand gating for this channel is shown in Figure 3.28.

A more powerful recent case study in the analysis of ligand-gated ion channels is that of the so-called olfactory cyclic nucleotide–gated CNGA2 ion channels. An example of the kind of important data to come from electrophysiology studies of these channels is shown in Figure 3.29. As a result of the ability to synthetically construct different combinations of mutant and wild-type subunits, there is a rich collection of mutant data carefully constructed to permit an investigation of the role of the different subunits in inducing channel gating, as

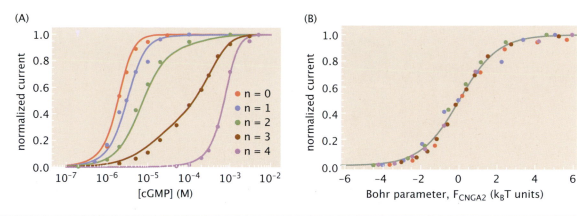

Figure 3.29

Normalized currents of CNGA2 ion channels with a varying number n of mutant subunits. (A) Normalized voltage-clamp currents of the ion channel with n wild-type subunits and $4 - n$ mutated subunits with weaker affinity for cGMP. Because the mutant subunits have lower affinity, the curves shift to the right as n decreases and the number of mutated subunits increases. Overlaid on the data are theoretical best-fit curves assuming an allosteric receptor with four ligand binding sites. These curves have the form given by equation 3.73 with the parameters $K_O = 1.2 \times 10^{-6}$ M, $K_C = 23 \times 10^{-6}$ M, $K_O^* = 510 \times 10^{-6}$ M, $K_I^* = 140 \times 10^{-3}$ M, and $\beta\varepsilon = -3.4$. (B) Data collapse for the different mutants. (A) adapted from Wongsamitkul et al. (2016).

shown in Figure 3.29. This is the first of many examples of the powerful insights garnered by using chimeric proteins.

Figure 3.30 gives a pictorial representation of the different states and weights that emerge as the fraction of wild-type and mutant subunits is varied. The left column recapitulates the states and weights we already determined for the wild-type channel, as shown in Figure 3.27. The second row in the figure tells us how to compute the statistical weights associated with the closed (left column) and open (right column) channels for the case in which there is a single mutant subunit. Each subsequent row shows the statistical weights as the number of mutant subunits is increased, until finally in the bottom row, we have the statistical weights for the fully mutant channel. Given a receptor with n unmutated subunits (with ligand affinities K_C and K_O for the closed and open states, respectively) and $4 - n$ mutated subunits (with ligand affinities K_C^* and K_O^*), the probability that this receptor will be in the active or open state can be found as usual by appealing to our states-and-weights diagrams. To compute the open probability, we then take the statistical weights from the right column of Figure 3.30 for the case of interest and divide by the sum of all the statistical weights.

In light of the states and weights given above, the open probability for a channel with n wild-type subunits and $4 - n$ mutant subunits is given by

$$p_{open}(c) = \frac{\left(1 + \frac{c}{K_O}\right)^n \left(1 + \frac{c}{K_O^*}\right)^{4-n}}{\left(1 + \frac{c}{K_O}\right)^n \left(1 + \frac{c}{K_O^*}\right)^{4-n} + e^{-\beta\varepsilon} \left(1 + \frac{c}{K_C}\right)^n \left(1 + \frac{c}{K_C^*}\right)^{4-n}} \quad (3.73)$$

or in the normalized form,

$$\text{normalized current} = \bar{p}_{open}(c) = \frac{p_{open}(c) - p_{open}(c \to 0)}{p_{open}(c \to \infty) - p_{open}(c \to 0)}. \quad (3.74)$$

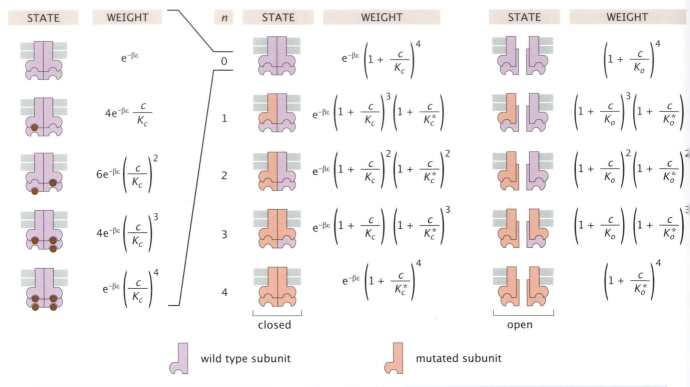

Figure 3.30

States and weights for a heteromeric MWC ion channel. The left column shows the states and weights for the wild-type channel. Each row in the diagram on the right corresponds to a different number (n) of mutant subunits. The statistical weights result from summing over *all* states of ligand occupancy for a given mutant, as shown in the left column for the case in which there are no mutant subunits.

We note that $p_{open}(c)$ carries even more information than the normalized current, since not only does the former allow direct computation of the latter, but $p_{open}(c)$ also describes the leakiness and dynamic range of the ion channel.

The limits of $p_{open}(c)$ allow us to directly compute the leakiness and dynamic range of the ion channels, as shown earlier with the results

$$\text{leakiness} = p_{open}^{min} = \frac{1}{1 + e^{-\beta\varepsilon}} \tag{3.75}$$

and

$$\text{dynamic range} = p_{open}^{max} - p_{open}^{min} = \frac{1}{1 + e^{-\beta\varepsilon}\left(\frac{K_O}{K_C}\right)^n\left(\frac{K_O^*}{K_C^*}\right)^{4-n}} - \frac{1}{1 + e^{-\beta\varepsilon}}. \tag{3.76}$$

Note that our MWC model of these channels provides us with a polarizing physical hypothesis about the leakiness. Equation 3.75 tells us that the leakiness depends only upon the energy difference between the closed and open states, and for the mutants considered here we have hypothesized that the only parameters altered by the mutations are the binding constants K_C and K_O. If this hypothesis is correct, then the leakiness will be indifferent to the mutations, an assertion that seems entirely consistent with the leakiness revealed by the data in Figure 3.29.

The most compelling feature of the curves shown in Figure 3.29 is the shifts in the EC_{50} for different mutants. Here we now seek a simple explanation of those shifts from the point of view of the MWC model. We start with the wild-type channel whose EC_{50} is given by

$$EC_{50}^{(n=4)} = \left(\left(2 + e^{-\beta\varepsilon}\right)^{1/4} - 1 \right) K_O \approx e^{-\beta\varepsilon/4} K_O. \tag{3.77}$$

Thus, the EC_{50} of the wild-type channel depends linearly on K_A but has a much slower dependence on ε owing to the 1/4 power. An ion channel with all subunits mutated $n = 0$ is given by this same expression but with $K_O \to K_O^*$,

$$EC_{50}^{(n=0)} \approx e^{-\beta\varepsilon/4} K_O^*. \tag{3.78}$$

We next consider the EC_{50} of the ion channel with two mutated subunits and two unmutated subunits. In this case, we find

$$EC_{50}^{(n=2)} = \frac{1}{2} \left(-K_O^* - K_O + \sqrt{\left(K_O^* + K_O\right)^2 + 4\left(\left(2 + e^{-\beta\varepsilon}\right)^{1/2} - 1\right) K_O^* K_O} \right)$$

$$\approx \frac{1}{2} \left(-K_O^* - K_O + \sqrt{\left(K_O^* + K_O\right)^2 + 4 e^{-\beta\varepsilon/2} K_O^* K_O} \right)$$

$$\approx \frac{e^{-\beta\varepsilon/2} K_O^* K_O}{2 \left(K_O^* + K_O\right)}, \tag{3.79}$$

where in the last step we expanded the square root assuming that the second term was significantly smaller than the first. For our particular parameters where $K_{A,I} \ll K_O^*$, the denominator can be written as $K_O^* + K_O \approx K_O^*$, so that $EC_{50}^{(n=2)}$ depends linearly on K_O, while its ε dependence is now mitigated only by a 1/2 power.

For the cyclic nucleotide–gated channels considered here, it is impressive that we can work out the biophysical phenotypic parameters of different mutant channels analytically. It is perhaps useful to have all those results in one place. Thus, for the situation with n wild-type subunits and $4 - n$ mutant subunits, we can summarize the phenotypic parameters as follows:

$$\text{leakiness} = \frac{1}{1 + e^{-\beta\epsilon}} \tag{3.80}$$

$$\text{dynamic range} = \frac{1}{1 + e^{-\beta\epsilon} \left(\frac{K_O}{K_C}\right)^n \left(\frac{K_O^*}{K_C^*}\right)^{4-n}} - \frac{1}{1 + e^{-\beta\epsilon}} \tag{3.81}$$

$$[EC_{50}] \approx \begin{cases} e^{-\beta\epsilon/4} K_O & n = 4 \\ e^{-2\beta\epsilon/4} \frac{K_O K_O^*}{K_O + K_O^*} & n = 2 \\ e^{-\beta\epsilon/4} K_O^* & n = 0 \end{cases} \tag{3.82}$$

$$h \approx \begin{cases} 4 & n = 4 \\ 2 & n = 2 \\ 4 & n = 0. \end{cases} \tag{3.83}$$

These equations are one of the very important messages of this book. My reason for making this claim is that these properties of the input-output response of a given molecule are an expression of its molecular phenotype and thus determine how the molecule functions. As a result, these parameters are critical evolutionary values that can be altered to give a system freedom to explore new functions. These equations are also important because they make very precise predictions about how these phenotypic properties scale as various knobs are tuned.

Now that we have these various phenotypic properties in hand, we can reconsider the data and the physical hypotheses they engender. One point of contact between theory and experiment that is a powerful measure of our understanding that goes beyond our treatment of the leakiness and EC_{50} described earlier is to explore data collapse using ideas like those introduced in equation 3.43 (p. 96). For the cyclic nucleotide–gated channels, the Bohr parameter is slightly more complicated and can be found by dividing numerator and denominator of equation 3.73 by

$$\left(1 + \frac{c}{K_O}\right)^n \left(1 + \frac{c}{K_O^*}\right)^{4-n} \tag{3.84}$$

with the result that

$$F_{Bohr} = \varepsilon - k_B T \ln \frac{\left(1 + \frac{c}{K_C}\right)^n \left(1 + \frac{c}{K_C^*}\right)^{4-n}}{\left(1 + \frac{c}{K_O}\right)^n \left(1 + \frac{c}{K_O^*}\right)^{4-n}}. \tag{3.85}$$

The predictive power of this full model is revealed in the data collapse offered by the Bohr parameter, as seen in Figure 3.29(B) (p. 109). This figure shows that for the entire set of mutants considered in this elegant work, if we describe the data in terms of the natural variables provided by the Bohr parameter instead of the pipettor's variables (i.e., the concentration of cGMP), all the data fall on a single master curve, suggesting the single, self-consistent view implied by the model.

3.8 Beyond the MWC Model in Ion Channelology

Often, the emergence of new technologies allows us to ask new questions about the world around us. One vision of the history of science is that it is a long series of episodes punctuated by successive approximations to the truth as new technologies emerge, providing new observations or measurements that challenge the way we have thought about the world before. This chapter is the first of many detailed case studies on allostery that will unfold throughout the book, taking us from ion channels to enzymes to hemoglobin to transcription to nucleosomes. As a result, in this final section of our first case study, it is important to start off on a good footing by being maximally self-critical. Much of this self-criticism is driven by a new generation of beautiful experiments that will be described further that force us to rethink the detailed molecular picture used in this chapter of a strictly concerted conformational change as dictated by the most literal interpretation of models of allostery.

Often in this book, our first contact with experimental data will come in the form of dose-response curves, such as the normalized current versus ligand concentration curves shown, for example, in Figure 3.10 (p. 88). Note that in thinking about the data for open probabilities (or channel currents) that emerge from experiments on ligand-gated channels, such as the nicotinic acetylcholine receptor or the cGMP-gated channels, it is possible to resort to phenomenological fits of the normalized current as a function of the ligand concentration, which take the form

$$I = I_{max} \frac{c^n}{c^n + K_d^n},$$

(3.86)

known conventionally as a Hill function and introduced in section 2.5 (p. 57). When such fits are made to the data for the cGMP-gated channels such as is shown in Figure 3.28, the best fits are found for the case in which $n \approx 3$, despite the fact that there are known to be four binding sites. Some might argue that there are fewer parameters in using the Hill function (two parameters) rather than the MWC function of equation 3.19 (three parameters—two binding energies and an energy difference) and that this gives preference to the Hill function. It is not clear that being able to fit the data in the absence of some insight to be gained from that fit is really a cause for celebration. The view I adopt here is that our theoretical ideas prove themselves not so much by how well they fit data but, rather, in how far we succeed in writing models that tell a story about our understanding of the system. Not only should these models unify our understanding of different phenomena by revealing that they are based on common mechanisms, but they should also suggest new experiments by making predictions about measurements that have not yet been done. A central argument of this book is that statistical mechanical models of allostery do just that.

Many of the themes of the book that first saw action in this chapter will play out again and again. We have already noted that the mathematical structure of the governing equations characterizing the relative probability of open/closed, active/inactive states all have a similar form. Examples of this framework in this chapter were given by equations 3.19 and 3.73 (p. 89 and 109, respectively). This equation shares the feature with the Hill function of being a ratio of polynomials. Recall that the particular ratio of polynomials implied by the classic MWC model is dictated by a specific molecular interpretation of the microscopic states accessible to the MWC molecule. Recent experiments, however, call our molecular assumptions into question, and in the remainder of this section we examine how the same statistical mechanical thinking used already can be generalized to account for the broader collection of states that appear to be required by this new generation of elegant experiments.

3.8.1   Conductance Substates and Conformational Kinetics

Throughout the book we will often use dose-response curves as our first reflection on a given problem, whether the ligand-gated ion channels considered here, the binding of oxygen to hemoglobin, the binding of inducers to transcription factors, or the binding of DNA to histones to form nucleosomes. However, we will see that dose-response curves by themselves are often insufficient to distinguish between competing models or even alternative parameter

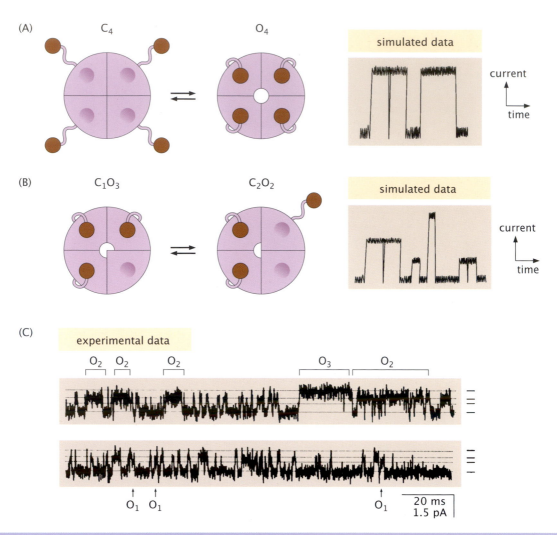

Figure 3.31

Ligand-tethering experiments for cyclic nucleotide–gated channels. (A) Closed (C_4) and open (O_4) states of the channel with four tethered ligands and no free ligands and hypothetical current traces showing what the current would look like as the channel switches between these two states. (B) Closed and open states of channel with three tethered ligands showing the hypothesized intermediate conformational states (C_1O_3 and C_2O_2) and their corresponding conductance substates. (C) Experimental data from a channel with three tethered ligands. (A) and (B) adapted from Miller (1997); (C) adapted from Ruiz and Karpen (1999).

sets within one type of model. Part of the challenge is that these dose-response curves emerge as averages over the different states of occupancy and activity. To that end, a brilliant experiment was conceived to test allosteric models of channel gating by making it so that only one state of occupancy was allowed. Figure 3.31 gives a schematic view of how these clever experiments work. The goal was to impose a specific state of ligand occupancy on CNG channels. Recall that there are four binding sites on these channels. The experiments shown in Figure 3.31 used flexible linkers to tether the ligands to the channels themselves in such a way that only one, two, or three ligands were tethered to the channel.

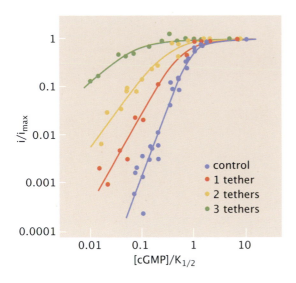

Figure 3.32

cGMP dose-response curves for different tethering constructs. The cGMP concentration is normalized for each curve by $K_{1/2}$, the concentration that results in half-maximal activation. The solid curves correspond to fits of the data using a Hill function. Adapted from Ruiz and Karpen (1997).

The resulting channel currents were then measured, as shown in the figure. The most important insight to come out of those experiments for our purposes is the existence of conductance substates, as shown in Figure 3.31(C). Stated simply, these conductance substates belie the all-or-nothing (i.e., "open" or "closed") mentality of the strictest interpretation of the MWC model.

Instead, these experiments hint at the possibility that there are states of partial opening that lead to current magnitudes lower than in the fully open state. For example, we can imagine, instead, a picture in which the current of a collection of such channels is given by

$$\langle I \rangle = \sum_n p(I_n) I_n, \tag{3.87}$$

where n labels the number of ligands bound, and $p(I_n)$ is the probability of such states. In the MWC model, I_n is the same for all states of occupancy in the open state. If we now consider molecular models in which states of partial opening are allowed, then within our model framework, the I_n are different, because each state of occupancy has its own conductances substate.

It is possible to take these experiments even further. In the absence of soluble ligands, the gating in channels with fewer than four tethered ligands is by definition determined by partial occupancy. But these channels can themselves then be subjected to a ligand titration with results like those shown in Figure 3.32. These results extend our opportunities to test our models. In the next subsection we consider several models in which these states of partial opening are allowed and explore how to think about them from the statistical mechanical perspective.

3.8.2   The Koshland-Némethy-Filmer Model Revealed

The experiments just described invite us to expand our theoretical repertoire. Specifically, we begin with a model that is the extreme opposite of the classic

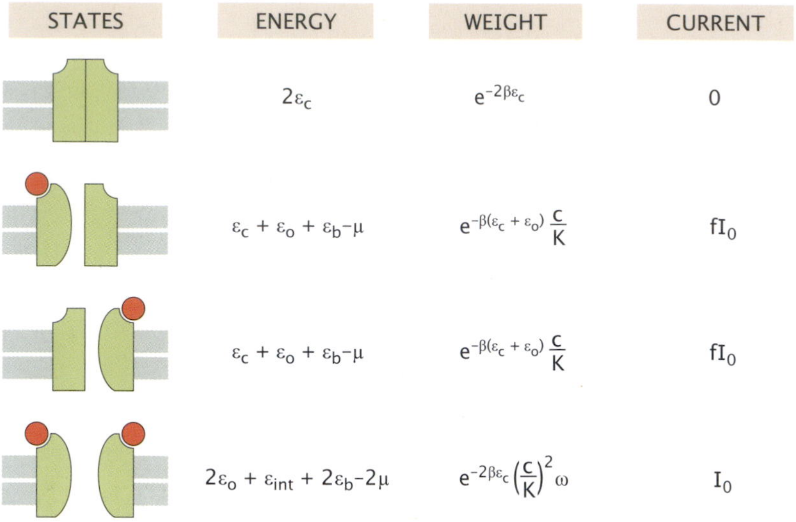

Figure 3.33

States and weights for a KNF model of ion-channel gating. For ease of interpretation, we consider a two-site channel such as the nicotinic acetylcholine receptor rather than the case of the four-binding site CNG channels considered in the experiments of Figure 3.31.

Figure 3.33

States and weights for a KNF model of ion-channel gating. For ease of interpretation, we consider a two-site channel such as the nicotinic acetylcholine receptor rather than the case of the four-binding site CNG channels considered in the experiments of Figure 3.31.

MWC model. The classic MWC model makes a molecular commitment to the idea that all subunits undergo the conformational transition together. Instead, the Koshland-Némethy-Filmer (KNF) model makes a similarly rigid molecular commitment, but this time to the idea that each subunit makes its own distinct transition when bound to its corresponding ligand, as shown in Figure 3.33. Note that rather than considering the full complexity of the four-site cyclic nucleotide–gated ion channels, we will begin instead by considering a toy model which has two binding sites, simply because then we can focus on the physical intuition and avoid the mathematical messiness that comes with the full four-site case.

For the two-site ion channel that is the basis for our explorations beyond the MWC model, we begin with the case with no tethered ligands, and in this case there are four states. Note that we attribute an energy ε_o and ε_c for open and closed subunits, respectively, and we introduce an interaction energy ε_{int} which explicitly allows the possibility of cooperative binding when there is more than one ligand. We can write the partition function for this case by summing over all the states shown in Figure 3.33, with the result

$$Z_{KNF} = e^{-2\beta\varepsilon_c} + 2e^{-\beta(\varepsilon_c+\varepsilon_o)}\frac{c}{K} + e^{-2\beta\varepsilon_o}\left(\frac{c}{K}\right)^2\omega, \tag{3.88}$$

where we have introduced the cooperativity parameter $\omega = e^{-\beta\varepsilon_{int}}$. In view of the states and weights and the resulting partition function, we can write the average current as a function of ligand concentration as

$$\langle I \rangle = fI_0 \frac{2e^{-\beta(\varepsilon_c+\varepsilon_o)}\frac{c}{K}}{Z_{KNF}} + I_0 \frac{e^{-2\beta\varepsilon_o}\left(\frac{c}{K}\right)^2\omega}{Z_{KNF}}, \tag{3.89}$$

now acknowledging two kinds of conductance substates found in the experiment analogous to those in Figure 3.31.

The ideas inherent in the KNF model give us enough flexibility to consider the tethering experiments. The reason for appealing to the KNF model

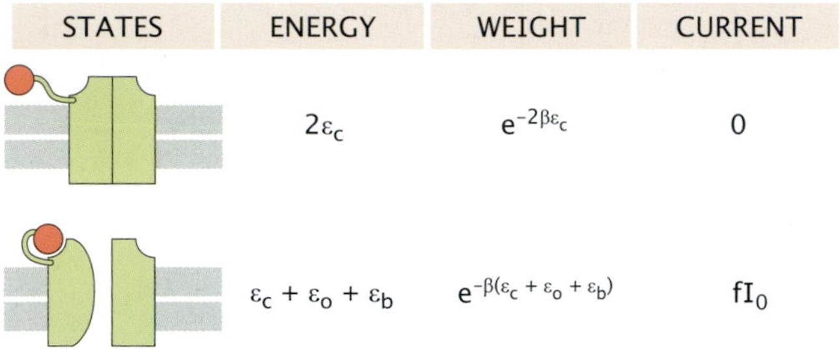

STATES	ENERGY	WEIGHT	CURRENT
	$2\varepsilon_c$	$e^{-2\beta\varepsilon_c}$	0
	$\varepsilon_c + \varepsilon_o + \varepsilon_b$	$e^{-\beta(\varepsilon_c + \varepsilon_o + \varepsilon_b)}$	fI_0

Figure 3.34

KNF model interpretation of the tethering experiments of Figure 3.31 in the absence of soluble ligand. In this case, there are only two states.

comes from the data shown in Figure 3.31, where we see that there are substate conductances smaller than the conductance when the channels are saturated with ligands. These data lead to the physical hypothesis of states of partial opening. For the case in which there is a single tethered ligand and no soluble ligands, the states and weights are given in Figure 3.34. Here we see that there is the closed state with the tethered ligand unbound, and the partially opened state with the tethered ligand bound. The point of showing the states and weights here is simply to give a sense of how we will proceed in the more complex case.

In light of these ideas about the role of the tethered ligand, we can now consider the presence of soluble ligands to complement those ligands that are tethered. To begin, we first consider a simpler picture in which there are no partial channel openings, again with the ambition of illustrating the key concepts and notation. Such models will allow us to consider the data of Figure 3.32, where we see that with increasing ligand concentration, the constructs with different numbers of tethered ligands have distinctive behaviors. For the case in which there is a single tethered ligand, as shown in Figure 3.35, the states and weights acknowledge different binding energies for the tethered ligands than for the free ligands. Thus, we can write the open probability as a function of ligand concentration in the form

$$p_{open} = \frac{e^{-\beta\varepsilon_o}(e^{-\beta\varepsilon_{T,F}^o} + e^{-\beta\varepsilon_{T,B}^o})(1 + \frac{c}{K_O})}{e^{-\beta\varepsilon_o}(e^{-\beta\varepsilon_{T,F}^o} + e^{-\beta\varepsilon_{T,B}^o})(1 + \frac{c}{K_O}) + e^{-\beta\varepsilon_c}(e^{-\beta\varepsilon_{T,F}^c} + e^{-\beta\varepsilon_{T,B}^c})(1 + \frac{c}{K_C})}.$$

(3.90)

From the point of view of Figure 3.32, the key result of this analysis is seen if we rewrite this result in the form

$$p_{open} = \frac{e^{-\beta\varepsilon_o^{eff}}(1 + \frac{c}{K_O})}{e^{-\beta\varepsilon_o^{eff}}(1 + \frac{c}{K_O}) + e^{-\beta\varepsilon_c^{eff}}(1 + \frac{c}{K_C})},$$

(3.91)

where we have introduced the effective energies

$$\varepsilon_o^{eff} = \varepsilon_o + k_B T \ln(e^{-\beta\varepsilon_{T,F}^o} + e^{-\beta\varepsilon_{T,B}^o})$$

(3.92)

and

$$\varepsilon_c^{eff} = \varepsilon_c + k_B T \ln(e^{-\beta\varepsilon_{T,F}^c} + e^{-\beta\varepsilon_{T,B}^c}).$$

(3.93)

Figure 3.35

MWC model interpretation of the tethering experiments of Figure 3.31 in the presence of soluble ligand.

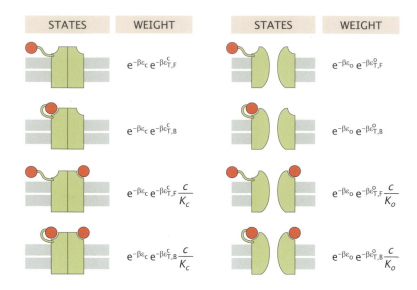

This insight demonstrates that this kind of model behaves in the same way as a ligand-gated ion channel with only a single ligand binding site, since this expression for the open probability has a Hill coefficient of 1. The reader is encouraged to now imitate this analysis for a four-site ion channel by considering in succession a single tethered ligand, then two tethered ligands, three tethered ligands, and four tethered ligands. The outcome will be expressions for the open probability with different effective Hill coefficients, as exhibited by the data in Figure 3.32.

Obviously, there are a number of subtle effects associated with considering the beautiful tethered-ligand experiments of Ruiz and Karpen. The purpose of the present section was not so much to carry out a detailed analysis of their data (or corresponding fits to the data) but, rather, to show the kinds of thinking that can be brought to bear on these problems and how they push us to expand the analysis that comes from the classic MWC model.

Like the MWC model, the KNF model is highly restrictive in the class of molecular states it admits. Let's now explore an even more general model of the full suite of molecular states available to our two-site ion channel.

3.8.3 Kinetic Proliferation

I have argued that one goal of this book is to show that we have statistical mechanical flexibility with respect to different molecular commitments about the classes of underlying states a given allosteric molecule might adopt. I don't want to patriotically fly the flag of one model over another. Since both the MWC and KNF models impose strict molecular conformations, the resulting statistical mechanical models are rather minimalistic. The ideas of those models can be further expanded by considering the Eigen model introduced in section 1.5.1 (p. 28). As shown in Figure 3.36, the Eigen model when applied to ion channels allows us to consider partially opened channels with all different allowed states of occupancy. Implementation of the states-and-weights philosophy will once again yield a curve for p_{open} that is a ratio of polynomials. It

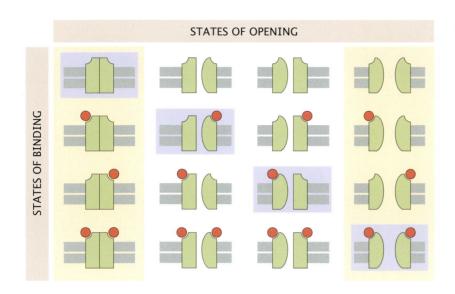

STATES OF OPENING

STATES OF BINDING

Figure 3.36

Eigen model for ligand-gated ion channel. States of a two-site ligand-gated ion channel when states of partial activity and partial occupancy are both allowed. Those states colored in yellow correspond to the MWC model, and those states shaded in blue correspond to the KNF model.

is left as an exercise for the reader to implement the full states and weights of the Eigen model for the four-site ion channel and then to explore how to restrict those states and weights to account only for tethered ligands, as shown in Figure 3.31.

As shown in section 3.6, we can go beyond equilibrium thinking to consider a fully kinetic treatment of the transitions between the different states of some allosteric molecule of interest such as the ion channels considered here. In particular, once we have committed to some set of microscopic states, such as those shown in Figure 3.36, we can ask how the probabilities of these different states change over time. Given an initial state of the ion-channel distribution (e.g., open-closed and occupancy), a model of the conformational dynamics of the ion channels should predict the probabilities of finding the channel in each of the possible states as a function of time, as already demonstrated in section 3.6. There are many dynamical models that can describe such a process, but we will restrict ourselves to so-called noiseless Markovian ("memory-less") dynamical models. Many systems are Markovian in that the state at time $t + \Delta t$ is determined only by the state at time t and not by any previous states for small enough Δt. For example, in the case of the two-site nicotinic acetylcholine receptor shown in Figure 3.7, we can write the chemical master equation in the form

$$p(x_i, t + \Delta t) = \sum_{x_j \in \{O_2, O_1, O_0, C_0, C_1, C_2\}} M(x_i, t + \Delta t | x_j, t) p(x_j, t), \qquad (3.94)$$

where $M(x_i, t + \Delta t | x_j, t)$ is the probability of transitioning to state x_i at time $t + \Delta t$ given that the system starts in state x_j at time t. The labels O_i and C_i refer to open and closed states with i bound ligands.

As our universe of kinetic states increases in passing from MWC to KNF to Eigen (and potentially beyond), it becomes less and less feasible to use the states-and-weights enumerations we have considered thus far. The language of linear algebra gives us the notational opportunity to reduce the complexity of our description in much the same way that Gibbs's version of vector calculus made it possible to write the unwieldy partial differential equations introduced by Maxwell to describe electromagnetic fields in a much more compact and pleasing form. Here we consider the two-site ion channel used throughout this section and introduce the notation

$$\mathbf{p}(t) = \begin{pmatrix} p(O_2,t) \\ p(O_1,t) \\ p(O_0,t) \\ p(C_0,t) \\ p(C_1,t) \\ p(C_2,t) \end{pmatrix} \tag{3.95}$$

to define our state vector $\mathbf{p}(t)$. Here, $p(O_1,t)$ is the probability that the receptor will be in state O_1 at time t, and the other probabilities are defined similarly. Thus, we can write the kinetics of our system in the form

$$\frac{d\mathbf{p}}{dt} = \mathbf{Mp}, \tag{3.96}$$

where we have introduced the transition matrix $\mathbf{M}$ as

$$M_{ij} = M(x_i, t + \Delta t | x_j, t). \tag{3.97}$$

M is usually called a transition matrix because it describes the probability of transitioning between states during the time interval Δt. There are a number of conditions on the matrix $\mathbf{M}$, and the reader is invited to see the suggestions in Further Reading to explore these questions more deeply.

Note that if the system is in equilibrium, then the system's state x_{eq} is given by the Boltzmann weights described earlier, that is,

$$x_{eq} = \frac{1}{Q} \begin{pmatrix} (\frac{c}{c_0 e^{\beta \Delta \varepsilon_o}})^2 \\ 2\frac{c}{c_0 e^{\beta \Delta \varepsilon_o}} \\ 1 \\ e^{-\beta \varepsilon} \\ 2e^{-\beta \varepsilon}\frac{c}{c_0 e^{\beta \Delta \varepsilon_c}} \\ e^{-\beta \varepsilon}(\frac{c}{c_0 e^{\beta \Delta \varepsilon_c}})^2 \end{pmatrix}, \quad Q = \left(1 + \frac{c}{c_0 e^{\beta \Delta \varepsilon_o}}\right)^2 + e^{-\beta \varepsilon}\left(1 + \frac{c}{c_0 e^{\beta \Delta \varepsilon_c}}\right)^2. \tag{3.98}$$

The full weight of this formalism can be brought to bear on the experiments of Figure 3.31.

3.8.4 The Question of Inactivation

In nearly all the chapters that follow, we will find that the specialists in the field, as shown in this section for the case of ion channels, have objections to the use of the simplest allosteric models. In the previous part of this section, we explored one of these objections, namely, that the conformational changes considered in the MWC and KNF frameworks are kinetically restrictive. A second key critique we will also return to repeatedly throughout the book is that the idea of only two classes of states (inactive and active) is too restrictive. We finish this chapter by pointing out one of the most important classes of such states in the context of ion channels. Here, not only are there the closed and open states, but kinetic studies have revealed the presence of another kind of closed state that is strictly inactive. This means that the channel is not only closed but also unresponsive to the binding of ligands.

These are well-founded objections and provide fertile ground for refinements in our thinking and in the design of precision measurements that will place higher demands on our models. Indeed, we will seek out these critiques at every opportunity. That said, in the arena of words-based biology, dismissal of these models with no better quantitative description strikes me as simply defeatist and unhelpful. The fact that mathematical descriptions force us to sharpen our thinking and thereby make dangerous predictions helps us understand the limits and validity of our current understanding. As shown in this section, the beautiful experiments of Ruiz and Karpen, which enforced a particular state of ligand occupancy, revealed conductance substates that make us stand up and take notice of the inadequacy of the minimal description, and by any measure I can think of, this is progress.

3.9   Summary

We used ion channels as our first concrete example to delve into a rigorous and thorough examination of the practice of allosteric thinking in general and the two-state formalism in particular. One of the reasons for this choice is the conceptual simplicity of the readout of the channel state in the form of the current passing through it, as measured using electrophysiology. After considering the physiological role of ion channels, we turned to our first case study, the nicotinic acetylcholine receptor. This gave us a chance to write the full states and weights, to compute the channel-open probability, and to use our model to compute the key phenotypic properties: leakiness, dynamic range, EC_{50}, and effective Hill coefficient. We also took the opportunity to introduce the method of data collapse, which invites us to seek the "natural variable" of a given problem, in this case the free-energy difference between the closed and open states, which we christened the Bohr parameter in honor of Christian Bohr and the effect that bears his name in the context of hemoglobin. With our equilibrium tools in hand, we then examined channel gating from the standpoint of chemical kinetics, allowing us to reconcile the equilibrium and kinetic viewpoints. We ended with an analysis of recent beautiful experiments on cyclic nucleotide–gated channels. All the ideas in this specific case study will serve as the intellectual basis for much of the work of coming chapters.

3.10   Further Reading

Auerbach, A. (2012) "Thinking in cycles: MWC is a good model for acetylcholine receptor-channels." *J. Physiol.* 590 (1):93–98. This short paper discusses many of the key issues necessary to the contemplation of how the MWC model can be used to think about ion channels.

Biskup, C., J. Kusch, E. Schulz, V. Nache, F. Schwede, F. Lehmann, V. Hagen, and K. Benndorf (2007) "Relating ligand binding to activation gating in CNGA2 channels." *Nature* 446:440–443.

Karlin, A. (1967) "On the application of 'a plausible model' of allosteric proteins to the receptor for acetylcholine." *J. Theor. Biol.* 16:306–320. This is one of the first papers to really unleash the full weight of the MWC model on the problem of ion channels.

Karpen, J. W., and M. L. Ruiz (2002) "Ion channels: Does each subunit do something on its own?" *Trends Biochem. Sci.* 27:402. This excellent article describes the shortcomings of the MWC and KNF models for interpreting their own clever experiments which impose different discrete states of ligand binding and shows the need for the more general Eigen model.

Marzen, S., H. G. Garcia, and R. Phillips (2013) "Statistical mechanics of Monod-Wyman-Changeux (MWC) models." *J. Mol. Biol.* 425(9): 1433–1460. This paper describes the matrix approach to MWC dynamics described in the final section of the chapter.

Zagotta, W. N., T. Hoshi, and R. W. Aldrich (1994) *Shaker* potassium channel gating. III: Evaluation of kinetic models for activation." *J. Gen. Physiol.* 103:321–362. This paper makes a systematic analysis of the space of microscopic states required by the kinetic behavior of channels.

3.11 REFERENCES

Einav, T., and R. Phillips (2017) "Monod-Wyman-Changeux analysis of ligand-gated ion channel mutants." *J. Phys. Chem. B* 121:3813–3824.

Fawcett, D. W., and R. P. Jensch (1997) *Bloom and Fawcett: Concise Histology.* London: Chapman & Hall.

Feynman, R. P. (1965) "The Development of the Space-Time View of Quantum Electrodynamics." Nobel Lecture, December 11, Stockholm.

Goulding, E. H., G. R. Tibbs, and S. A. Siegelbaum (1994) "Molecular mechanism of cyclic-nucleotide-gated channel activation." *Nature* 372:369–374.

Guggenheim, E. A. (1945) "The principle of corresponding states." *J. Chem. Phys.* 13:253.

Jackson, M. B. (1986) "Kinetics of unliganded acetylcholine receptor channel gating." *Biophys. J.* 49(3):663–672.

Labarca, C., M. W. Nowak, H. Zhang, L. Tang, P. Deshpande, and H. A. Lester (1995) "Channel gating governed symmetrically by conserved leucine residues in the M2 domain of nicotinic receptors." *Nature* 376:514–516.

Miller, C. (1997) "Cuddling up to channel activation." *Nature* 389:328–329.

Ruiz, M. L., and J. W. Karpen (1997) "Single cyclic nucleotide–gated channels locked in different ligand-bound states." *Nature* 389:389–392.

Ruiz, M. L., and J. W. Karpen (1999) "Opening mechanism of a cyclic nucleotide–gated channel based on analysis of single channels locked in each liganded state." *J. Gen. Physiol.* 113:873–895.

Wongsamitkul, N., V. Nache, T. Eick, S. Hummert, E. Schulz, R. Schmauder, J. Schirmeyer, T. Zimmer, and K. Benndorf (2016) "Quantifying the cooperative subunit action in a multimeric membrane receptor." *Sci. Rep.* 6:20974.

4

HOW BACTERIA NAVIGATE THE WORLD AROUND THEM

> We may find illustrations of the highest doctrines of science in games and
> gymnastics, in travelling by land and by water, in storms of the air and of
> the sea, and wherever there is matter in motion.
>
> —James Clerk Maxwell

4.1 Bacterial Information Processing

As best we can tell, bacteria have been dominant players on our planet for nearly
the entirety of its more than 4 billion years of existence. Current estimates hold
that there are roughly 10^{30} bacteria on our planet, constituting the second most
abundant source of biomass, with plants assuming the dominant place. Part of
the reason for their great success is that bacteria are sophisticated information-
processing machines with a host of different molecular circuits designed to take
stock of the environment around them and to use that information to make
good choices. In this chapter, we explore several case studies that illustrate the
intimate connection between cellular decision making and the allostery concept
that forms the centerpiece of this book.

4.1.1 Engelmann's Experiment and Bacterial Aerotaxis

Cells have preferences of all types. A classic early study that quantified such
preferences was performed by Theodor Wilhelm Engelmann. The logic of his
experiment is shown in Figure 4.1. Algal cells were illuminated with different
wavelengths of light, resulting in photosynthesis occurring with different efficiency at different points along the algal cell. Engelmann used bacterial cells to
"read out" the efficiency of photosynthesis, since they tended to move to regions
of higher oxygen concentration, as shown in Figure 4.2.

The interpretation of this experiment in the context of bacterial decision
making is that the bacterial cells are engaged in a form of chemotaxis known
as *aerotaxis* (flagellar-based swimming in response to oxygen). Specifically,
in Engelmann's experiment, cells piled up preferentially where there was the

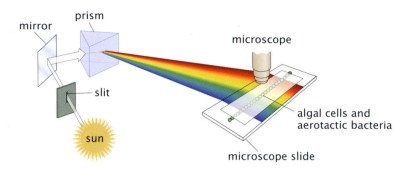

Figure 4.1

Schematic of apparatus used in classic experiments of Engelmann. This experiment revealed wavelength dependence of photosynthesis in *Spirogyra*, with more photosynthesis occurring near red and blue wavelengths. Bacteria "read out" the efficiency of photosynthesis in the algae by moving to regions of higher oxygen concentration in the process of aerotaxis.

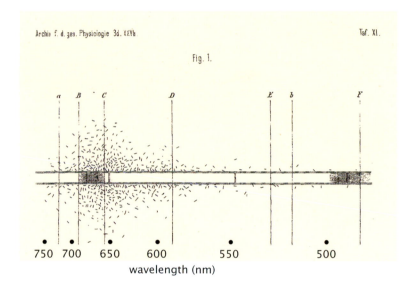

Figure 4.2

Classic experiments of Engelmann showed that the production of oxygen due to photosynthesis depended upon the wavelength of the light incident upon the algae. This effect was revealed by the preferences of aerotactic bacteria for some regions over others. Large bacteria (3–4 μm) from the Rhine River swam at speeds of 20–40 μm/s and used aerotaxis to move up the concentration gradient to regions of higher oxygen concentration. Maximum oxygen production was between the Fraunhofer lines shown as B and C. Adapted from Engelmann (1882).

highest concentration of oxygen, which was in turn dictated by regions that were the most photosynthetically active. The oxygen production itself was controlled by illuminating the thylakoid membranes of unicellular algae with different wavelengths of light using a specially constructed device constructed by Carl Zeiss, famous to this day as an innovative manufacturer of optical devices for biologists (and many others). These experiments left in their wake a great challenge. How can we understand the way in which cells detect, interpret, and then act on such environmental signals? This chapter attempts to answer that question. Further, cells don't just make individual choices such as where to go; they also engage in collective behaviors that are also dependent upon sophisticated signaling pathways.

4.1.2 Love Thy Neighbors: Signaling between Bacteria

A deeply surprising discovery that helped redefine our view of the microbial world was that bacterial cells communicate with one another. This insight emerged from the realization that the bioluminescence of certain bacteria depends upon their density. Though this discovery was initially met with great

Figure 4.3

The autoinduction phenomenon. Cell density and bioluminescence in *Photobacterium fischeri* were measured as a function of time. Adapted from Nealson, Platt, and Hastings (1970).

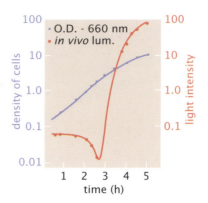

skepticism, a subsequent decade of intensive research in the 1960s led to seminal work on the mechanism of density-dependent microbial control. Stated simply, cells express their genes differently depending on their density. Some of the signature achievements of that era were the development of a quantitative bioassay for the density-dependent activity, the purification of the autoinducer molecules secreted by bacterial cells that signal their neighbors, beautiful chemostat experiments that made it possible to induce the phenomenon even when cells were at low densities and would otherwise not exhibit bioluminescence, and the cloning and expression of the genes responsible for regulating the phenomenon.

To be more concrete about the autoinduction phenomenon itself, the idea is that depending upon whether there are few or many bacteria present, the cells exhibit different patterns of gene expression, with palpable changes to their behavior, such as the emission of light when cells are crowded together. Figure 4.3 shows data from one of the early experiments that established the phenomenon, though the definitive causal link was established in chemostat experiments to be discussed later in the chapter (Figure 4.36). In particular, we see that over the course of several hours, the number of cells increased by more than 100-fold, resulting in a concomitant increase in the bioluminescence of the cells. Interestingly, this cell–cell communication was later discovered to take place not only within a given species but also between different species.

The original discovery of the autoinduction phenomenon (i.e., an enzyme apparently switching from an inactive to an active state in the absence of an externally prescribed inducer) was made in the context of Gram-negative luminous bacteria that are symbionts of squids and fish. Figure 4.4 provides a schematic of current thinking about the physiological implications of quorum sensing in the context of those symbionts. The argument goes that organisms such as the Hawaiian bobtail squid shown in the figure use the light generated by their bacterial partners as a kind of sophisticated way of preventing them from casting dangerous underwater shadows to be exploited by swimming predators. Though provocative, these hypotheses raise questions of their own, such as the fact that the starry background does not move, though presumably the bobtail squid do. Later in the chapter, we will explore the mechanisms that control the light-giving properties of these bacterial symbionts and describe how these mechanisms can be viewed through the prism of allostery.

In this chapter, we explore the two classic case studies in bacterial decision making introduced in this section, namely, bacterial chemotaxis and quorum

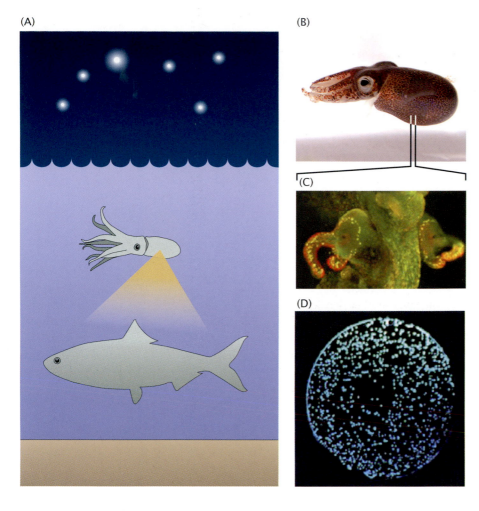

(A)

(B)

(C)

(D)

Figure 4.4

The physiological significance of quorum sensing. (A) The hypothesized mechanism of luminous camouflage used by squids to avoid casting a shadow at night. (B) The bobtail squid *Euprymna scolopes*. (C) Close-up view of the light organ of the squid harboring the bioluminescent bacteria. (D) Colonies of luminescent *Vibrio fischeri* bacteria on a petri dish. When at high densities, these bacteria give off light! (B) and (C) courtesy of Margaret McFall-Ngai. (D) courtesy of J. W. Hastings, Harvard University, through E. G. Ruby, University of Hawaii.

sensing. In both cases, we will find that the quantitative description of these phenomena takes us right back to the important property of molecular switching. Here, switching will be seen to take place at many molecular nodes in the signaling pathways that mediate these processes, but we begin with an analysis of how membrane-bound receptors switch their activity in response to the absence and presence of chemoattractants. Once we have completed our analysis of this single-cell phenomenon, we then turn to the case of collective bacterial decision making in the form of the quorum-sensing problem just described.

4.2 Bacterial Chemotaxis

4.2.1 The Chemotaxis Phenomenon

As noted in the previous section, one of the most well studied of signaling phenomena is bacterial chemotaxis, the directed motion of bacteria up a chemoattractant gradient. Stated simply, bacteria such as *E. coli* are propelled forward by the spinning of their flagella. But this forward motion known as "runs" is punctuated intermittently by "tumbles," in which the flagellar

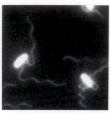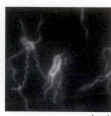

2 μm

Figure 4.5

Fluorescence microscopy images of bacteria engaged in tumbles. Individual flagella are fluorescently labeled. When undergoing a run, the flagella form a single bundle, leading to directed motility, while during tumbles as shown here, the flagella splay apart, leading to erratic cell motion. From Phillips, R., J. Kondev, J. Theriot, and H. Garcia (2013), *Physical Biology of the Cell*, 2nd ed. Reproduced by permission of Taylor & Francis LLC, a division of Informa plc. Adapted from Turner, L., W. S. Ryu, and H. C. Berg (2000), *J. Bacteriol.* 182:2793, fig. 2. Amended with permission from American Society for Microbiology.

motors spin in the opposite direction and the bacterium reorients as shown in Figure 4.5.

The modern biophysical understanding of the bacterial chemotaxis phenomenon owes much to early experiments like those shown in Figure 4.6. This elegant series of experiments on bacterial chemotaxis revealed the preferences of bacteria for chemoattractant by providing these chemoattractants in a micropipette and then watching the enrichment (or exclusion) of bacteria from the vicinity of the pipette, as shown in Figure 4.6. Such experiments hinted at the idea that the bacteria express preferences for particular kinds of molecules, which result in behaviors such as directed motility.

A more precise quantitative analysis of this phenomenon was made using a capillary assay with an experimental setup like that shown in Figure 4.7(A). Capillary tubes harboring different concentrations of chemoattractant were incubated with the bacterial suspension and then removed after a certain amount of time, and the total number of cells within the capillary was measured by plating those cells and counting colonies. This apparatus and protocol allowed for measuring the average number of bacteria that moved into the capillary and the dependence of this number upon the magnitude of the chemoattractant concentration, as shown in Figure 4.7(B). As seen in the figure, the number of bacteria found in the capillary increased with increasing concentration of chemoattractant. As the concentration of chemoattractant increased, the bacteria increasingly accumulated in the capillary. Such experiments raise mechanistic questions about key points such as how cells detect chemoattractants and with what precision, and given such detection, how do they implement their decisions about whether to continue their run or to tumble?

Figure 4.6

Bacterial response to chemoattractants. Chemoattractants are provided, and bacteria clearly accumulate in the vicinity of the pipette tip (dense black cloud). Republished with permission of Annual Reviews, Inc., from Adler, J. (2011), "My life with nature," *Ann. Rev. Biochem.* 80:4270; permission conveyed through Copyright Clearance Center.

(A)

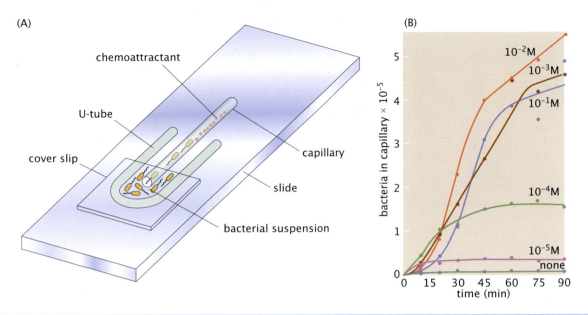

(B)

Figure 4.7

Accumulation of bacteria in chemotaxis capillary assay. (A) Schematic of the apparatus showing how bacteria move up the capillary in response to the presence of chemoattractants. (B) Number of bacteria in the capillary as a function of time. Each curve corresponds to a different concentration of chemoattractant. Adapted from Adler (1973).

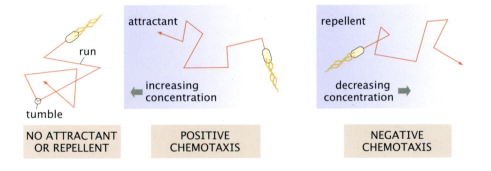

Figure 4.8

Prokaryotic chemotaxis. The frequency of tumbling depends upon whether chemoattractants or chemorepellents are present. As seen in the middle panel, bacteria move up a concentration gradient of chemoattractant. Similarly, they will move down the gradient if faced with a chemorepellent.

4.2.2 Wiring Up Chemotaxis through Molecular Switching

Given the facts described, we are left to wonder, how does a bacterium work its way up a concentration gradient? If instead of focusing on the bulk behavior exhibited in Figures 4.6 and 4.7 we instead watch the path taken by individual cells, these different experimental conditions reveal dynamical trajectories like those shown in Figure 4.8. In the absence of chemoattractant or repellent, the bacterium undergoes an unbiased random walk. By way of contrast, in the presence of chemoattractant, the bacterium still undergoes a series of runs and tumbles, but now we observe that the frequency of tumbling is reduced when the cell's motion takes it up the chemoattractant gradient. Early experiments done

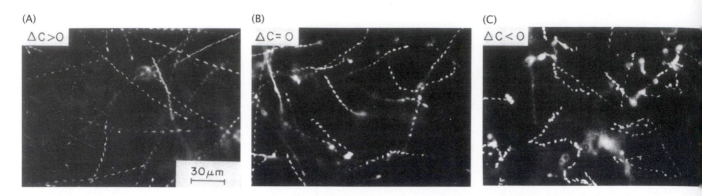

Figure 4.9

Run-and-tumble behavior and chemoattractant concentration. A stroboscopic lamp was used to measure the motility tracks of *S. typhimurium* with a frequency of five pulses per second. They measured bacterial velocities of roughly 30 μm/s. (A) Motion after a sudden increase in serine concentration ($\Delta c > 0$). (B) Control experiment in the absence of serine. (C) Motion after a sudden reduction in the concentration of serine ($\Delta c < 0$). Adapted with permission from Macnab, R. M. and D. E. Koshland, *Jr., Proc. Nat. Acad. Sci.,* 69:2509–2512.

with primitive but elegant time-lapse microscopy revealed these behaviors, as shown in Figure 4.9. The idea of the experiment shown in Figure 4.9(A) is that the medium is quickly changed from low to high concentrations of serine. In this case, we see that the trajectories are long and smooth, corresponding to longer runs without a tumble, since the bacteria interpret the sudden shift in concentration as resulting from swimming up the gradient. Figure 4.9(B) is a control experiment in which there is no change in chemoattractant concentration and thus shows the tumble frequency of cells in a uniform concentration field. Figure 4.9(C) shows the result when the chemoattractant concentration is instantaneously reduced, which bacteria interpret as resulting from moving down the chemoattractant gradient, thus creating a much higher frequency of tumbling, as seen by the erratic trajectories.

How is this behavior implemented at the molecular level? The answer is a signal transduction pathway that alters the frequency of motor switching depending upon the concentration of chemoattractant in the surrounding medium. As seen in Figure 4.10, the run motions are characterized by the flagella working in harmony by rotating in a counterclockwise direction. Because of fluid dynamic coupling, during counterclockwise rotation, the flagella bundle together. When the motors spin in the opposite (clockwise) direction, the flagella splay apart and no longer work together, thus resulting in reorientation of the cells so that they can swim off in a new direction. By controlling the frequency of motor switching, the cells can tune the length of their runs, increasing the run lengths when conditions are improving.

We now aim to see how the phenomenon of chemotaxis makes contact with the MWC concept that is our central preoccupation. To begin to answer that question, we first need to explore the molecular circuitry that makes the chemotaxis behavior work in bacteria. Figure 4.11 shows a conceptual description of the underlying molecular circuitry, dividing it into three modules: the sensor module, the transduction module, and the actuator module.

The sensor module. The chemotaxis process begins in the sensor module with the chemoreceptors that bind the chemoattractants. These receptors act

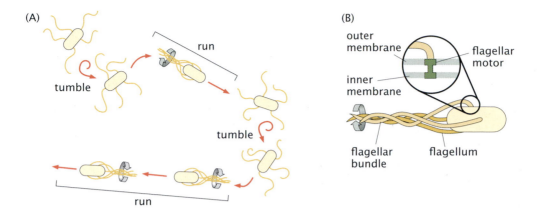

Figure 4.10

Prokaryotic chemotaxis mechanism. (A) Run-and-tumble motion. During counterclockwise rotation, the flagella bundle together. Rotation in the clockwise direction leads to splaying of the flagella and reorientation of the bacterium. (B) The flagellar motor protrudes from the cell membrane and attaches to the flagella. Depending upon the direction of motor rotation, the flagella either bundle (counterclockwise) or splay apart (clockwise).

as a switch, since in the absence of chemoattractant, they are in their active state, engaged in the phosphorylation reactions that lead to motor reorientation. When chemoattractant binds to the receptor, the probability of being in the inactive state is increased, as we will show using statistical mechanics later in the chapter.

As illustrated in Figure 4.11, there are multiple types of chemoreceptors on the cell surface, each detecting different chemoattractants. For example, the Tar receptor (also known as aspartate receptor) allows bacteria to move toward both aspartate and maltose while mediating repulsion away from metals such as nickel and cobalt. An example of such a receptor is shown in Figure 4.12. These receptors consist of a periplasmic ligand-binding domain, a transmembrane domain shown in the figure as the region threading through the inner membrane (gray-shaded region), and a cytoplasmic domain. Besides the Tar receptor, in *E. coli* there is also the Trg receptor, which is sensitive to ribose and galactose, the Tsr receptor sensitive to serine and leading to chemotaxis away from leucine, the Tap receptor for peptides, and the Aer receptor, related to the kind of aerotaxis studied by Engelmann and presented in Figure 4.2 (p. 125).

The transduction module. Once the sensor module has detected the presence or absence of chemoattractant, the next layer in the chemotaxis molecular circuitry that responds has been dubbed the transduction module. As seen in Figure 4.11, the chemoreceptors bind to a sensor kinase CheA that can phosphorylate the response regulator known as CheY. When phosphorylated, the CheY-P molecule interacts with the motor in the actuator module, causing the motor to switch its direction of rotation. Specifically, as we see in Figure 4.12, the cytoplasmic part of the chemoreceptors have a CheA binding region that allows them to bind the sensor kinase (CheW/CheA) that mediates the addition of a phosphate group to the response regulator CheY. Note also the presence on the receptors (see Figure 4.12) of an adaptation region, where the chemoreceptor itself can be "edited" by posttranslational methylation reactions which allow the

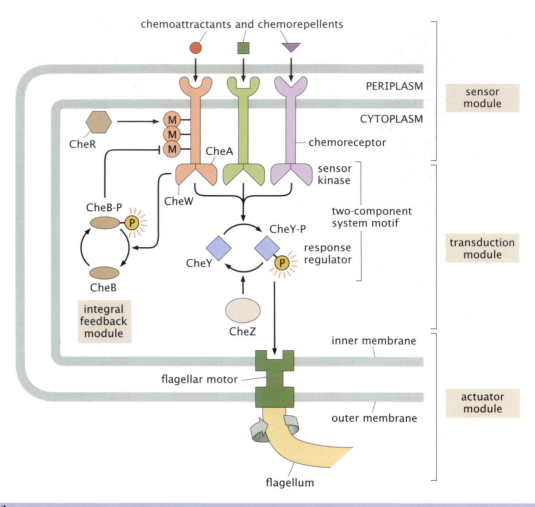

Figure 4.11

The molecular circuit that drives chemotaxis. The circuit can be conceptually divided into three modules: the sensor module, the transduction module, and the actuator module (top to bottom). In the sensor module, chemoreceptors at the cell membrane detect chemoattractants. The binding of chemoattractants sets off a signaling cascade in the transduction module that culminates in the phosphorylation of the messenger molecule CheY. In the actuator module, the interaction of CheY-P with the flagellar motor alters the frequency of the change in rotation direction of the flagellar motor. Adapted from McAdams, Srinivasan, and Arkin (2004).

chemotaxis detection system to reset its concentration setpoint in the process of adaptation.

One of the most important aspects of the chemotaxis circuitry is that nearly all the molecular players are subject to posttranslational modifications, as shown in Figure 4.13. The chemoreceptors themselves, as part of the critical process of adaptation, have multiple methylation sites that, as we will see, when modified by the addition or removal of a methyl group can be thought of as modifying one of the key parameters ($\Delta\epsilon$) in the MWC description of these molecules, thus changing the relative equilibrium of the inactive and active states of the receptor. These methyl groups are added to the chemoreceptors by the enzyme CheR. The soluble response regulator CheY has different levels

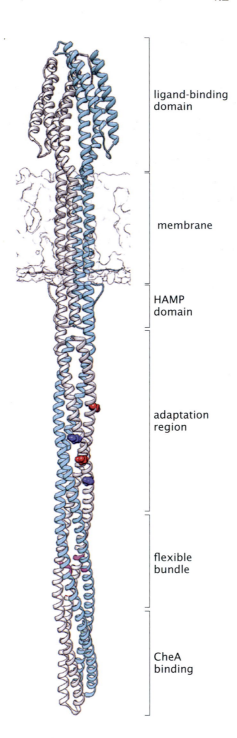

ligand-binding
domain

membrane

HAMP
domain

adaptation
region

flexible
bundle

CheA
binding

Figure 4.12

Chemotaxis receptor structure. The chemoreceptor has a periplasmic ligand-binding domain, a transmembrane domain shown threading the shaded membrane region, and a cytoplasmic region. Adapted from Yang et al. (2019).

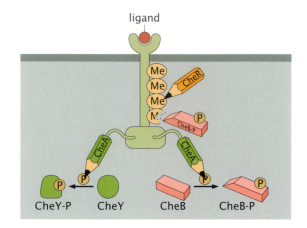

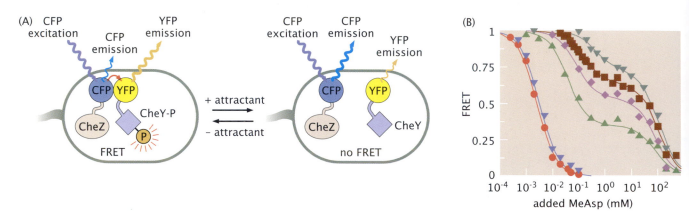

Figure 4.14

Activity of chemotaxis receptors. (A) Schematic of the FRET experiment used to measure chemoreceptor activity. When CheZ actively
dephosphorylates CheY-P, the proximity of these two molecules leads to FRET. (B) Measured activity curves as a function of chemoattractant concentration for different chemoreceptor mutants designed to mimic different states of adaptation of the chemoreceptors.
Adapted from Sourjik and Berg (2002); see Further Reading.

of activity (similar to the chemoreceptors themselves) in the chemotaxis circuit controlled by phosphorylation, with the phosphorylated state (CheY-P)
active to interact with the motor and switch its direction of rotation. Also,
the protein CheB responsible for demethylating the chemoreceptors in the
process of adaptation is itself subject to phosphorylation. Clearly, posttranslational modifications are the basis for a lot of feedback and control within the
chemotaxis signal transduction module, tuning the relative activities of several
different molecular players in the chemotaxis circuitry

Beautiful experiments using the technique of fluorescence resonance energy
transfer (FRET) make it possible to peer into the mechanics of signal transduction during chemotaxis. As shown in Figure 4.14(A), the idea of these FRET
experiments is to look for spatial proximity of the enzyme CheZ, which dephosphorylates CheY-P, to its substrate CheY-P. In the absence of chemoattractant,
CheY will constantly be phosphorylated, which in turn will be constantly
dephosphorylated through the action of CheZ. However, in the presence of

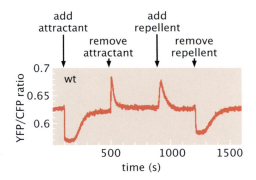

Figure 4.15

Adaptation in bacterial chemotaxis. When chemoattractant is added ("+attr"), the FRET ratio changes temporarily. However, after an adaptation time scale, the FRET ratio returns to the same value, a process known as adaptation. Adapted from Sourjik and Berg (2002).

chemoattractant, CheY is phosphorylated less, meaning that the rate of CheZ dephosphorylation of CheY-P is reduced. Hence, the proximity of CheZ and CheY-P is used as a stand-in for chemoreceptor activity itself. As seen in Figure 4.14(B), the activity of the chemoreceptors depends upon chemoattractant concentration, such that when those concentrations are low, the activity of the receptor is high. By way of contrast, at high chemoattractant concentrations, the activity of the chemoreceptors is reduced. The family of different activity curves will serve as a key conceptual challenge to our modeling efforts, since they are a series of different mutants that shed light on the process of chemotactic adaptation, as shown in Figure 4.15. The reader is invited in Further Reading to see how single-cell FRET experiments have refined our picture of the chemotaxis transduction module.

The actuator module. The final step in the signal transduction process is in the actuator module, shown in Figure 4.11. In this module, CheY-P leads to tumbling behavior when it interacts with the protein FliM on the flagellar motor. As seen in Figure 4.10, the interaction of CheY-P with FliM leads to a change from counterclockwise to clockwise rotation of the flagellar motor, which disrupts the bundling of the flagella and results in reorientation of the cells through tumbling.

4.3   MWC Models of Chemotactic Response

To confront the full complexity of the experimental data on chemotaxis receptor activity shown in Figure 4.14, we will proceed in increasing levels of model sophistication, as shown in Figure 4.16. We begin by thinking about a single chemotaxis receptor without adaptation, as shown in Figure 4.17. True to the allosteric philosophy already espoused throughout the book, we assume that the receptors can exist in two states of activity: inactive and active. Each of those two states can themselves exist in two states of ligand occupancy. The fraction of time the receptor spends in the inactive and active states is altered by the binding of chemoattractant molecules to the receptor. As seen in Figure 4.16(B), at the next level in the modeling hierarchy, we generalize the first model to include the fact that chemotaxis receptors cluster together. With those results in hand, we then generalize even further to the case shown in Figure 4.16(C), where we see that the chemotaxis clusters are heterogeneous, with more than one type of receptor per cluster. At the final level in the hierarchy (see Figure 4.16(D)), we

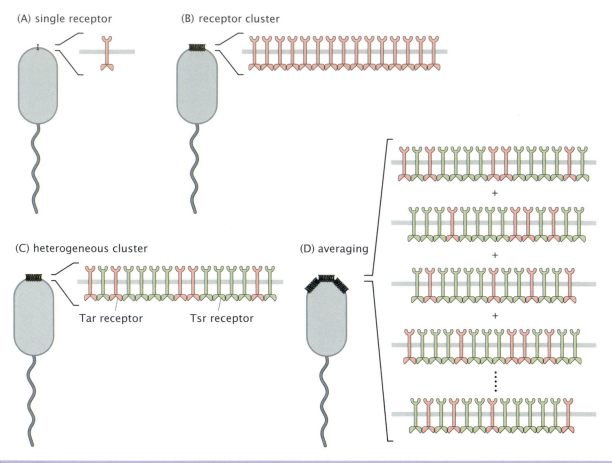

Figure 4.16

Hierarchy of increasingly sophisticated models of chemoreceptor activity. (A) Model of a single chemoreceptor. (B) Model of a homogeneous chemoreceptor cluster. (C) Model of a heterogeneous chemoreceptor cluster, with two types of receptor in the cluster. (D) Model which averages over both cluster composition and size.

average over both cluster composition and cluster size to attempt to respond to the data of Figure 4.14.

We begin with the activity of a single chemotaxis receptor. The states shown in Figure 4.17 represent an extremely coarse-grained view of the molecular essences of chemotaxis receptor activity. As with the example of the previous chapter on ion channels (p. 77), it is now time to exploit the statistical mechanical protocol to evaluate the statistical weights of the different states shown in Figure 4.17.

At this point, our goal is to compute the activity of the chemoreceptors as a function of the concentration of chemoattractant. In particular, the quantity p_{active} measures the ability of the receptor to produce phosphorylated CheY, resulting in a change in the motor's direction of rotation. The statistical mechanics of this system can be examined by appealing to a states-and-weights diagram like that shown in Figure 4.18. The probability that the receptor will be active is obtained by constructing the ratio of the statistical weights for the active state to the sum over the statistical weights of all states. Using the states and weights

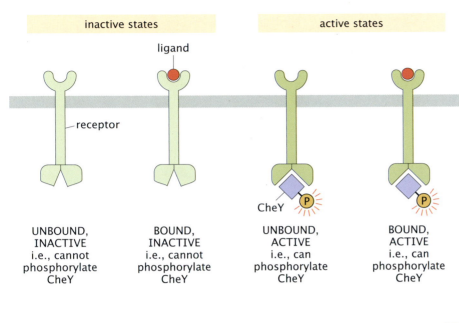

Figure 4.17

Molecular states of a chemotaxis receptor. The two states on the left show an inactive chemoreceptor in two different states of chemoattractant occupancy. The two states on the right show a chemoreceptor that is active and can phosphorylate CheY.

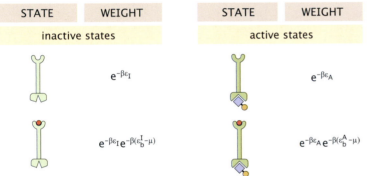

Figure 4.18

States and weights for MWC model of chemotactic receptors.

of Figure 4.18 tells us

$$p_{active} = \frac{e^{-\beta\varepsilon_A}(1 + e^{-\beta(\varepsilon_b^A - \mu)})}{e^{-\beta\varepsilon_A}(1 + e^{-\beta(\varepsilon_b^A - \mu)}) + e^{-\beta\varepsilon_I}(1 + e^{-\beta(\varepsilon_b^I - \mu)})}. \tag{4.1}$$

Invoking equation 2.51 (p. 52) which tells us that the chemical potential is of the form $\mu = \varepsilon_{sol} + k_B T \ln(c/c_0)$, we can rewrite this as

$$p_{active}(c) = \frac{e^{-\beta\varepsilon_A}(1 + \frac{c}{c_0}e^{-\beta(\varepsilon_b^A - \varepsilon_{sol})})}{e^{-\beta\varepsilon_A}(1 + \frac{c}{c_0}e^{-\beta(\varepsilon_b^A - \varepsilon_{sol})}) + e^{-\beta\varepsilon_I}(1 + \frac{c}{c_0}e^{-\beta(\varepsilon_b^I - \varepsilon_{sol})})}. \tag{4.2}$$

Recalling that we are often interested in interconverting between the statistical mechanical and thermodynamic language, we now use the fact that

$$K_A = c_0 e^{\beta(\varepsilon_b^A - \varepsilon_{sol})} \tag{4.3}$$

Figure 4.19

Figure 4.19

Chemoreceptor activity as a function of chemoattractant concentration. Parameters used are $K_I = 10^{-6}$ M, $K_A = 10^{-4}$ M, and $\varepsilon_I - \varepsilon_A = 2\,k_B T$.

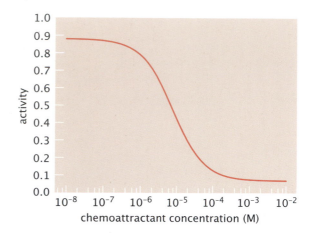

and

$$K_I = c_0 e^{\beta(\varepsilon_b^I - \varepsilon_{sol})} \tag{4.4}$$

with the result that

$$p_{active}(c) = \frac{(1 + \frac{c}{K_A})}{(1 + \frac{c}{K_A}) + e^{-\beta \Delta \varepsilon}(1 + \frac{c}{K_I})}, \tag{4.5}$$

where we have defined $\Delta \varepsilon = \varepsilon_I - \varepsilon_A$.

This formula tells us that in the context of this statistical mechanical model the probability of the active state of the chemoreceptor depends on a few biologically important variables: the energy difference between the active and inactive states of the receptor in the absence of ligand ($\Delta \varepsilon = \varepsilon_I - \varepsilon_A$), the affinities of the ligand for the active state (ε_b^A) and the inactive state (ε_b^I) of the receptor, and the amount of ligand itself. For chemoattractants, binding of the ligand will tend to favor the inactive state (where CheY is not phosphorylated), that is, $K_I < K_A$. We will find it convenient later in the context of exploring data collapse to rewrite this equation in the form

$$p_{active} = \frac{1}{1 + e^{-\beta(\varepsilon_I - \varepsilon_A)}\dfrac{1 + (c/K_I)}{1 + (c/K_A)}}. \tag{4.6}$$

Let's consider the implications of this result. In the absence of ligand (if $c = 0$), the equation simplifies to the familiar result for a two-state system such as an ion channel with the active and inactive states controlled by the relative values of ε_I and ε_A. Specifically, we have the leakiness given by

$$\text{leakiness} = p_{active}(c=0) = \frac{1}{1 + e^{-\beta(\varepsilon_I - \varepsilon_A)}}. \tag{4.7}$$

Since in the absence of ligand the receptor is active for phosphorylation, we know that ε_I is larger than ε_A, thus favoring the active state. On the other hand, we expect that with increasing ligand concentration, the inactive state will predominate. This means within this model that $K_I < K_A$. These ideas are shown in graphical form in Figure 4.19. Note that the key element of the

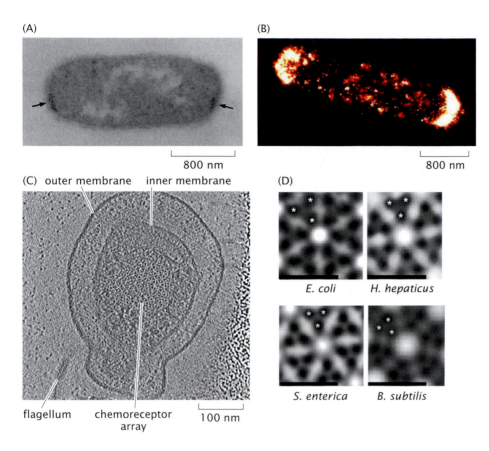

(A)

(C) outer membrane   inner membrane

flagellum   chemoreceptor
               array

800 nm

(B)

(D)

E. coli   *H. hepaticus*

S. enterica   *B. subtilis*

800 nm

100 nm

Figure 4.20

Spatial organization of chemotaxis receptor clusters. (A)
Localization of chemotaxis
receptors in *E. coli* as shown by
immunogold labeling in a thin-
section electron micrograph.
Two clusters are apparent, one at
each pole of the cell, indicated
by arrows. (B) High-resolution
fluorescence image of *E. coli*
chemotaxis receptors. Here it
is clear that there are many
small clusters along the sides
of the bacterium, as well as the
two major ones at the poles.
(C) Cryo-electron microscopy
showing a tomographic slice
through the pole of *Salmonella
enterica* (a pathogen closely
related to *E. coli*). At this resolution, it becomes apparent that the
receptors are packed into regular
hexagonal arrays. (D) Averaged
tomograms from four different
species of bacteria show similar
receptor packing. Each asterisk
indicates a single receptor dimer.
These group together to form
triangular trimers of dimers,
which then pack hexagonally.
From Phillips, R., J. Kondev, J.
Theriot, and H. Garcia (2013),
Physical Biology of the Cell, 2nd
ed. Reproduced by permission of
Taylor & Francis LLC, a division
of Informa plc. (A) adapted from
Maddock, J. R. and L. Shapiro
1993), "Polar location of the
chemoreceptor complex in the
Escherichia coli cell," *Science*
259:1717. Reprinted with permission from AAAS. (B) adapted
from Greenfield et al. (2009), CC
BY-NC 4.0. (C) and (D) adapted
from Briegel, A., et al. (2012),
"Bacterial chemoreceptor arrays
are hexagonally packed trimers
of receptor dimers networked
by rings of kinase and coupling
proteins," *Proc. Natl. Acad. Sci.
USA* 109:3766, fig. 1. Used with
permission of PNAS.

response is that at low chemoattractant concentrations, the chemoreceptors are active, while at high chemoattractant concentrations, the chemoreceptors are inactive. To fully investigate whether these ideas are consonant with the measured behavior of chemotaxis highlighted in Figure 4.14 (p. 134), we have to consider several other features of the chemoreceptors, such as their clustering together.

4.3.1 MWC Model of Chemotaxis Receptor Clusters

A twist in the plot is that experiments have shown that structurally and functionally, the chemotaxis receptors do not act in isolation. As shown in Figure 4.20, chemoreceptors form clusters preferentially near the cell poles. This preferential polar localization has been seen using both electron and fluorescence microscopy, and has been observed in many different species of bacteria. How might such clustering affect the behavior of the chemoreceptors? Conceptually, the idea is that in the spirit of the MWC model, binding of chemoattractant to one of the chemoreceptors in the cluster will have the effect of shifting the equilibrium of all the chemoreceptors in the cluster. In fact, though the names have changed, the statistical mechanics we used in the previous chapter to consider ligand-gated ion channels that bind more than one site on the channel tells us exactly what we need to know to handle the chemotaxis cluster problem.

Figure 4.21

States and weights for MWC model of chemotactic receptors. The cluster of chemoreceptors is of size N. As a result, there are $N + 1$ different states of occupancy, ranging from $n = 0$ sites occupied to $n = N$ sites occupied by ligand.

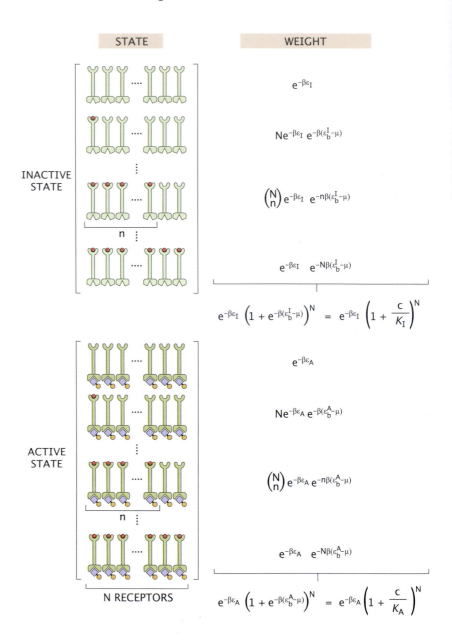

STATE	WEIGHT

INACTIVE STATE

$e^{-\beta \varepsilon_I}$

$N e^{-\beta \varepsilon_I} e^{-\beta(\varepsilon_b^I - \mu)}$

$\binom{N}{n} e^{-\beta \varepsilon_I} e^{-n\beta(\varepsilon_b^I - \mu)}$

n

$e^{-\beta \varepsilon_I} e^{-N\beta(\varepsilon_b^I - \mu)}$

$$e^{-\beta \varepsilon_I} \left(1 + e^{-\beta(\varepsilon_b^I - \mu)}\right)^N = e^{-\beta \varepsilon_I} \left(1 + \frac{c}{K_I}\right)^N$$

ACTIVE STATE

$e^{-\beta \varepsilon_A}$

$N e^{-\beta \varepsilon_A} e^{-\beta(\varepsilon_b^A - \mu)}$

$\binom{N}{n} e^{-\beta \varepsilon_A} e^{-n\beta(\varepsilon_b^A - \mu)}$

n

$e^{-\beta \varepsilon_A} e^{-N\beta(\varepsilon_b^A - \mu)}$

N RECEPTORS

$$e^{-\beta \varepsilon_A} \left(1 + e^{-\beta(\varepsilon_b^A - \mu)}\right)^N = e^{-\beta \varepsilon_A} \left(1 + \frac{c}{K_A}\right)^N$$

Our calculations so far using only a single chemoreceptor illustrate the key ideas, but they will not suffice to capture the full complexity of chemotactic behavior as revealed in Figure 4.14(B). In addition to the precise adaptation phenomenon seen in Figure 4.15 and which will be the subject of section 4.4.1, the system exhibits a high degree of cooperativity. To account for cooperativity, we now need to acknowledge that the chemoreceptors come in clusters and see how we should amend the states and weights to account for this.

To illustrate how this result emerges, in Figure 4.21 we resort to our usual states-and-weights procedure in which we imagine a cluster of N receptors. The fate of the one is the fate of the many: either all receptors are inactive, or all are active. As in the usual MWC framework, the relative energies of the inactive and active states are different, and the K_d for the binding of ligands depends upon which of the two states the receptors are in. To see how the statistical weight of the active state arises, note that the number of *bound* ligands can be anything between 0 and N. The generic statistical weight for the active state when it has n ligands bound is of the form

$$w_n = e^{-\beta \varepsilon_A} \frac{N!}{(N-n)!n!} e^{-n\beta(\varepsilon_b^A - \mu)}. \tag{4.8}$$

Conceptually, this term accounts for the many different ways of binding n ligands to N receptors, as captured by the term $\binom{N}{n}$. The factor $e^{-n\beta(\varepsilon_b^A - \mu)}$ acknowledges that we have taken n ligands out of solution, and they are now bound to the chemoreceptors. As a result, we note that equation 4.8 gives us the nth term in a binomial of order N (except for the prefactor $e^{-\beta \varepsilon_A}$).

Using the states and weights presented in Figure 4.21, we can compute the partial partition function for the active state as

$$Z_{active} = \sum_{n=0}^{N} e^{-\beta \varepsilon_A} \frac{N!}{n!(N-n)!} e^{-n\beta(\varepsilon_b^A - \mu)} \tag{4.9}$$

Recall that the binomial theorem tells us that

$$(x+y)^N = \sum_{n=0}^{N} \frac{N!}{n!(N-n)!} x^n y^{N-n}. \tag{4.10}$$

For our case, if we take

$$x = e^{-\beta(\varepsilon_b^A - \mu)} \tag{4.11}$$

and

$$y = 1, \tag{4.12}$$

we are left with

$$Z_{active} = e^{-\beta \varepsilon_A} (1 + e^{-\beta(\varepsilon_b^A - \mu)})^N. \tag{4.13}$$

By symmetry, we can similarly determine the partial partition function for the inactive states as

$$Z_{inactive} = e^{-\beta \varepsilon_I} (1 + e^{-\beta(\varepsilon_b^I - \mu)})^N \tag{4.14}$$

by following precisely the same kind of argument we used to obtain Z_{active}. Given the partial partition functions, we can then write the probability that the chemoreceptor cluster is active as

$$p_{active} = \frac{Z_{active}}{Z_{active} + Z_{inactive}}. \tag{4.15}$$

We can rewrite this more explicitly as

$$p_{active}(c) = \frac{e^{-\beta \varepsilon_A} (1 + e^{-\beta(\varepsilon_b^A - \mu)})^N}{e^{-\beta \varepsilon_A} (1 + e^{-\beta(\varepsilon_b^A - \mu)})^N + e^{-\beta \varepsilon_I} (1 + e^{-\beta(\varepsilon_b^I - \mu)})^N}. \tag{4.16}$$

Figure 4.22

Active probability versus chemoattractant concentration. The different curves correspond to different sizes of chemoreceptor clusters, the exponent N in equation 4.18. Parameters used are $K_I = 10^{-6}$ M, $K_A = 10^{-4}$ M, and $\varepsilon_I - \varepsilon_A = 2\,k_B T$.

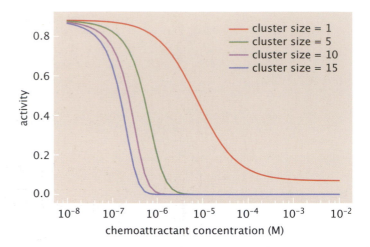

Invoking equation 2.51 (p. 52) which tells us that the chemical potential is of the form $\mu = \varepsilon_{sol} + k_B T \ln(c/c_0)$, we can rewrite this as

$$p_{active}(c) = \frac{e^{-\beta\varepsilon_A}(1 + \frac{c}{c_0}e^{-\beta(\varepsilon_b^A - \varepsilon_{sol})})^N}{e^{-\beta\varepsilon_A}(1 + \frac{c}{c_0}e^{-\beta(\varepsilon_b^A - \varepsilon_{sol})})^N + e^{-\beta\varepsilon_I}(1 + \frac{c}{c_0}e^{-\beta(\varepsilon_b^I - \varepsilon_{sol})})^N}. \quad (4.17)$$

As usual, we pass between the statistical mechanical and thermodynamic language using equations 4.3 and 4.4 with the result that

$$p_{active}(c) = \frac{(1 + \frac{c}{K_A})^N}{(1 + \frac{c}{K_A})^N + e^{-\beta(\varepsilon_I - \varepsilon_A)}(1 + \frac{c}{K_I})^N}. \quad (4.18)$$

We can rewrite this result as

$$p_{active} = \frac{1}{1 + e^{-\beta(\varepsilon_I - \varepsilon_A)}\dfrac{\left(1 + (c/K_I)\right)^N}{\left(1 + (c/K_A)\right)^N}}, \quad (4.19)$$

where the only difference relative to equation 4.6 is that the cluster size N shows up as an exponent for the terms involving the ligand concentration.

One of the simplest questions we might now ask is, how do the activity curves depend upon the cluster size? Figure 4.22 shows the activity curves for different number of receptors. Note that the leakiness does not depend upon the number of chemoreceptors in the cluster, while the saturation value (i.e., $\lim_{c\to\infty} p_{active}(c)$) clearly does, with the value

$$\text{saturation} = \lim_{c\to\infty} p_{active}(c) = \frac{1}{1 + e^{-\beta\Delta\varepsilon}(\frac{K_A}{K_I})^N}. \quad (4.20)$$

STATE	WEIGHT
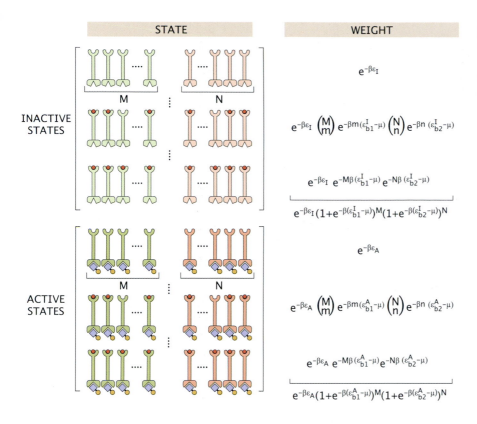	

Figure 4.23

States and weights for heterogeneous receptor clusters. There are M receptors of one type and N receptors of the second type in each cluster.

The weights shown in the figure:

$e^{-\beta \varepsilon_I}$

$e^{-\beta \varepsilon_I} \binom{M}{m} e^{-\beta m (\varepsilon_{b1}^I - \mu)} \binom{N}{n} e^{-\beta n (\varepsilon_{b2}^I - \mu)}$

$e^{-\beta \varepsilon_I} e^{-M \beta (\varepsilon_{b1}^I - \mu)} e^{-N \beta (\varepsilon_{b2}^I - \mu)}$

$e^{-\beta \varepsilon_I}(1 + e^{-\beta(\varepsilon_{b1}^I - \mu)})^M (1 + e^{-\beta(\varepsilon_{b2}^I - \mu)})^N$

$e^{-\beta \varepsilon_A}$

$e^{-\beta \varepsilon_A} \binom{M}{m} e^{-\beta m (\varepsilon_{b1}^A - \mu)} \binom{N}{n} e^{-\beta n (\varepsilon_{b2}^A - \mu)}$

$e^{-\beta \varepsilon_A} e^{-M \beta (\varepsilon_{b1}^A - \mu)} e^{-N \beta (\varepsilon_{b2}^A - \mu)}$

$e^{-\beta \varepsilon_A}(1 + e^{-\beta(\varepsilon_{b1}^A - \mu)})^M (1 + e^{-\beta(\varepsilon_{b2}^A - \mu)})^N$

4.3.2 Heterogenous Clustering

The result of the dialogue between experiment and theory in the analysis of chemoreceptor activity is the conclusion that the chemoreceptor clusters are not built up from only one kind of chemoreceptor but, instead, are hetero-clusters with mixtures of the different kinds of chemoreceptors, as shown in Figure 4.16(C). That is, the family of activity curves revealed in Figure 4.14 is best explained on the assumption that chemoreceptors do not act in isolation but, rather, are clustered together heterogeneously with more than one chemoreceptor type per cluster. To see how the MWC framework can be extended to that setting, we now explore the statistical mechanics of such heterogeneous clusters.

To begin, we consider a chemoreceptor cluster with two types of chemoreceptors. We imagine there are M copies of the first type of chemoreceptor per cluster and N copies of the second type of chemoreceptor per cluster, for a total of $M + N$ receptors per cluster, as shown in Figure 4.23. In particular, we begin by thinking about the case in which there are m ligands bound to the first class of receptors and n ligands bound to the second class of receptors. In this case, the contribution to the inactive state grand partition function is given by

$$Z_{inactive}(m, n) = e^{-M \beta \varepsilon_I^{(1)}} \frac{M!}{m!(M-m)!} e^{-\beta m (\varepsilon_b^I(1) - \mu)}$$

$$\times e^{-N \beta \varepsilon_I^{(2)}} \frac{N!}{n!(N-n)!} e^{-\beta n (\varepsilon_b^I(2) - \mu)}, \qquad (4.21)$$

where we have defined the binding energy to receptor type 1 and type 2 as $\varepsilon_b^I(1)$ and $\varepsilon_b^I(2)$, respectively. Note that the model has now taken another step in complexity, since we treat the energy difference between the inactive and active states as a linear superposition of the inactive energies of each receptor type, $\varepsilon_I = M\varepsilon_I^{(1)} + N\varepsilon_I^{(2)}$. It is useful to build up the grand partition function step by step by considering its constituent elementary binding configurations. Given the preceding expression, we can find the contribution of all the inactive states to the grand partition function by summing over all the possible states of occupancy as

$$Z_{inactive} = \sum_{m=0}^{M} \sum_{n=0}^{N} Z_{inactive}(m, n). \tag{4.22}$$

However, we also see that the two sets of sums are independent, and hence we can rewrite this as

$$Z_{inactive} = e^{-M\beta\varepsilon_I^{(1)}} \sum_{m=0}^{M} \frac{M!}{m!(M-m)!} x_1^m y_1^{M-m}$$

$$\times e^{-N\beta\varepsilon_I^{(2)}} \sum_{n=0}^{N} \frac{N!}{n!(N-n)!} x_2^n y_2^{N-n}, \tag{4.23}$$

where we define $x_1 = e^{-\beta(\varepsilon_b^I(1)-\mu)}$, $y_1 = 1$, $x_2 = e^{-\beta(\varepsilon_b^I(2)-\mu)}$, and $y_2 = 1$. In light of these definitions, we now recognize that each of these sums is nothing more than a conventional binomial, and hence the inactive states contribute to the grand partition function as

$$Z_{inactive} = e^{-M\beta\varepsilon_I^{(1)}} e^{-N\beta\varepsilon_I^{(2)}} (1+x_1)^M (1+x_2)^N. \tag{4.24}$$

The calculation goes through identically for the case in which we interest ourselves in Z_{active}, except that now the subscripts and superscripts I are replaced by A to indicate "active" rather than "inactive". Given these results, we can now construct the active probability itself as

$$p_{active} = \frac{Z_{active}}{Z_{active} + Z_{inactive}}. \tag{4.25}$$

We can rewrite this in a more explicit form by appealing to equation 4.24 and the analogous formula for the active state and remembering the expression for the chemical potential in a dilute solution, namely, $\mu = \mu_0 + k_B T \ln \frac{c}{c_0}$, resulting in

$$p_{active} = \frac{e^{-M\beta\varepsilon_A^{(1)}} e^{-N\beta\varepsilon_A^{(2)}} \left(1 + \frac{c}{K_A(1)}\right)^M \left(1 + \frac{c}{K_A(2)}\right)^N}{\left(e^{-M\beta\varepsilon_A^{(1)}} e^{-N\beta\varepsilon_A^{(2)}} \left(1 + \frac{c}{K_A(1)}\right)^M \left(1 + \frac{c}{K_A(2)}\right)^N + e^{-M\beta\varepsilon_I^{(1)}} e^{-N\beta\varepsilon_I^{(2)}} \left(1 + \frac{c}{K_I(1)}\right)^M \left(1 + \frac{c}{K_I(2)}\right)^N\right)}. \tag{4.26}$$

This equation reduces to our previous results in the limits when the cluster is composed of only one type of receptor or the other. Using this equation, we can examine the activity of the chemoreceptors as a function of chemoattractant concentration, as shown in Figure 4.24, revealing how the activity depends upon the cluster composition.

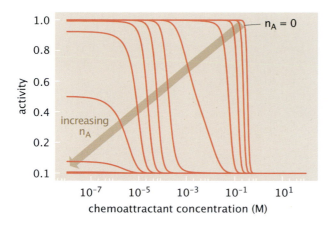

Figure 4.24

Chemoreceptor activity versus cluster composition. All curves are for the case of a heterogeneous cluster of size $N = 15$. The rightmost curve gives the activity as a function of chemoattractant concentration for a cluster composed entirely of Tsr receptors, $n_A = 0$. Each curve moving from right to left gives the activity as a function of chemoattractant concentration for clusters having one more Tar receptor than the previous one.

4.3.3   Putting It All Together by Averaging

Figure 4.16 (p. 136) alerted us to the possibility of constructing a series of increasingly sophisticated models that will make it possible to take stock of the beautiful and provocative quantitative FRET data available on chemotaxis. We have now reached the final step in the progression of models, which asks us to average over the heterogeneous clusters we considered in the previous subsection. The reason for such averaging is that there is a distribution of different cluster sizes and compositions, even at the level of a single cell. To perform such averaging then, first, we need to consider the fact that the concentration of different types of receptors within a given cluster can vary. Second, we need to average over the different possible cluster sizes.

For example, let's consider the case in which the ratio of Tar receptors to Tsr receptors is 1:2. In this case, if we randomly choose a receptor, we have a $p_a = 1/3$ chance that receptor will be a Tar receptor. For a cluster of size N, the probability the cluster will contain n Tar receptors is given by the binomial distribution as

$$p(n, N) = \frac{N!}{n!(N-n)!} p_a^n (1 - p_a)^{N-n}. \tag{4.27}$$

Thus, we can average over all different cluster compositions as

$$\langle p_{active}^{(N)} \rangle \text{composition} = \sum_{n=0}^{N} \frac{N!}{n!(N-n)!} p_a^n (1 - p_a)^{N-n} p_{active}(n; N), \tag{4.28}$$

where $p_{active}(n; N)$ is given by equation 4.26, where we replace M with n and N with $N - n$.

The second step in our analysis is to average over cluster sizes (i.e., different N). In the work culminating in Figure 4.25, the hypothesis was advanced that the clusters were equally likely to contain 14, 15, or 16 chemoreceptors, thus demanding the average

$$\langle p_{active} \rangle \text{cluster size} = \frac{1}{3} (p_{active}^{(14)} + p_{active}^{(15)} + p_{active}^{(16)}). \tag{4.29}$$

Wild-type chemotaxis response. (A) The activity of wild-type cells as a function of chemoattractant concentration was measured using FRET experiments. (B) The theory was carried out using the analysis of equation 4.29. Adapted from Keymer et al. (2006); see Further Reading.

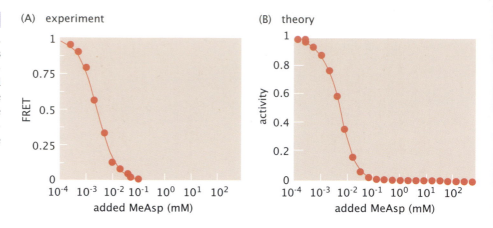

Figure 4.25 shows the activity as a function of chemoattractant concentration that results from the full model presented here and compares it with the wild-type response of *E. coli* cells.

Now that we have taken our MWC modeling effort all the way to the point that we can make contact with experiments, we are ready to consider the full set of mutants leading to the data in Figure 4.14. This brilliant set of experiments took this theory to the next level by creating a series of mutants that mimicked the phenomenon of adaptation shown in Figure 4.15. These experiments were based on mutants that were incapable of adaptation. Each mutant was explicitly constructed to correspond to a different level of adaptation by tuning the relative number of glutamate (E) and glutamine (Q) residues at the methylation sites. This made it possible within the MWC framework described here to understand which parameters are responsible for real-time physiological adaptation.

4.4   The Amazing Phenomenon of Physiological Adaptation

So far, our theoretical treatment of the chemotaxis phenomenon has missed one of its most important and beautiful characteristics, namely, adaptation. The phenomenon of adaptation is familiar to us all through our sense of vision. Adaptation gives us the ability to see stars on a moonless night on a deserted tropical island in the middle of the Pacific Ocean or to read a map while standing on a glacier in the middle of the day in summer, as shown in Figure 4.26(A). Other examples abound. The great American physiologist Walter Bradford Cannon dedicated much of his career to exploring how animals maintain a state of homeostasis through connected networks of chemical reactions (see Figure 4.26(B)). Cellular physiology is predicated upon similar mechanisms, with the profile of genes being transcribed or proteins being subjected to posttranslational modifications giving rise to functionally different chemistries in response to the environment, as shown in Figure 4.26(C). In each of these cases, the molecular machinery of living organisms has to be

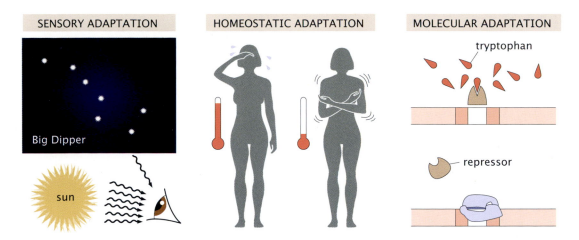

Figure 4.26

The concept of adaptation across biology. (A) Adaptation in vision. Under the guidance of the Big Dipper, the human eye can see the North Star, even though in that dim light, photons are roughly a kilometer apart. The eye can also comfortably accommodate to the bright illumination of a sunny day. (B) The human body maintains temperature homeostasis by sweating and shivering. (C) Adaptation in the processes of the central dogma. A transcription factor controlling tryptophan biosynthesis "measures" the amount of tryptophan present and represses that gene if there is sufficient tryptophan present.

able to change its response function in response to some new environmental conditions. This section explores the depth of our understanding of such phenomena within the framework of the MWC model.

One of the best-understood examples of adaptation is exhibited by bacteria in response to varying concentrations of chemoattractant. Figure 4.15 (p. 135) shows data on one such experiment. In that experiment, the ambient concentration was changed from one level to another, as shown in Figure 4.27. Initially, the bacterium recognizes the change in environmental conditions and changes its tumble frequency. However, over time as the bacterium "realizes" that it is simply in a new but spatially homogeneous environment, it adapts through changes in the methylation state of the chemoreceptors so as to restore the original tumble frequency. More sophisticated experiments in which bacteria are subjected to time-dependent concentrations have made it possible to dissect the adaptive response in exquisite detail (see Further Reading).

Figure 4.28 explores the concept of adaptation by showing four different concentration gradients of the form

$$c(x) = c_0 + kx, \qquad (4.30)$$

where c_0 indicates the ambient concentration, and k expresses the steepness of the gradient. Adaptation makes it possible for bacterial cells to detect an underlying gradient (dc/dx) across quite disparate background concentrations. The idea of adaptation is that for each background concentration, there is a different input-output function, as illustrated in Figure 4.28(B). Mechanistically, these new input-output responses are effectively the response of a new chemoreceptor, because there is a different methylation state, which leads to a shift in the

Figure 4.27

Response to a jump in the ambient concentration of chemoattractant. (A) The bacterium has a steady-state tumble frequency in the presence of chemoattractant at concentration c_1. (B) After a jump to chemoattractant concentration c_2, the bacterium's response is dictated by a different part of the input-output response curve. (C) A change in the methylation state of the receptor leads to a change of shape of the input-output curve so that the response is sensitive at the new ambient concentration.

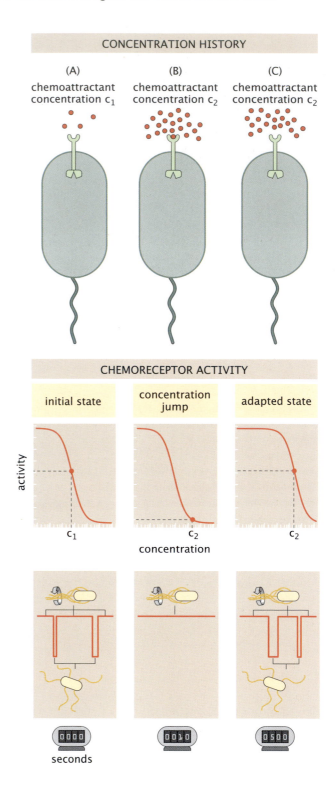

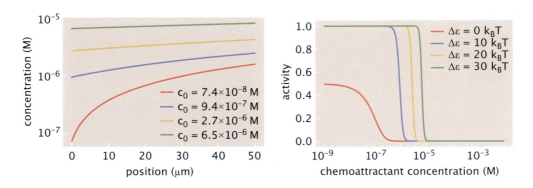

Figure 4.28

The concept of adaptation in bacterial chemotaxis. (A) Four different concentration profiles, all showing the *same* gradient in eqn. 4.30 ($k = dc/dx = 0.03 \, \mu M/\mu m$) against very different background concentrations c_0. (B) Activity curves as a function of concentration for four different background chemoattractant concentrations. The parameter that mediates the shift in the activity curves is the energy difference between the inactive and active states $\Delta\varepsilon = \varepsilon_I - \varepsilon_A$.

EC_{50} of the chemoreceptor. Our task in the pages that follow is to work out how the chemotaxis receptors can be modified through posttranslational modifications (methylation in particular) to center their response functions at the correct chemoattractant concentration "sweet spot."

To convey a sense of the molecular machinery that mediates these adaptation processes, Figure 4.29 shows how the enzyme CheR adds a methyl group to the chemotaxis receptor, and the phosphorylated version of the enzyme CheB removes a methyl group. From a structural perspective, the geometry of the adaptation apparatus was hinted at in Figure 4.12, where we see the "adaptation region" of the chemotaxis receptor, an idea codified in our chemoreceptor schematics by the addition of the lollipop-like structure with the label "M" for methylation. Though at the level of the chemoreceptor this physiological adaptation is due to the addition of a molecular group, in the context of the statistical mechanical model, this adaptation is hypothesized to be due to a change in the parameter $\Delta\varepsilon = \varepsilon_I - \varepsilon_A$, the energy difference between inactive and active states, with the dissociation constants K_I and K_A unchanged by the mutations. The method of data collapse to be highlighted later in the chapter makes it possible to test this hypothesis.

Interestingly, to achieve the full dynamic range exhibited by the chemotaxis receptors, chemoreceptors have multiple methylation sites. As seen in Figure 4.30, there are four such sites on each chemoreceptor. Figure 4.31 gives a qualitative sense of how the allosteric model of chemoreceptor clusters we have worked out in this chapter treats the adaptation process. At the molecular level, adaptation corresponds to addition and removal of methyl groups. From a statistical mechanical perspective, these molecular changes are attended by shifts in the value of the MWC parameter $\Delta\varepsilon$, the energy difference between the active and inactive states. In the figure, we plot the activity using equation 4.18 (p. 142), with cluster size $N = 15$ and for a family of different $\Delta\varepsilon = \varepsilon_I - \varepsilon_A$ chosen to be roughly consistent with those used in the full theory for heterogeneous clusters. Note that the treatment given here is not quantitatively correct, since we consider chemoreceptor clusters with receptors of only one

Figure 4.29

Simple model of adaptation due to methylation. CheR adds methyl groups when the chemoreceptor is inactive. Phosphorylated CheB (CheB-P) removes methyl groups when the chemoreceptor is active.

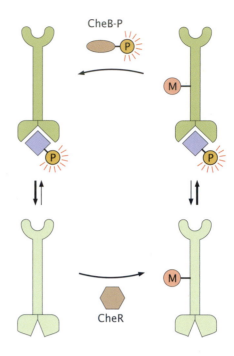

Figure 4.30

Adaptation in chemotaxis through successive methylation and demethylation events. Each chemoreceptor has four methylation sites, and each such methylation event creates a larger shift in the MWC parameter $\Delta\varepsilon$. Adapted from Nelson (2017).

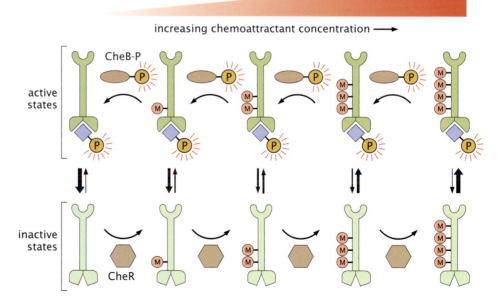

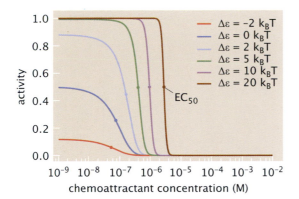

Figure 4.31

Shifts in activity curves with chemore-ceptor methylation. Parameters used are those for a homogeneous cluster with $N = 15$ chemoreceptors and $K_I = 10^{-6}$ M and $K_A = 10^{-4}$ M. The energy difference between the inactive and active states is chosen to roughly mimic the parameters used by Keymer et al. in their original work, though the full treatment requires heterogeneous clusters, as we do in the text.

type. To actually jibe with the experimental data, the model must consider chemoreceptor clusters with more than one species of chemoreceptor present. Further, we ignore the fact that the chemoreceptors form dimers, thus resulting in eight states of methylation. The point of this exercise before launching into the full treatment is simply to show how the energy difference between active and inactive chemoreceptors dictates the position of the EC_{50}, thus permitting the chemoreceptors to reset their "sweet spot" for sensitivity to changes in the current background concentration. As $\Delta\varepsilon$ increases, the midpoint of the activity curve shifts to the right, meaning that the chemoreceptors are responsive at higher concentrations.

4.4.1   Adaptation by Hand

How can we test all these ideas about adaptation? To explore the activity curves for different states of methylation, a series of strains were constructed in which the enzymes that mediate methylation and demethylation were removed ($\Delta CheRCheB$), and then the effective state of methylation was varied by changing the relative number of glutamate (E) and glutamine (Q) residues at these methylation sites. For the purposes of these experiments, it is argued that glutamines (Q) are physiologically equivalent to glutamates (E) that have been methylated. The mutants considered were EEEE, QEEE, QEQE, and QEQQ. Effectively, the experiment amounts to hardwiring the adaptations by hand, as seen in Figure 4.32. To be concrete, each chemotaxis receptor has four amino acids that in the wild-type cells can be methylated. Since the methylation and demethylation enzymes are knocked out in the mutants, this means that those four amino acids are unchanged. Hence, switching between different combinations of glutamates (E) and glutamines (Q) causes the receptors to behave as if they are methylated to varying extents. The activity curves for these different strains were shown in Figure 4.14 (p. 134). From an MWC perspective, each one of these mutants has a different value of $\Delta\varepsilon = \varepsilon_I - \varepsilon_A$ and hence a different midpoint in its activity curve.

We see, then, that these different mutants have different corresponding activity curves with several key features. First, the EC_{50} of these different receptors is shifted, giving a sense of how the methylation process leads to adaptation by shifting the concentration for the midpoint activity. From the point of view of

Figure 4.32

Mutants that explore the space of adaptations. Strains lacking the enzymes for methylation and demethylation are replaced by mutants with specific arrangements of "methylation," as shown in the top part of the figure. The FRET pair shown in each cell is used to measure the chemoreceptor activity, as shown in the bottom of the figure. As the number of methyl groups on the chemoreceptors increases, the activity curves shift to the right, implying that it takes more chemoattractant to render them inactive.

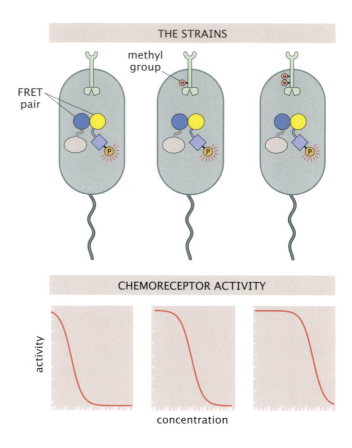

the MWC model, these shifts come from changes in a single parameter, namely, $\Delta\epsilon$, the energy difference between the inactive and active states in the absence of ligand.

We can get an analytic handle on how the EC_{50} shifts with $\Delta\varepsilon$ by making several reasonable approximations. First, we note that $K_A >> K_I$, which means that at the concentration $c = EC_{50}$, $EC_{50}/K_A << 1$, resulting in

$$p_{active} \approx \frac{1}{1 + e^{-\beta\varepsilon}(1 + \frac{c}{K_I})^N}. \tag{4.31}$$

The second critical approximation is to assume that for very small concentrations, $p_{active} \rightarrow 0$, while for large concentrations $p_{active} \rightarrow 1$. This means that at the EC_{50} we have $p_{active} \approx 1/2$, resulting in the condition

$$\frac{1}{1 + e^{-\beta\Delta\varepsilon}(1 + \frac{EC_{50}}{K_I})^N} = \frac{1}{2}. \tag{4.32}$$

To satisfy this condition, the denominator needs to equal 2, or equivalently,

$$e^{-\beta\Delta\varepsilon}(1 + \frac{EC_{50}}{K_I})^N = 1. \tag{4.33}$$

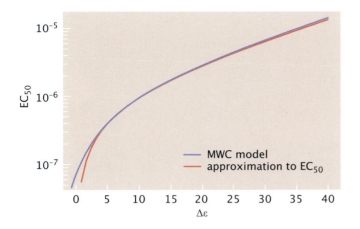

Figure 4.33

Shift in EC_{50} with methylation. The curves compare the position of the EC_{50} as computed using the full MWC model and using the approximation of equation 4.34.

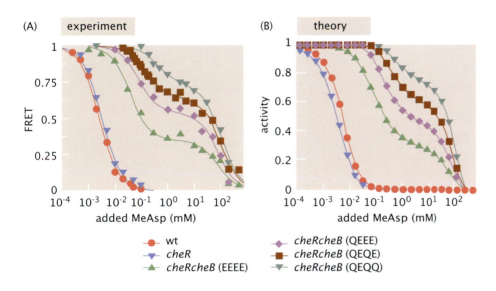

Figure 4.34

Chemoreceptor cluster activity as a function of chemoattractant concentration for heterogeneous clusters of chemoreceptors. (A) Measured activity curves. (B) Theoretical results for activity curves. Adapted from Keymer et al. (2006); see Further Reading.

Using a few lines of algebra we can write the EC_{50} analytically as

$$EC_{50} = K_I(e^{\beta \Delta \varepsilon / N} - 1). \qquad (4.34)$$

4.4.2   Data Collapse in Chemotaxis

We have argued repeatedly throughout the book that one of the signatures of our understanding of a given problem is our ability to unify disparate data. Specifically, when a family of measurements has been made as a function of some control parameter such as the concentration of chemoattractant, the question is whether we can find a way to understand all those data at once without a new set of fit parameters and new stories and explanations for each dataset. Data collapse of different datasets provides one of the most satisfying demonstrations of such understanding. Here, we explore how the allosteric model described thus far in the chapter can be used to bring all the mutant data from Sourjik and Berg under one roof.

Our starting point is to rewrite equation 4.19 for a homogeneous cluster of chemoreceptors of size N here as

$$p_{active} = \cfrac{1}{1 + e^{-\beta(\varepsilon_I - \varepsilon_A)} \cfrac{\left(1 + (c/K_I)\right)^N}{\left(1 + (c/K_A)\right)^N}}. \tag{4.35}$$

This equation gives us the possibility of learning the natural variable of the problem. Specifically, since we are writing p_{active} in the form

$$p_{active} = \frac{1}{1 + e^{-\beta F_{Bohr}}}, \tag{4.36}$$

whenever two receptor clusters have achieved the same F_{Bohr}, they will have achieved the same level of activity. Hence, if we plot our data with respect to F_{Bohr} rather than with respect to chemoattractant concentration, by definition, all data will collapse onto the same curve. Note that in light of equation 4.19, we have

$$F_{Bohr} = (\varepsilon_I - \varepsilon_A) - k_B T \ln \frac{\left(1 + (c/K_I)\right)^N}{\left(1 + (c/K_A)\right)^N}. \tag{4.37}$$

We can adopt this same strategy for the case in which we have heterogenous clusters. For example, if we have M receptors of type 1 and N receptors of type 2, then using equation 4.26 (p. 144), we will have the generalized Bohr parameter

$$F_{Bohr} = M(\varepsilon_I^{(1)} - \varepsilon_A^{(1)}) + N(\varepsilon_I^{(2)} - \varepsilon_A^{(2)}) - k_B T \ln \frac{(1 + \frac{c}{K_I(1)})^M (1 + \frac{c}{K_I(2)})^N}{(1 + \frac{c}{K_A(1)})^M (1 + \frac{c}{K_A(2)})^N}$$

$$\tag{4.38}$$

After all this hard work, we are finally ready to see how the MWC theory works in the context of the data of Sourjik and Berg. Figure 4.35 shows the response of wild-type cells to steps in chemoattractant concentration when adapted to different ambient concentrations. Each ambient concentration corresponds to a different energy difference between the inactive and active states, but the dissociation constants remain the same. Figure 4.35 also shows how using the data-collapse strategy highlighted here causes all the data to fall onto one master curve. As we can see from the raw data (left panel), the activity either goes up in response to a jump decrease in chemoattractant or goes down in response to a jump increase in chemoattractant. Note that from the theoretical perspective, the only differences in the curves arise from changes to the energy difference $\Delta\varepsilon$, which characterizes the difference in energy between the inactive and active states in the absence of chemoattractant. The right panel shows the corresponding data collapse according to the theoretical ideas developed thus far in the chapter.

In summary, this first part of the chapter on bacterial signaling has focused on the revealing and important case study of chemotaxis. This subject is characterized by precision measurements on well-conceived mutants described in turn by well-formulated theoretical ideas designed precisely to respond to those experiments. We see that the quantitative story holds together without any

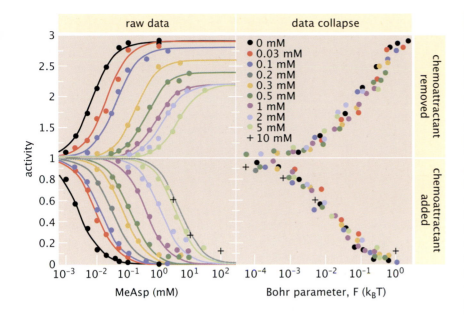

Figure 4.35

Response measured by FRET to steps of chemoattractant (MeAsp). The left panels show the chemotactic response of wild-type cells adapted to a particular background chemoattractant concentration after a step in MeASP concentration. The activity is measured by FRET, as illustrated in Figure 4.14. The top panel shows the response when chemoattractant is removed and the bottom panel shows the response when chemoattractant is added. The legend shows the ambient concentration when the cells are subjected to the MeASP concentration step. The right panels show the result of replotting the data (parameter free) with respect to the natural variable of the Bohr parameter (i.e., free-energy difference). Adapted from Keymer et al. (2006).

major blemishes and serves as a model for the kind of theory-experiment dialogue this book celebrates.

4.5   Beyond the MWC Model in Bacterial Chemotaxis

Thus far, our discussion of chemotaxis has focused on the successes of the MWC framework. As in the previous chapter in our discussion of ion channels, we end our discussion of bacterial chemotaxis on a more sober note in the hope that the reader recognizes that our models are incomplete caricatures of complex phenomena and that many questions remain unanswered.

In this brief section, we broach two key ways in which researchers have moved beyond the simplest MWC framework in the description of bacterial chemotaxis. First, a whole body of work (reviewed in Tu 2013; see Further Reading) has used one of the other powerhouse models of statistical physics, namely, the Ising model, to describe the function of clusters of chemotaxis clusters. Though our treatment in section 4.3.1 culminated in an ability to fit the data, it also featured a phenomenological fit to the cluster sizes. Alternative models based upon the Ising model provide a more natural interpretation of the effective cluster size and make it possible to abandon the strict all-or-none thinking of the MWC framework. Future work is needed to further uncover the physics of cooperative action of chemotaxis receptors.

The second key point that needs to be raised is our reliance on equilibrium thinking. The MWC equilibrium framework not only has been exploited for thinking about the activity of the chemotaxis receptors but also has served as a null model for the switching behavior of the bacterial flagellar motor. This motor switches between clockwise and counterclockwise rotation, as described earlier in the chapter, as a result of the binding of CheY-P to the FliM part of the motor. The distribution of duration times in the counterclockwise rotation state of the motor appears to defy description in terms of the MWC model and Ising-type conformational spread model, instead demanding that the system operate out of equilibrium with constant energy dissipation. Nonequilibrium effects will be broached more deeply in chapter 12. Indeed, the emergence of such nonequilibrium effects where the MWC framework breaks down represents one of the most exciting research frontiers for thinking about the function of allosteric molecules in the context of living cells.

4.6   The Ecology and Physiology of Quorum Sensing

Chemotaxis as we described it in the first part of this chapter is essentially a single-cell phenomenon. Individual cells take stock of their external environment's chemical attributes and make rapid decisions about where to go. But bacteria also participate in collective behaviors. As seen in Figure 4.4 (p. 127), bacteria such as *Vibrio fischeri* live in symbiosis with the Hawaiian bobtail squid, exhibiting bioluminescence when cell densities become sufficiently high. Though originally dubbed "autoinduction" by its discoverers (see Hastings in Further Reading), this phenomenon is now known by the more familiar name quorum sensing.

The molecules that induce this response were discovered in a series of hard-won and clever experiments, ultimately owing their discovery to the existence of bacterial mutants that showed less bioluminescence. The key point is that these mutant cells did not produce any substantial luminescence when at densities at which wild-type cells would be brightly glowing. But when these very same dim mutants were placed in media from cells with a fully functioning quorum sensing system, these mutants themselves became luminescent, as seen in Figure 4.3 (p.126).

Figure 4.36 shows the critical data in which a chemostat was used to control the cellular density. As seen in the figure, as the cell density was reduced the corresponding bioluminescence declined precipitously. However, when cells at densities normally associated with no (or very low) bioluminescence were bathed in media previously inhabited by cells at high cell density, they then became luminescent. These experiments led to the hypothesis that these mutant cells lacked functional genes to produce the relevant factor, later isolated and named AI (for autoinducer). This factor is produced within the cells and then released into the medium, and when concentrations become sufficiently high, the receptors that detect AI are activated, leading to a cascade of downstream molecular consequences culminating in the readout of luminescence.

The next step down the road to fully establishing the phenomenology of the quorum sensing phenomenon that will animate the remainder of this chapter was the determination of the molecules responsible for chemical signaling itself. As already described with reference to Figure 4.36, some molecule or molecules were present in the medium of those cells participating in the quorum sensing process that told cells they were in the presence of others. Using high-performance liquid chromatography, the autoinducer was purified and its structure was determined using a combination of nuclear magnetic resonance and other spectroscopies. Figure 4.37 gives a rogue's gallery of both the maiden version of an autoinducer molecule, as well as several others that have been

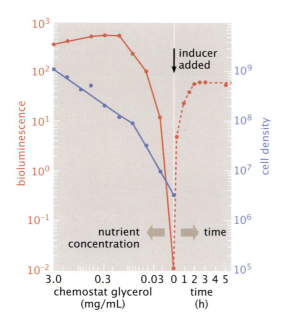

Figure 4.36

Discovery of the autoinducer. Bioluminescence and cell density of *Vibrio fischeri* were monitored as a function of the glycerol concentration in a chemostat, as shown in the left part of the figure. Using the chemostat, the density of bacteria could be continuously decreased, thus switching off bioluminescence. Once the cells reached a sufficiently low density, exogenously purified autoinducer was added, and the bioluminescence was monitored as a function of time (right part of the figure). Addition of autoinducer restores bioluminescence at low cell density. Adapted from Rosson and Nealson (1981).

Vibrio fischeri/LuxI

Pseudomonas aeruginosa/LasI

Pseudomonas aeruginosa/RhlI

Agrobacterium tumefaciens/TraI

Vibrio harveyi/LuxLM

Figure 4.37

Structures of autoinducer molecules from different species of Gram-negative bacteria. Each molecule has the name of the microorganism and the corresponding autoinducer. Adapted from Ng and Bassler (2009).

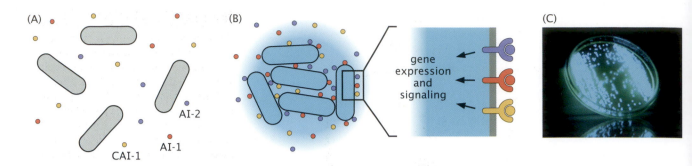

Figure 4.38

Schematic of the quorum sensing process in *Vibrio harveyi*. (A) Bacteria secrete autoinducer molecules of different types, AI-1, AI-2 and CAI-1. (B) Highly simplified view of the molecular circuit giving rise to the quorum sensing response. When the receptors are occupied by autoinducer, genes are turned on that give rise to bioluminescence. (C) Petri dish of cells at high densities exhibiting bioluminescence. (C) From Edward A. Quinto (2006), "Bioluminescent Plate Culture of the marine luminous microbe: Vibrio fischeri."

identified in other species. As seen in the figure, an autoinducer is a member of a class of molecules known as homeserine lactones, sometimes abbreviated as AHL which more precisely stands for *N*-acyl homoserine lactones. The concentration needed to induce luminescence was found to be $\approx 10^{-9}$ M, and interestingly, the molecule was also synthesized from scratch and shown to confer the expected biological activity. But what is the molecular mechanism associated with the presence of these autoinducer molecules in the medium?

4.6.1   Wiring Up Quorum Sensing

Stated simply, the consequence of the presence of high concentrations of autoinducer is ultimately the expression of critical genes that give rise to luminescence. Figure 4.38 summarizes in schematic fashion the quorum sensing process in *Vibrio harveyi*, the case study that will dominate our analysis in this chapter, because it has been so successfully investigated using the allosteric ideas that form the core of this book. The key point is that at high cell densities, autoinducer concentrations are sufficiently high that all the cells turn on their bioluminescence genes, as seen in Figure 4.38(C).

The original explorations and discoveries on the molecular circuitry in quorum sensing were focused on the Gram-negative bacterium *Vibrio fischeri*. In this case, the molecular circuit for the luciferase system that gives rise to bioluminescence is encoded on an operon known as the *lux* operon. In *Vibrio fischeri*, there are five such genes that are required for the production of bioluminescence, known as *luxCDABE*. This operon is regulated by LuxI and LuxR. LuxI is the synthase enzyme that produces the autoinducer. LuxR, in turn, is the transcriptional activator.

A more detailed molecular view of the underlying molecular circuits is given in Figure 4.39. As just noted, the LuxI enzyme is responsible for synthesis of the autoinducer (AHL). When at sufficiently high densities, this autoinducer binds the transcriptional activator LuxR, inducing in it the kind of allosteric change that forms the core substance of this book, thus licensing it to bind to DNA and to activate the quorum sensing genes. Though we will see that more complex

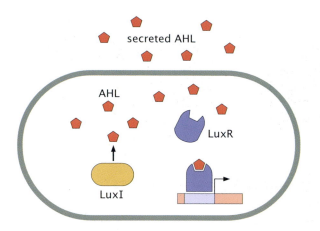

Figure 4.39

Allosteric nodes in the quorum sensing pathway in the Gram-negative bacterium *Vibrio fischeri*. The autoinducer molecule (AHL) is synthesized in the bacterium by LuxI. LuxR is an activator that adopts two different states depending upon whether it is bound to AHL or not, suggesting that it is another example of the allosteric phenomenon. Adapted from Ng and Bassler (2009).

quorum sensing architectures have been discovered in other organisms such as *Vibrio harveyi*, the allosteric behavior exhibited in Figure 4.39 serves as a call to arms for thinking about quorum sensing from an allosteric point of view.

The molecular circuit in *Vibrio fischeri* featured in Figure 4.39 involves the soluble transcriptional activator LuxR as its principal allosteric actor. By way of contrast, in the model system *Vibrio harveyi*, the receptor and regulator are distinct molecules, as seen in Figure 4.40. We see that there are three separate membrane receptors in the inner membrane of these quorum sensing bacteria. Depending upon whether the autoinduction signal is present, these receptors will be either inactive or active, with the result that only when active will these cells perform the reaction resulting in the emission of light. In this case, the molecular circuitry is quite analogous to that we saw earlier in the chapter in the context of chemotaxis. Binding of autoinducers to the histidine kinase quorum sensing receptor leads to phosphorylation of the response regulator, which is then licensed to activate its target gene.

Thus far, we have considered the wiring of the quorum sensing apparatus in the context of Gram-negative bacteria. An equally impressive array of work has gone into uncovering the mechanisms of quorum sensing in Gram-positive bacteria, as shown in Figure 4.41. One of the first observations we can make in this case is that the autoinducers are genetically encoded peptides. These peptides have a very low permeability across the cell membrane and thus, as seen in the figure, are carried across the membrane by transporters. The downstream signal transduction network is an example of a two-component signaling system like those shown in Figures 1.11 and 1.12 (p. 13). Here the peptide autoinducer binds the receptor histidine-kinase, which in its active form autophosphorylates. The receptor is then poised to transfer its phosphate group to a conserved histidine residue on the target response regulator, in this case a transcription factor.

The allosteric ideas developed so far in the book are relevant at multiple points in the cellular response to autoinducer molecules, as shown in Figures 4.39, Figure 4.40, and 4.41. For our purposes, we will consider two distinct kinds of allosteric interactions in these pathways. First, as seen in Figures 4.40 and 4.41, we will consider the way in which an autoinducer controls the probability that the membrane receptor will be in either the inactive or active state. Second, in our later chapter on transcription (chap. 8, p. 272)

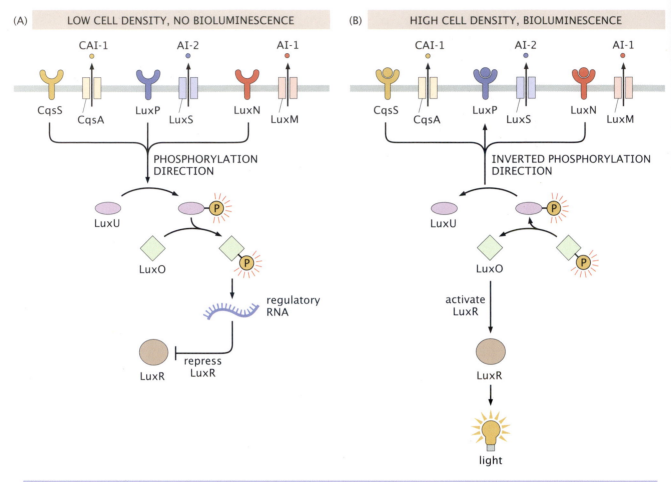

Figure 4.40

Molecular circuit involved in quorum sensing in the Gram-negative bacterium *Vibrio harveyi*. (A) There are three distinct surface receptors responsible for detecting different autoinducers. When autoinducer is absent, a regulatory RNA is produced which represses LuxR, with the result being no bioluminescence. (B) When autoinducer is present, LuxR is present and leads to transcription of the genes responsible for light generation. The "flow" of phosphorylation is in different directions in the two cases. Adapted from Ng and Bassler (2009).

we will explore how transcription factors like that shown in Figure 4.39 change their activity as a function of the presence of inducer molecules, though our focus there will not be on LuxR, since cleaner induction data exist for other transcription factors.

4.6.2   Dose-Response Curves in Quorum Sensing

Experimentally, one of the ways we characterize the operation of a system like the quorum sensing circuit just described is by appealing to the so-called dose-response curve. The idea of such a curve is to measure the magnitude of the response of the system as a function of the magnitude of the stimulus. In the quorum sensing context, this would mean examining how much the

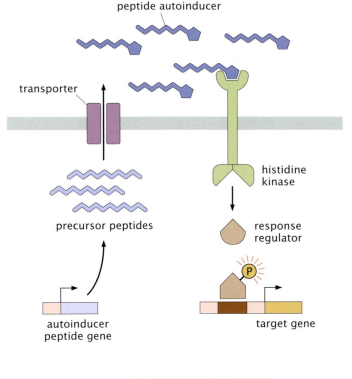

peptide autoinducer

transporter

histidine
kinase

precursor peptides

response
regulator

P

autoinducer
peptide gene

target gene

Figure 4.41

Schematic of membrane-receptor-mediated quorum sensing in Gram-positive bacteria. The peptide autoinducer is under genetic control and is actively transported out of the cell by a membrane-bound transporter. Binding of the peptide autoinducer to the quorum sensing receptor licenses the receptor to phosphorylate a histidine residue on the response regulator (transcription factor). Adapted from Ng and Bassler (2009).

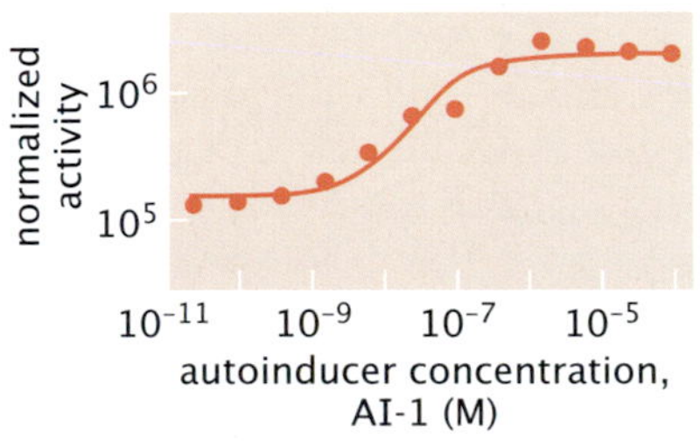

normalized activity

10^6

10^5

10^{-11} 10^{-9} 10^{-7} 10^{-5}

autoinducer concentration,
AI-1 (M)

Figure 4.42

Dose-response curve for quorum sensing receptor. The readout is light produced as a function of the AI-1 concentration for wild-type cells. Adapted from Swem et al. (2008); see Further Reading.

downstream genes in the quorum sensing circuit are expressed as a function of the concentration of autoinducer. However, in the interest of developing a detailed understanding of the system, one can focus on individual molecular components of the system and ask how those individual parts respond to their specific inputs. For example, in the context of quorum sensing, one might ask about the activity of the membrane receptors as a function of AI concentration.

An example of the wild-type response is shown in Figure 4.42. The readout of the response is the amount of light produced per cell (obtained by normalizing by the optical density) as a function of the autoinducer concentration. The dose-response curve has the characteristic sigmoidal shape that has already appeared multiple times throughout the book. Later in this section, we will see that entire families of dose-response curves were measured for wild-type cells and various mutants.

To get a sense of the way that the quorum sensing receptors work, consider Figure 4.43. In particular, what that figure shows us schematically is how the kinetics of switching between the active and inactive states depends upon the concentration of autoinducer, as indicated by the sizes of the arrows depicting

Figure 4.43

Kinetics of a quorum sensing receptor. The receptor can be in either the inactive or the active state depending on the concentration of autoinducer. For this simple model, we consider a monomeric receptor in the presence of AI-1. (A) At low concentrations of autoinducer, the active state is favored, as seen by the size of the arrows showing the magnitude of the rate constants. (B) At high autoinducer concentrations, the inactive state is favored.

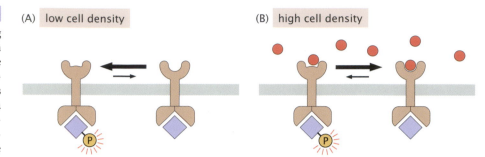
(A) low cell density   (B) high cell density

these different kinetics steps. An equilibrium statistical mechanical model of the relative probabilities of these different states uses the Boltzmann distribution to compute the statistical weight of each of the four competing states at each different autoinducer concentration. Then, by using these statistical weights, we can compute the probability of being active (p_{active}) as a function of autoinducer concentration. The challenge we now face in the remainder of the chapter is to implement our statistical mechanical understanding of the quorum sensing receptors using the MWC model.

4.6.3 Statistical Mechanics of Membrane Receptors

By now, we have begun to see the pattern in the statistical mechanical description of different kinds of MWC molecules. As usual, we begin by acknowledging the distinct states of the system and their corresponding statistical weights, as shown in Figure 4.44. Just as we did for the case of the chemotaxis receptors, we consider a quorum sensing receptor with only a single binding site for its cognate ligand, and we imagine that the receptor can be in either the active or the inactive state. As a result, as shown in the figure, the receptor can be in any one of four states, with two states of occupancy for both the inactive and active states.

It is convenient to sum the statistical weights of subsets of the states to access the statistical weights of the inactive and active states, respectively. To that end, by inspection of the states on the left column of Figure 4.44, we compute the partial partition function of the inactive states, resulting in

$$Z_{inactive} = e^{-\beta \varepsilon_I}(1 + e^{-\beta(\varepsilon_b^I - \mu)}), \quad (4.39)$$

and similarly, we compute the sum of the statistical weights associated with the active state by summing over the states on the right column of Figure 4.44, resulting in

$$Z_{active} = e^{-\beta \varepsilon_A}(1 + e^{-\beta(\varepsilon_b^A - \mu)}). \quad (4.40)$$

Together, the sum over these two partial partition functions gives the total partition function, namely,

$$Z_{tot} = Z_{active} + Z_{inactive}. \quad (4.41)$$

We can now compute the probability that the receptor will be active as a function of ligand concentration as

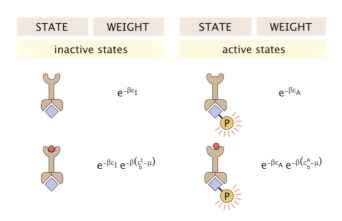

STATE	WEIGHT	STATE	WEIGHT
inactive states		active states	

$$p_{active} = \frac{Z_{active}}{Z_{active} + Z_{inactive}}, \qquad (4.42)$$

resulting in

$$p_{active} = \frac{e^{-\beta \varepsilon_A}(1 + e^{-\beta(\varepsilon_b^A - \mu)})}{e^{-\beta \varepsilon_A}(1 + e^{-\beta(\varepsilon_b^A - \mu)}) + e^{-\beta \varepsilon_I}(1 + e^{-\beta(\varepsilon_b^I - \mu)})}. \qquad (4.43)$$

Note that effectively we have done nothing more than recapitulate precisely the same equation that was advertised as the one equation to rule them all (see equation 1.2, p. 26), seen earlier in the context of ion channels and chemoreceptors. An equivalent way of writing this same result is to once again appeal to the chemical potential $\mu = \mu_0 + k_B T \ln c/c_0$. If we make this substitution, we then find that the probability of the active state can be written as

$$p_{active} = \frac{e^{-\beta \varepsilon_A}(1 + \frac{c_A}{K_A})}{e^{-\beta \varepsilon_A}(1 + \frac{c_A}{K_A}) + e^{-\beta \varepsilon_I}(1 + \frac{c_A}{K_I})}. \qquad (4.44)$$

This equation provides a framework for computing the dose-response curves as a function of the concentration of autoinducer. Indeed, the curve shown in Figure 4.42 is precisely the result of this model.

The theory just developed really reveals its power only when confronted with the very interesting question of quorum sensing mutants. Through genetic screens, "dark" mutants were identified; that is, they did not produce any bioluminescence. A representative example of a mutant dose-response phenotype is shown in Figure 4.45. There are multiple competing hypotheses for what is going on with those mutants. As noted in the original paper by Swem et al. (2008), "We had reasoned that the dark phenotypes of our LuxN mutants could stem from (1) a defect in the ability to bind AI-1, (2) a bias favoring the kinase state, (3) a defect in signaling, or (4) some combination of the above. The method of data collapse provides us with a powerful tool to distinguish among these possibilities." We now work our way up to an analysis of the data collapse in this problem that mirrors our earlier work to understand the data collapse in the chemotaxis mutants presented in Figure 4.35 (p.155).

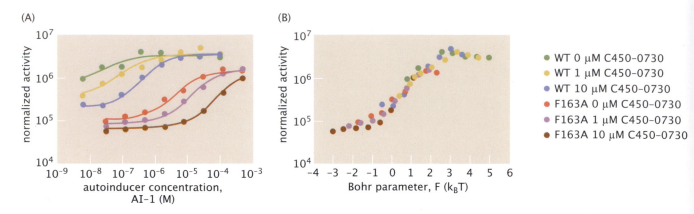

Figure 4.45

Dose-response curves for wild-type and mutant receptors in the presence of inhibitor molecules. (A) Activity as a function of AI-1 concentration for both wild-type and mutant LuxN receptors in the presence of the inhibitor C450-0730. (B) Data collapse for the quorum sensing system. The same data set as that of part (A) in which the activity is plotted as a function of the Bohr parameter. Adapted from Swem et al. (2008).

4.6.4 Statistical Mechanics of Membrane Receptors with Inhibitors

The dose-response curves of Figure 4.45 were also measured in the presence of molecules that inhibit LuxN signaling. For this more complicated case in which there is a competitor ligand, the determination of the total partition function requires us to sum not only over the states in which our ligand of interest is bound but also those states for which the competitor is bound. The states and weights for this case are shown in Figure 4.46. In particular, for the sum over all inactive states we have

$$Z_{inactive} = e^{-\beta \varepsilon_I}(1 + e^{-\beta(\varepsilon_b^I - \mu_a)} + e^{-\beta(E_b^I - \mu_c)}), \tag{4.45}$$

where we have defined ε_b^I as the binding energy for the ligand of interest and E_b^I as the binding energy for the competitor ligand. We have also defined the chemical potential μ_a for the autoinducer and μ_c for the competitor. Stated differently, there are two ligand reservoirs, one for the autoinducer and another for the competitor.

We follow similar analysis for the active states. Imitating our notation for the inactive state, we have

$$Z_{active} = e^{-\beta \varepsilon_A}(1 + e^{-\beta(\varepsilon_b^A - \mu_a)} + e^{-\beta(E_b^A - \mu_c)}). \tag{4.46}$$

With these results in hand, we can now compute the probability that the system will be in the active state as

$$p_{active} = \frac{e^{-\beta \varepsilon_A}(1 + e^{-\beta(\varepsilon_b^A - \mu_a)} + e^{-\beta(E_b^A - \mu_c)})}{e^{-\beta \varepsilon_A}(1 + e^{-\beta(\varepsilon_b^A - \mu_a)} + e^{-\beta(E_b^A - \mu_c)}) + e^{-\beta \varepsilon_I}(1 + e^{-\beta(\varepsilon_b^I - \mu_a)} + e^{-\beta(E_b^I - \mu_c)})}. \tag{4.47}$$

As we did previously, we can use the chemical potential for chemical species i as $\mu_i = \mu_0 + k_B T \ln c_i/c_0$ to simplify our expression. If we make this substitution,

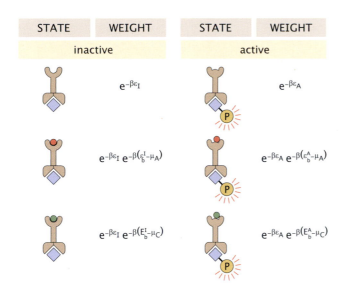

Figure 4.46

States and weights for a quorum sensing receptor in the presence of a competitor. The receptor can be in either the inactive or the active state. For this simple model, we consider a monomeric receptor in the presence of both AI-1 and a competitor molecule.

we then find that the active state can be written as

$$p_{active} = \frac{e^{-\beta \varepsilon_A}(1 + \frac{c_A}{K_A} + \frac{c_C}{L_A})}{e^{-\beta \epsilon_A}(1 + \frac{c_A}{K_A} + \frac{c_C}{L_A}) + e^{-\beta \varepsilon_I}(1 + \frac{c_A}{K_I} + \frac{c_C}{L_I})}, \quad (4.48)$$

where we have introduced the notation L_I and L_A as the dissociation constants for the competitor when the receptor is in the inactive and active states, respectively. This equation is the basis of the interpretation of the mutant data shown in Figure 4.45.

4.6.5   Data Collapse in Quorum Sensing

The same theorists who considered the chemotaxis problem earlier in the chapter turned their attention to the quorum sensing problem. Intriguingly, their labors went a long way to illustrating the broad reach of the MWC implementation of the allostery concept. Following in the footsteps of their own earlier work on data collapse in chemotaxis (see sec. 4.4.2 p. 153), they found the method of data collapse a useful tool for exploring the mechanisms of the different quorum sensing mutants.

To see how the data collapse works here, we can simplify equation 4.48 even further and put into a functional form convenient for exploring the data collapse by dividing top and bottom by the numerator, resulting in

$$p_{active} = \frac{1}{1 + e^{-\beta \Delta \epsilon_I} \frac{\left(1 + \frac{c_A}{K_I} + \frac{c_C}{L_I}\right)}{\left(1 + \frac{c_A}{K_A} + \frac{c_C}{L_A}\right)}}. \quad (4.49)$$

This finally implies that we can write p_{active} in the form

$$p_{active} = \frac{1}{1 + e^{-\beta F_{Bohr}}}, \quad (4.50)$$

where we define the Bohr parameter as

$$F_{Bohr} = \Delta\epsilon + k_B T \ln \frac{\left(1 + \frac{c_A}{K_I} + \frac{c_C}{L_I}\right)}{\left(1 + \frac{c_A}{K_A} + \frac{c_C}{L_A}\right)}. \tag{4.51}$$

To be specific about the implications of these calculations for thinking about the activity data, the key point is to plot p_{active} versus F_{Bohr} as opposed to as a function of the concentration of ligand. Figure 4.45(B) shows how when plotted with respect to the Bohr parameter the entirety of the data on quorum sensing considered here (and even more) reduces to a simple one-parameter family. The insight to emerge from this data collapse was that this one-parameter family of curves corresponds only to changes in the energy difference $\Delta\varepsilon = \varepsilon_I - \varepsilon_A$, providing a mechanistic interpretation of the nature of the mutants. The fact that the dissociation constants were unchanged in the mutants reveals that the mutant phenotype did not result from an inability of the receptors to bind AI-1, one of the competing hypotheses for the nature of these mutants.

4.7 Summary

This long chapter has delved deeply into the features of bacterial signaling. We have seen that bacteria are sophisticated information-processing machines with a host of different molecular circuits designed to take stock of the environment around them and to use that information to make good choices. One of the examples we focused on is the individualistic behavior of bacteria as they take stock of the possible food sources around them. Our second example considered the opposite extreme of social behavior in the context of quorum sensing. In both cases, we found that the statistical mechanics of allosteric molecules provided a powerful conceptual framework for thinking about these problems. The analysis reached its high point in the ability to develop data collapse for entire families of mutants. In the next chapter we continue with this theme of membrane-mediated signaling, this time focusing on eukaryotic membranes with their large and diverse collection of G proteins and their associated receptors.

4.8 Further Reading and Viewing

Berg H. C. (1993) *Random Walks in Biology*. Princeton: NJ: Princeton University Press. A modern classic that should be read and reread by anyone interested in physical biology.

Drews, G. (2005) "Contributions of Theodor Wilhelm Engelmann on phototaxis, chemotaxis, and photosynthesis." *Photosynth. Res.* 83:2534. A very interesting article on a fascinating scientist in the history of cell sensing and motility.

Endres, R. G. (2013) *Physical Principles in Sensing and Signaling: With an Introduction to Modeling in Biology*. Oxford: Oxford University Press. Excellent book covering all facets of the chemotaxis problem.

Hastings, W. iBioSeminars. "The discovery of quorum sensing." This great brief talk describes the work in the 1960s when Nealson, Hastings, and their colleagues made a series of discoveries culminating in the discovery of autoinduction, now called quorum sensing. This story is of great personal interest to me, since the beginnings for Hastings and Nealson were in the physiology course at the Marine Biological Laboratory, home to perhaps my greatest memories in my life in science.

Keegstra, J. M., K. Kamino, F. Anquez, M. D. Lazova, T. Emonet, and T. S. Shimizu (2017) "Phenotypic diversity and temporal variability in a bacterial signaling network revealed by single-cell FRET." *eLife* 6:e27455. This paper gives an example of the new insights garnered from single-cell FRET experiments on chemotaxis.

Keymer, J. E., R. G. Endres, M. Skoge, Y. Meir, and N. S. Wingreen (2006) "Chemosensing in *Escherichia coli*: Two regimes of two-state receptors. *Proc. Natl. Acad. Sci.* 103:1786–1791. This excellent paper cleanly shows how the MWC model can be used in the context of chemotaxis.

Mello, B. A., and Y. Tu (2005) "An allosteric model for heterogeneous receptor complexes: Understanding bacterial chemotaxis responses to multiple stimuli." *Proc. Natl. Acad. Sci.* 102:17354–17359. This paper gives a very clear discussion of how the MWC model can be generalized to include more than one type of input in the context of chemotaxis. The Supplemental Information has an excellent treatment giving intuitive (and rigorous) discussions of the phenotypic parameters relevant to MWC models, with special reference to the effective Hill coefficient.

Ng, W.-L., and B. L. Bassler (2009) "Bacterial quorum-sensing network architectures." *Ann. Rev. Genetics* 41:197 (2009). This excellent article describes how quorum sensing works in both Gram-negative and Gram-positive bacteria.

Shimizu, T. S., Y. Tu, and H. C. Berg (2010) "A modular gradient-sensing network for chemotaxis in *Escherichia coli* revealed by responses to time-varying stimuli." *Mol. Syst. Biol.* 6:382. This article goes beyond the discussion in this chapter to consider fascinating experiments in which the concentration presented to the bacterium is varied over time. Imposing time-dependent concentrations allows the adaptation network to be understood more deeply.

Sourjik, V., and H. Berg (2002) "Receptor sensitivity in bacterial chemotaxis." *Proc. Natl. Acad. Sci.* 99:123–127 and (2002) Binding of the *Escherichia coli* response regulator CheY to its target measured in vivo by fluorescence resonance energy transfer." *Proc. Natl. Acad. Sci.* 99:12669–12674. Classic experiments of the field of physical biology that show the kind of experimental design and data it takes to have a productive dialogue between theory and experiment, in this case driving much of the development of thinking in the theory of bacterial chemotaxis.

Swem, L. R., D. L. Swem, N. S. Wingreen, and B. L. Bassler (2008) "Deducing receptor signaling parameters from in vivo analysis: LuxN/AI-1 quorum sensing in *Vibrio harveyi*." *Cell* 134:461–473. This important paper shows a wonderful synergy between beautiful experiments and well-crafted theory. It is the basis of the entirety of my section on quorum sensing.

Tu, Y. (2013) "Quantitative modeling of bacterial chemotaxis: signal amplification and accurate adaptation." *Annu. Rev. Biophys* 42:337–359. This article gives a balanced treatment of the bacterial chemotaxis problem from the point of view of the MWC treatment described here but also shows how the Ising model has been invoked to describe the behavior of chemotaxis clusters. Much work still remains to be done to fully describe the bacterial chemotaxis problem.

Tu, Y. (2008) "The nonequilibrium mechanism for ultrasensitivity in a biological switch: Sensing by Maxwell's demons." *Proc. Natl. Acad. Sci.* 105:11737–11741. This paper shows how nonequilibrium effects have to be invoked to explain the temporal behavior of the bacterial flagellar motor, thus illustrating the shortcomings of the strictly equilibrium MWC framework, which forms the centerpiece of the chapter.

4.9 REFERENCES

Adler, J. (1973) "A method for measuring chemotaxis and use of the method to determine optimum conditions for chemotaxis by *Escherichia coli.*" *J. Gen. Microbiol.* 74:77–91.

Adler, J. (2011) "My life with nature." *Annu. Rev. Biochem.* 80:4270.

Briegel, A., D. R. Ortega, E. I. Tocheva, K. Wuichet, Z. Li, S. Chen, A. Müller, C. V. Iancu, G. E. Murphy, M. J. Dobro, I. B. Zhulin, and G. J. Jensen (2009) "Universal Architecture of Bacterial Chemoreceptor Arrays." *Proc. Natl. Acad. Sci. USA* 106:17181.

Engelmann, T. W. (1882) "Über Sauerstoffausscheidung von Pflanzenzellen im Mikrospektrum." *Pflüger, Arch.* 27:485–489, https://doi.org/10.1007/BF 01802976.

Greenfield, D., A. L. McEvoy, H. Shroff, G. E. Crooks, N. S. Wingreen, E. Betzig, and J. Liphardt (2009) "Self-organization of the *Escherichia coli* chemotaxis network imaged with super-resolution light microscopy." *PLoS Biol.* 7:e1000137.

Macnab, R. M., and D. E. Koshland Jr. (1972) "The gradient-sensing mechanism in bacterial chemotaxis." *Proc. Natl. Acad. Sci.* 69:2509–2512.

Maddock, J. R., and K. Shapiro (1993) "Polar location of the chemoreceptor complex in the *Escherichia coli* cell." *Science* 259:1717.

McAdams, H. H., B. Srinivasan, and A. P. Arkin (2004) "The evolution of genetic regulatory systems in bacteria." *Nat. Rev. Genetics* 5:169–178.

Nealson, K. H., T. Platt, and J. W. Hastings (1970) "Cellular control of the synthesis and activity of the bacterial luminescent system." *J. Bact.* 104:312–322.

Nelson, P. (2017) *From Photon to Neuron: Light, Imaging, Vision.* Princeton, NJ: Princeton University Press.

Ng, W.-L., and B. H. Bassler (2009) "Bacterial quorum-sensing network architectures." *Ann. Rev. Genetics* 41:197.

Rosson, R. A., and K. H. Nealson (1981) "Autoinduction of bacterial bioluminescence in a carbon limited chemostat." *Arch. Microbiol.* 129:299–304.

Turner, L.W.S., W. S. Ryu, and H. C. Berg (2000) "Real-time imaging of
fluorescent flagellar filaments." *J. Bacteriol.* 182:2793.

Yang, W., C. K. Cassidy, P. Ames, Diebolder C. A., Schulten K, Luthey-Schulten
Z, Parkinson JS, and A. Briegel (2019). "In situ conformational changes of
the *Escherichia* coli serine chemoreceptor in different signaling states. mBio
10:e00973-19.

5

THE WONDERFUL WORLD OF
G PROTEINS AND G PROTEIN–COUPLED
RECEPTORS

> Ask not what mathematics can do for biology, Ask what biology can do for mathematics.
>
> —Stanislaw Ulam

Cell signaling often begins at membranes. Chapter 3 (p. 77) showed us how ion channels open and close in response to environmental stimuli such as the presence of a specific ligand. This channel gating is the basis of a myriad of physiological responses ranging from muscle contraction to vision. Chapter 4 (p. 124) offered a second example of membrane proteins engaged in signaling with the case studies of bacterial chemotaxis and quorum sensing. In both of these cases, membrane receptors have the capacity to bind ligands and to deliver that message to the cellular interior.

In the present chapter, we consider perhaps the most important and ubiquitous membrane-bound signaling molecules, the so-called G protein–coupled receptors (GPCR). The G proteins are able to transduce signals initiated by the very large family of G protein–coupled receptors, which include the opsins but also a huge number of other transmembrane proteins (almost always with seven transmembrane helices) that are involved in a wide variety of physiological functions. For example, the receptor that detects the hormone adrenaline, which floods our bodies when we are excited or frightened, is a G protein–coupled receptor. It has been estimated that between 1 and 3 percent of the proteins encoded in animal genomes are members of this family. Their downstream effectors, the heterotrimeric G proteins, also come in a variety of forms.

These proteins can exist in either inactive or active conformations, rendering them completely amenable to the same kind of statistical mechanical thinking we have used to describe ligand-gated ion channels and the receptors that mediate bacterial chemotaxis and quorum sensing. In the pages that follow, we begin with a biological story that illustrates one of the many roles of GPCRs. This is followed by a general description of the GPCR concept, namely, how they switch between inactive and active conformations and what the consequences are for their partner G proteins. With those ideas in hand, we then examine a few more case studies, focusing on the all-important model GPCRs, the

Figure 5.1

Batesian mimicry in butterflies. The butterflies on the left are poisonous. Those on the right are palatable but through Batesian mimicry give the same visual appearance as their poisonous counterparts. Adapted from Barton et al. (2007). Courtesy of Nipam H. Patel.

β-adrenergic receptor and rhodopsin, and finish with a physiological story that links GPCRs and ion channels.

5.1   The Biology of Color

Modern biology is built on the travails of the great naturalists who went out into an unexplored and dangerous world, and brought back incredible and unexpected samples and ideas. Perhaps none of these intrepid travelers inspires me more than Alfred Russel Wallace, who spent four years in South America, nearly died after his ship sank off Bermuda, and was rescued to return to England so he could do it all over again during an eight-year odyssey in modern-day Indonesia. During his trip to South America, Wallace was accompanied by another naturalist, Henry Walter Bates, with whom he intended to "gather facts towards solving the problem of the origin of species," as Wallace put it in a letter to Bates. Bates is now known to science, among other things, for Batesian mimicry, the evolutionary trick in which a palatable species mimics the appearance of a second, noxious species, as shown in Figure 5.1.

So impressed was he by Bates's insights, that Darwin could not contain himself, writing in chapter 14 of *On the Origin of Species*, "Why then, we are naturally eager to know, has one butterfly or moth so often assumed the dress of another quite distinct form; why to the perplexity of naturalists has Nature condescended to the tricks of the stage?" He then goes on to posit the mechanistic underpinnings of this dress by noting, "Now if a member of one of these persecuted and rare groups were to assume a dress so like that of a well-protected species it would often deceive predacious birds and insects, and thus escape entire annihilation. This we believe is the true explanation of all this mockery."

Color and evolution is one of those great biological mysteries that is constantly before our eyes and yet often taken for granted. In this section, I tell the story of the humble field mouse and how it has shed light not only on the evolution of coloration but also into the vast role of G protein–coupled receptors in cell signaling.

5.1.1   Crypsis in Field Mice

In a classic experiment bolstering the hypothesis of the fitness benefits of coat coloration, Kaufman (1974; see Further Reading) used two different

Figure 5.2

Different coat colors within the same species of mice. These coat colors provide camouflage in different habitats in Florida, varying from "oldfields which are vegetated and have dark loamy soil, and coastal sand dunes which have little vegetation and brilliant white sand." As seen in the figure, the mice in these habitats have colors that blend more with their corresponding habitats. Adapted from Hoekstra (2010); see Further Reading. Courtesy of Hopi E. Hoekstra. Habitat photos courtesy of Sacha Vignieri.

phenotypes of the field mouse *Peromyscus polionotus* (see Figure 5.2) as prey for owls. The key point of these experiments was that of the two colors of mice, one was more conspicuous than the other in each of the sets of conditions, and hence it was possible to measure their rate of capture and how it depended upon the relative colors of mouse and soil. Consistent with our intuition, but at the same time with the humility to recognize how often such intuitions are wrong, when coat color matched the surroundings, the mice were less likely to be captured by owls.

A modern version of this experiment was done in the wild, not with living mice but, rather, with plastic mouse models. The "coat colors" of the plastic mice mimicked the natural colors seen in mainland and beach environments, as shown in Figure 5.2. In this case, the painted mouse models were "released" in the different habitats and then collected after having been attacked by a variety of disappointed predators. By examining the nature of the injuries to these unfortunate plastic mice, in some cases researchers could even determine whether the predators came from the sky or the land. Perhaps more important, these experiments measured the correlation between survival rate and coat color in much the same way that Kaufman had done earlier, with the result shown in Figure 5.3, where it is clear that coat color confers a survival advantage when it is matched to the surrounding environment.

One aspect of the story of coat coloration in field mice that makes it so appealing intellectually is that the researchers who made these discoveries made it a multiscale intellectual voyage running the gamut from field ecology experiments like those just described all the way to the molecules of the cell such as the individual membrane proteins that confer coat color. To find those molecular players required a dedicated genetic study involving repeated rounds of crosses of mice with different coat colors and investigation of the genetic loci responsible for those differences. Figure 5.4 shows the F, F_1, and F_2 generations of these crosses with corresponding schematics of the chromosomal regions responsible for differences in coat color. The interested reader is invited to consult the suggestions in Further Reading at the end of the chapter to learn more about these impressive and fascinating experiments. With regret at missing the opportunity

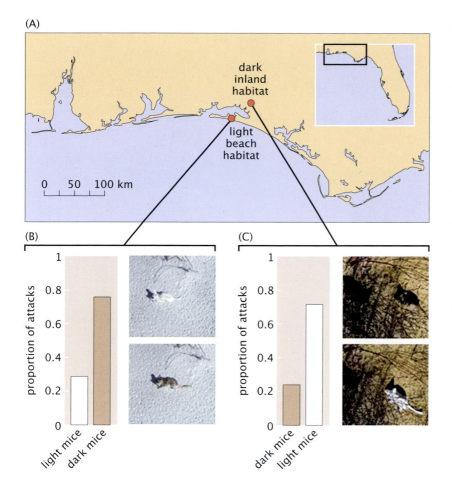

Figure 5.3

Survival and coat color. (A) Plastic mouse models were used to measure the fraction of "mice" with different coat colors that were attacked in two different environments. When coat color matched the environment, the fraction of mice that were attacked was reduced significantly. (B) Proportion of attacks on light mice and dark mice in a light beach habitat. (C) Proportion of attacks on light mice and dark mice in a dark inland habitat. Adapted from Vignieri, S. N., J. G. Larson, and H. E. Hoekstra (2010), "The selective advantage of crypsis in mice." *Evolution* 64, 2153–2158, fig. 1. © 2010 S. N. Vignieri, J. G. Larson, H. E. Hoekstra. Journal compilation; © 2010 The Society for the Study of Evolution. Used with permission of the publisher, John Wiley & Sons.

to tell such an interesting story, we turn instead to the story of one of the key molecules that emerged from these genetic studies, the G protein–coupled receptor known as the melanocortin-1 receptor (Mc1r).

5.1.2 Coat Color and GPCRs

As a result of crosses like those shown in Figure 5.4 and the genetic analyses that followed, several key molecular candidates presented themselves as driving differences in coat color. One of the most important such candidate is a G protein–coupled receptor known as the melanocortin-1 receptor (Mc1r). The G protein–coupled receptors (GPCR) are a huge class of membrane-bound receptors characterized by a particular seven transmembrane α-helix structure, as indicated in Figure 5.5 and will be described in more detail in section 5.2 (p. 177). The GPCR protein can bind a specific ligand, which has the effect of switching it from an inactive to an active conformation. When activated, the GPCR can interact with an inactive G protein (see Figure 5.5(A)), which is then activated.

Mc1r is a powerful example of a GPCR molecule that acts as a switch, rendering the decision as to whether dark eumelanin or light phaeomelanin is produced in the pigmentation-producing melanocyte cells. The melanocytes

Crosses of mice with different coat colors. The mice in the parental generation have different coat colors. The F_1 mice all have the same color, but in the second generation, interesting variations in coat color arise. Beneath each mouse is a schematic representation of their chromosomes showing different combinations of genetic regions from their parents. Adapted from Hoekstra (2010). Courtesy of Hopi E. Hoekstra.

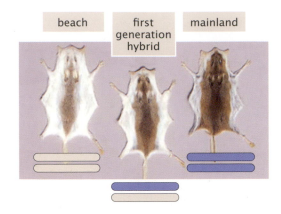

beach | first generation hybrid | mainland

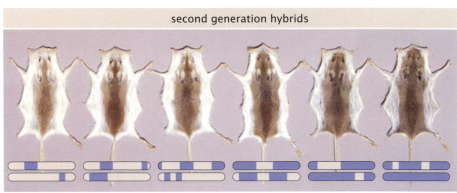

second generation hybrids

The G protein–coupled receptor paradigm. (A) The G protein–coupled receptor (GPCR) is a seven-transmembrane helix protein that can bind a ligand externally and that interacts with the G protein on the cytoplasmic side of the cell. (B) Structure of G protein–coupled receptor. (C) Structure of a G protein.

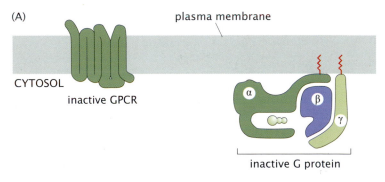

(A) plasma membrane

CYTOSOL

inactive GPCR

α β γ

inactive G protein

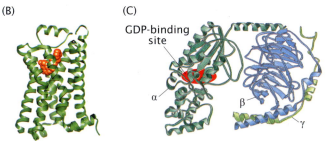

(B)

(C)

GDP-binding site

α β γ

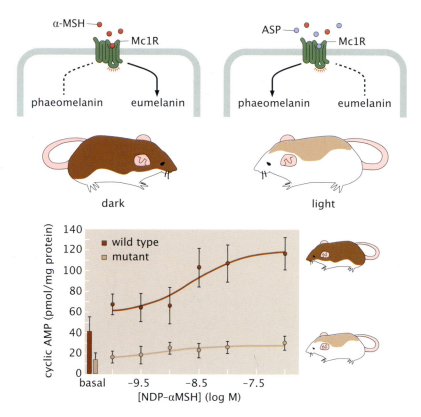

Figure 5.6

The coat color switch. Binding of α-MSH to the Mc1r protein results in the production of the darker eumelanin. When ASP is present, it competes for binding to Mc1r, resulting in the phaeomelanin phenotype.

Figure 5.7

Activity of the Mc1r receptor. The activity of the receptor is read out on the basis of cAMP accumulation. The wild-type receptor has a characteristic sigmoidal response with increasing concentration of α-MSH. By way of contrast, the mutant receptor binds the α-MSH weakly, thereby resulting in low activity. Adapted from Hoekstra et al. (2006).

are found in the skin's epidermis and produce melanin that confers coat and skin color. The Mc1r receptors present in the cells of the epidermis have several ligands—α-MSH (alpha-melanocyte-stimulating hormone) and ASP (Agouti signaling peptide)—which turn the switch either on or off. When α-MSH is present, the melanocyte expresses dark pigment, as shown in Figure 5.6. When ASP is present, it competes with α-MSH for the binding to the same site. If ASP is at sufficiently high concentrations to compete off α-MSH, then the light pigment is produced.

As with our earlier study of a myriad of other examples ranging from ligand-gated ion channels to quorum sensing receptors, our functional window onto the behavior of these molecules is often through the dose-response curve. The activity of the melanocortin-1 receptor has been tested as a function of the α-MSH ligand, as shown in Figure 5.7 for both wild-type and mutant versions of the receptor. Functionally, the binding of α-MSH leads to the activation of the Mc1r and its downstream consequences in the form of pigmentation.

How are these proteins linked to the light and dark phenotypes? The genetic screen described earlier discovered a particular mutation in the Mc1r protein that effectively inactivates the protein. As seen in Figure 5.8, this mutation results in an arginine at position 65 being replaced with a cysteine. As mentioned in the context of Figure 5.6, the mutation leads to reduced binding of α-MSH. When the switch is "off," instead of the dark eumelanin, the light phaeomelanin is produced. This reduced activity was revealed in Figure 5.7 and is seen for several species of mice in Figure 5.8(B). Note that for the Atlantic Coast field mouse, the Arg[65]Cys mutation is *not* found. On the other hand,

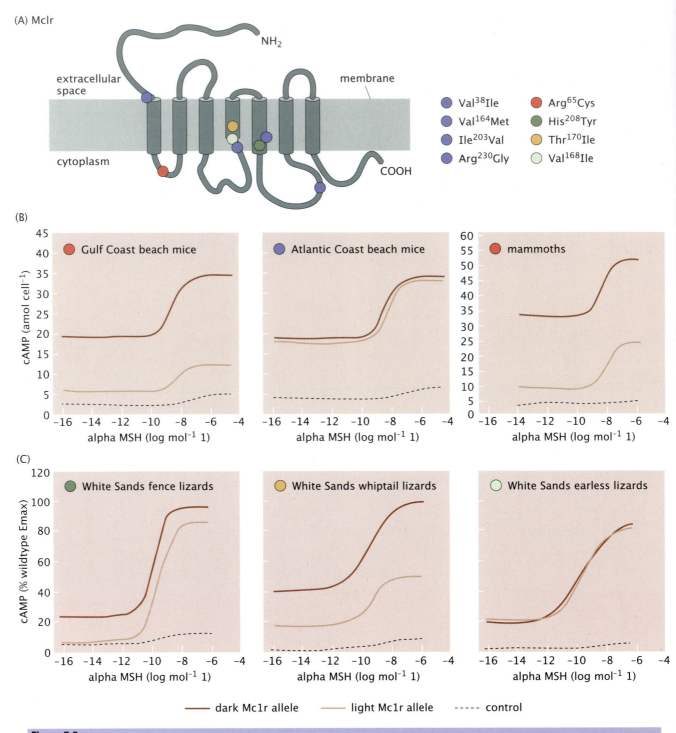

Figure 5.8

Mutations in Mc1r. (A) The seven-transmembrane domain receptors have a particular transmembrane topology. This schematic illustrates some of the key mutations found in mice, mammoths, and lizards that are thought to be responsible for differences in coat color. (B) Functional test of mutant Mc1r proteins. The dose-response curves shed light on the functional response of these different mutants. None of the mutations (blue) in the Atlantic Coast beach mice showed a measurable difference in receptor activity, so only one graph is shown that is typical of all four mutations. (C) Functional test of mutant Mc1r proteins in lizards. Adapted from Manceau et al. (2010).

interestingly, in a case of convergent evolution, precisely the same $Arg^{65}Cys$ mutation was discovered in the woolly mammoth, as shown in Figure 5.8(B).

This very interesting case study demonstrates how in the context of one particular GPCR, evolutionarily important differences in pigmentation confer fitness effects. That said, we need to finish on a note of caution. Though I enthusiastically embraced this particular example to make a point about allostery and GPCRs, pigmentation evolution can occur for very different mechanistic reasons, even when the mutations are in the same protein, as shown in Figure 5.8. For example, in the fascinating case of lizards that live in the dunes of White Sands, New Mexico, as shown in Figure 5.9, some mutations in the Mc1r gene rather than leading to an abrogation in the direct signaling capabilities of the protein instead affect its ability to integrate into the membrane in the first place. But both mutations result in a change of coat color.

5.2   The G Protein–Coupled Receptor Paradigm

As we have now seen repeatedly throughout the book, the question of how cells interact with signals from their environment often starts with a membrane protein whose job it is to detect some kind of signal. The story of control of pigmentation in vertebrates by the melanocortin-1 GPCR gave us a hint of the importance of these proteins in biology. Now we undertake a brief introduction to the mechanistic properties of these proteins and their G protein partners.

Figure 5.10 give us a schematic of the kinetic progression of signals through the G protein–coupled receptor signaling cascade. As shown in the top panel, in the absence of the signaling ligand, both the GPCR and the G protein are inactive. Note that both the α and γ subunits of the G protein are tethered to the lipid bilayer membrane. Binding of a ligand to the GPCR unleashes a succession of molecular processes. First, in keeping with the general allosteric theme of the book, the probability that the GPCR is in the active state is greatly increased by virtue of this ligand binding. When it is in this active state, we can then think of the GPCR itself as the regulatory ligand for the G protein, as shown in the middle panel of Figure 5.10. When the G protein is activated, the α subunit releases GDP. Parenthetically, we note that the very name G protein refers to the fact that these proteins bind guanine nucleotides in the form of both GDP and GTP. When the GDP has been released from the α subunit, it can then bind a GTP molecule, activating it in turn, which now dissociates it from its $\beta\gamma$ partner, which is also activated. As shown in the bottom panel of Figure 5.10, both the α and $\beta\gamma$ subunits are now in an active state, ready to perform their own jobs. With this brief explanation, we are now in a position to explore some of the most famous examples of GPCRs that have led the way to our modern understanding of these important proteins.

5.3   Paradigmatic Examples of GPCRs

Several GPCRs have been paradigmatic in catapulting the field forward, and we consider two of them here. First, we consider the β-adrenergic receptor, as shown in Figure 5.11. The second example is rhodopsin, shown in Figure 5.12.

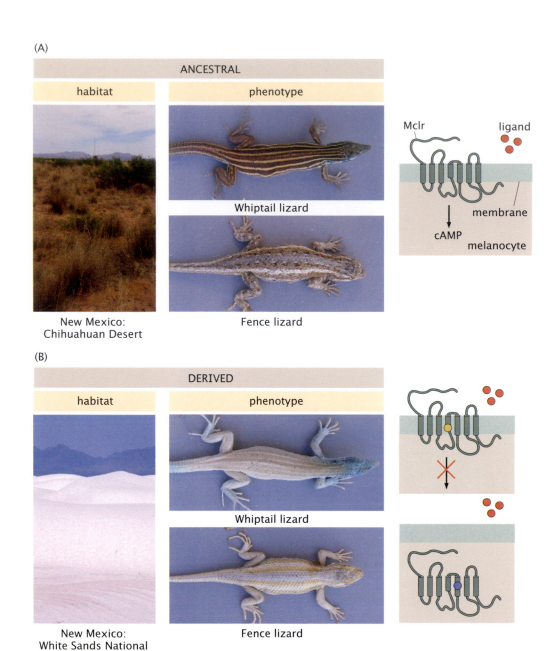

Figure 5.9

Lizard color and Mc1r. (A) Natural habitat in the New Mexico desert and corresponding phenotypes. (B) For those lizards living in the White Sands environment, there are several distinct mutations, one of which alters the function of the protein and the other of which negatively impacts its ability to incorporate into the membrane. Adapted from Manceau, M., V. S. Domingues, C. R. Linnen, et al. (2010), "Convergence in pigmentation at multiple levels: mutations, genes and function," *Phil. Trans. R. Soc.* 365:2445, fig. 4. Reproduced with permission of the Royal Society via Copyright Clearance Center.

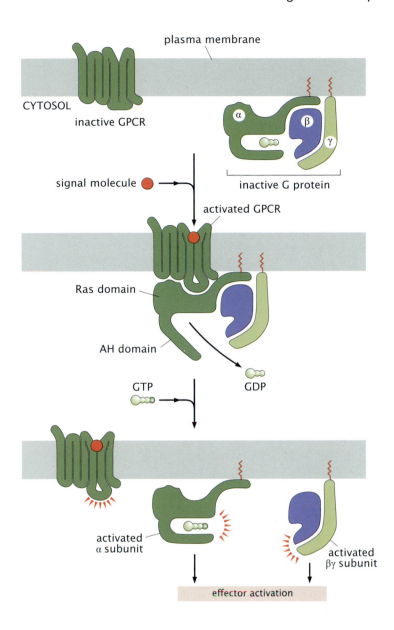

Figure 5.10
The G protein–coupled receptor paradigm. In the top panel, the GPCR is inactive, as is the G protein itself. Binding of a
signaling ligand to the GPCR activates it,
which in turn activates the G protein by
inducing release of GDP. Binding of GTP
activates the G protein, which dissociates
into an activated α subunit and an activated $\beta\gamma$ subunit. Adapted from Alberts
et al. (2016); see Further Reading.

5.3.1   The β-Adrenergic Receptor

The long history of how GPCRs acquired such a central place in our understanding of modern biology is a detective story that led researchers down many
blind alleys. To get a sense of the tortuous path to our modern thinking on these
iconic molecules, the reader is highly encouraged to see the lectures of Robert
Lefkowitz cited in the Further Reading section at the end of the chapter.

   In the beginning of the modern era in molecular biology, many were skep-
tical as to whether GPCRs even existed. In his Nobel Lecture, which serves as
the template for much of the story described here, Robert Lefkowitz (2012) tells
us of the views of noted pharmacologist Raymond Ahlquist, who on the subject of what are now known as GPCRs said: "This would be true if I were so

Figure 5.11

Structure of the β-adrenergic receptor. (A) Inactive state of the molecule. (B) Active state of the molecule.

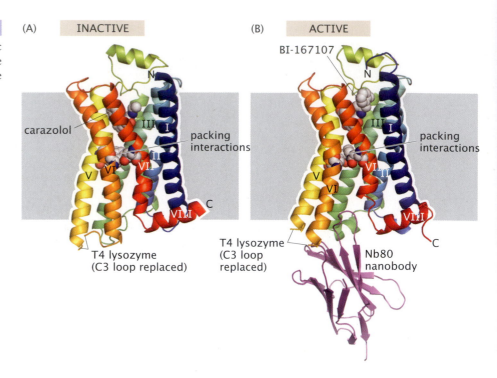

(A)   INACTIVE

(B)   ACTIVE

presumptuous as to believe that α and β receptors really did exist. There are those that think so and even propose to describe their intimate structure. To me they are an abstract concept conceived to explain observed responses of tissues produced by chemicals of various structure." This reminds me of the way the concept of the gene also started as an abstraction in the hands of Mendel and had to await the labors of Thomas Hunt Morgan and his gene hunters to be promoted from an abstraction to a particular molecular reality.

As a result of this skepticism regarding the existence of GPCRs, the burden of proof was high, requiring a series of experiments of increasing sophistication, each of which addressed some lingering doubts. As is always the case in science where skepticism is rooted out by careful and unequivocal experiments, these experiments end up providing the foundations for a subject that in the end can be dignified with the title "knowledge." The first step in this progression of experiments was purifying the receptors by fishing them out using a radioligand binding assay. This general idea of following the radioactivity has been used repeatedly in biochemistry in incredibly clever ways. For example, Calvin and Benson (1948) entitled their classic studies the "path of carbon" in photosynthesis, and this path was followed by measuring which molecules were radioactive as a function of time. The binding studies for the β-adrenergic receptor also began to reveal the allosteric character of the GPCRs through analysis of competition assays with both agonists and antagonists.

But questions remained. Are these proteins that bind the relevant ligands able to activate the allied biological processes such as the stimulation of adenylate cyclase in the case of the β-adrenergic receptor? To find out, functional assays were developed in which erythrocytes from *Xenopus laevis* were fused with lipid vesicles containing purified β-adrenergic receptor proteins. The

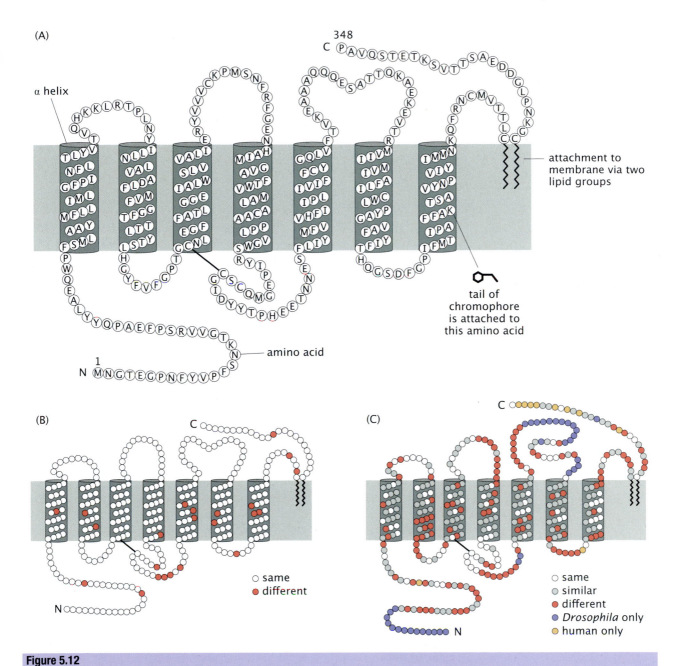

Figure 5.12

Sequence and structure of rhodopsin. (A) 348 amino acid sequence of human rhodopsin. (B) Comparison of sequence of human and bovine rhodopsin with differences in a particular amino acid indicated by red color. (C) Comparison of sequence of human and *Drosophila* rhodopsin. Compared with (B), there are many more differences between the rhodopsins. Adapted from Rodieck (1998); see Further Reading.

Xenopus cells lacked these specific GPCRs, but through the process of membrane fusion with the vesicles, the cells thereby acquired a membrane complement of these receptors. As a readout of functionality, it was shown that cAMP accumulated in the *Xenopus* cells when they were exposed to the specific β-adrenergic drugs that interacted with the reconstituted GPCR.

Once it had been proven that the receptors exist and that they had the capacity to bind ligands and induce downstream molecular processes, the next key phase in the creation of the modern understanding of GPCRs was developing the tools to clone these proteins. Because scientists had highly purified receptor protein, they could obtain small stretches of the amino acid sequence, thus providing the clues needed to design nucleotide probes that would make it possible to amplify the gene for β-adrenergic receptor. An outcome of this stage in the research was the realization that the seven-transmembrane structure is a generic feature of GPCRs. The cloned sequence of the gene revealed a key hallmark of these proteins, namely, the hydrophobic regions associated with the seven-transmembrane helixes. I find it interesting that we now take these seven-transmembrane α-helices for granted, but at the time they came as a surprise. Lefkowitz notes that until the time his research group cloned these proteins, those canonical seven-transmembrane structures had been seen only in rhodopsin and bacteriorhodopsin, providing a false lead indicating that somehow those structures were a feature only of light-sensitive proteins, not GPCRs in general. Work on the adrenergic receptors changed all that.

Once the β-adrenergic receptor was understood, many others followed. This paved the way for creating chimeric proteins that would allow mixing and matching of subunits from different GPCRs, making it possible to dissect the functional roles of these different parts. Just as the achievements of biochemistry are deeply impressive, so too, are the advances born of modern genetics and DNA science. The reason the chimeras were so important is that in the era before the structures were even known, it was possible to figure out which parts of the proteins were responsible for the "core" functions of ligand binding and activating G proteins. One set of chimeras focused on the α_2- and β_2-adrenergic receptors, which harbor similar sequence and structure but have totally different functions. The α_2-adrenergic receptor inhibits adenylate cyclase activity, while the β_2-adrenergic receptor stimulates this activity. Impressive and definitive experiments showed that the α_2-adrenergic receptor could be converted from an inhibitor to an activator by replacing one cytoplasmic loop with the corresponding loop from the β_2-adrenergic receptor. Amazing! I have loved similar stories in the context of transcription factors (see Ptashne and Gann 2002 in Further Reading), where DNA-binding domain and activation domain chimeras were similarly explored, turning transcription factors for one gene into transcription factors for another gene by swapping the DNA-binding domain. Ultimately, these examples are a monument to my personal favorite kind of science, namely, indirect work, in which major insights are inferred rather than observed directly. For example, the existence of atoms was inferred long before scanning tunneling microscopes could manipulate individual atoms on surfaces. In the case of the key structural features of the GPCRs, this was a kind of structural biology without the atoms, much like the genomics without sequences in the context of hemoglobin, we discussed in section 7.1.3 (p. 236). The last big step in this progression was determining the structures of GPCRs themselves, with examples like the β-adrenergic receptor shown in Figure 5.11 providing some of the first critical direct insights into the connection between structure and function.

To my mind, it is very important to tell stories like the one described by Lefkowitz that I have paraphrased here. Specifically, this story represents a classic example of the systematic, detailed, tedious work of actually knowing things. Speaking of one of his polyacrylamide gels, Lefkowitz notes: "This figure represents approximately one decade of work by a number of devoted students and postdocs," and I for one am deeply grateful for these kinds of labors. Ultimately work like this has as its ambition building the edifice of knowledge, in contrast with the frustrating current insistence that the only truly exciting and important results at any given time are those that are viewed as surprising.

5.3.2   Vision, Rhodopsin, and Signal Transduction

Despite all the anatomical variety of animal eyes, the fundamental light-driven chemical reaction that provides vision for all animals is conserved, namely, the isomerization of retinal. For vision, the protein that acts as the receptor for photons is called opsin and is well-conserved at the level of amino acid sequence among all animal species. Figure 5.12 shows the amino acid sequence for the human rhodopsin as well as the differences between that of humans and cows, and humans and flies. As seen in the figure, the opsin associated with animal vision is a transmembrane protein with seven helices that which holds the retinal molecule covalently attached to a lysine residue. Figure 5.13 shows how the close packing of the amino acid side chains surrounding the retinal constrain it in a bent configuration. Absorption of a photon by the 11-*cis* retinal in an opsin causes it to flip into the all-trans form, leading to a protein conformational change that initiates a complex signal transduction cascade, eventually resulting in a change in the transmembrane potential for the cell.

Indeed, rhodopsin is another supremely important example of a GPCR that has been paradigmatic in unearthing the structure and function of GPCRs and G proteins more generally. An appealing aspect of the rhodopsin case study is that in many ways it forces us to expand our view of allostery and regulatory ligands, because in this case, the ligand is a photon. And that ligand interacts with a molecule of retinal, shifting it from one conformational state to another.

In an animal eye, many copies of the opsin protein and its associated light-detecting retinal are expressed in specialized photoreceptor cells, whose ultimate function is to convert the informational signal received by absorbing photons into an electrical signal that can be communicated to neurons in the brain. Vertebrates that are able to distinguish different wavelengths of light and perceive them as different colors typically have at least two kinds of photoreceptor cells, called rods and cones. The rods and cones in a salamander eye are shown in Figure 5.14(A). Rods are very sensitive to low levels of light, being capable of perceiving the presence of a single photon. Cones function well only at higher levels of light but are usually specialized for detecting light of different colors.

How is it possible for a rod cell to detect a single photon? The cell packs many copies of the opsin protein into a series of stacked membrane discs that are orthogonal to the direction of incident light. A small section of a rod cell is shown in Figure 5.14(B), illustrating the extremely dense and regular packing

Figure 5.13

Light as ligand and the structure of rhodopsin. (A) The inactive state of the rhodopsin molecule. (B) The light activated structure of rhodopsin.

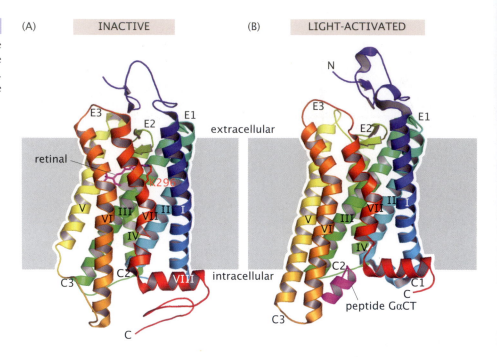

(A) INACTIVE   (B) LIGHT-ACTIVATED

of the discs. Thus, an incoming photon that does not happen to collide with an opsin in the first disc has another chance to be absorbed when passing through the second disc and up to several thousand chances for the longest rod cells found in vertebrates. To make these claims more specific, let's estimate the number of rhodopsin molecules in a rod outer segment.

ESTIMATE

Estimate: Number of Rhodopsin Molecules in a Rod Outer Segment

To get a sense of the possibility that a photon will encounter a rhodopsin molecule in this estimate we further explore the geometry of the rod cells shown in Figure 5.14. As seen in the figure, the diameter of a rod cell is between 1 and 2 μm, while its length is of order 10 of its diameters. The spacing between the membrane discs in the rod cell is roughly 25 nm. Hence, for a 20 μm long rod cell, we would estimate there are roughly 1000 such membrane discs. In a given membrane disc, if we assume the rhodopsin molecules are packed at a modest density with a mean spacing of roughly 10 nm (or 25,000 rhodopsins/μm^2), this still implies that there are over 100,000 rhodopsin molecules in every disc, leading to an estimate of in excess of 10^8 rhodopsin molecules in every rod cell.

The central problem in animal vision is translating the conformational change of the retinal molecule that has absorbed a photon into some kind of signal that can be perceived by the brain of the animal, and that is where G proteins and GPCRs come into the story. The informational transactions in animal nervous systems are mostly carried out by alterations in the transmembrane ion gradient, resulting in electrical signals. Vision is no exception. In the visual

(A)   rods

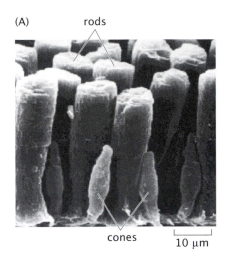

cones   10 µm

(B)

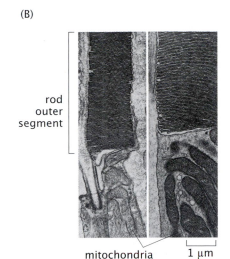

rod
outer
segment

mitochondria   1 µm

Figure 5.14

Animal photoreceptor cells. (A) This scanning electron micrograph shows the organization of rods and cones in the retina of a salamander. The relative sizes and numbers of rods and cones varies for different vertebrate animal species. (B) Electron micrographs of thin sections through a mammalian rod cell (left) and cone cell (right) reveal the close packing of the membrane discs within the outer segments that hold the rhodopsin. Below the outer segments, mitochondria pack the inner segments to provide the large amounts of ATP necessary for phototransduction. Note that these rod and cone cells are substantially smaller than those for salamanders as revealed in part (A). (A) from S. Mittman and D. R. Copenhagen, courtesy of S. Mittman; (B) adapted, courtesy of the Estate of Toichiro Kuwabara.

system, however, the effect of light-induced retinal isomerization on transmembrane ion gradients is much more indirect. The reason for this is quantitative: in bacteriorhodopsin, the absorption of one photon results in the transmembrane transport of one hydrogen ion (i.e., one charge equivalent). This amount of charge difference is orders of magnitude too low to be detected as a voltage change. For a rod cell, a change in membrane potential of 1 mV would require transport of some millions of ions. Furthermore, the rhodopsin in a rod cell is not located in the plasma membrane, where the relevant voltage change must take place but, rather, in internal disc membranes. How can the isomerization of one retinal cause such a large change in cell behavior on a different membrane compartment?

The overall scheme used in vertebrate photoreceptors is shown in Figure 5.15. The signal generated by the retinal conformational change is passed through several important intermediate signaling proteins before ultimately affecting the conductance of ion channels in the plasma membrane. The resting membrane potential in most cells is established by ATP-consuming ion pumps that export sodium ions while importing potassium ions, coupled with potassium leak channels that allow some potassium ions back outside, resulting in a net transmembrane potential that is negative on the inside. In rod cells kept in the dark, a specific cation channel also remains open, allowing sodium and calcium ions to flow back into the cell, down their electrochemical gradients. Because of this extra leakiness, the resting potential of a rod cell membrane is only about −30 mV, much closer to neutral than the resting potential of a typical nerve or muscle cell. Critically, though, the channel allowing reentry of sodium and potassium is a ligand-gated channel, remaining open only when its ligand is present, an intracellular signaling molecule called cyclic GMP (cGMP) (see sec. 3.7, p. 106, where we discussed these channels in detail), which is generated from GTP by a cyclase enzyme. When the concentration of cGMP drops, the leaking channels close, causing the membrane to slightly hyperpolarize (i.e., the potential becomes more negative).

So, how does rhodopsin activation alter the intracellular concentration of cGMP? The answer to that question forces us to harken back to the general

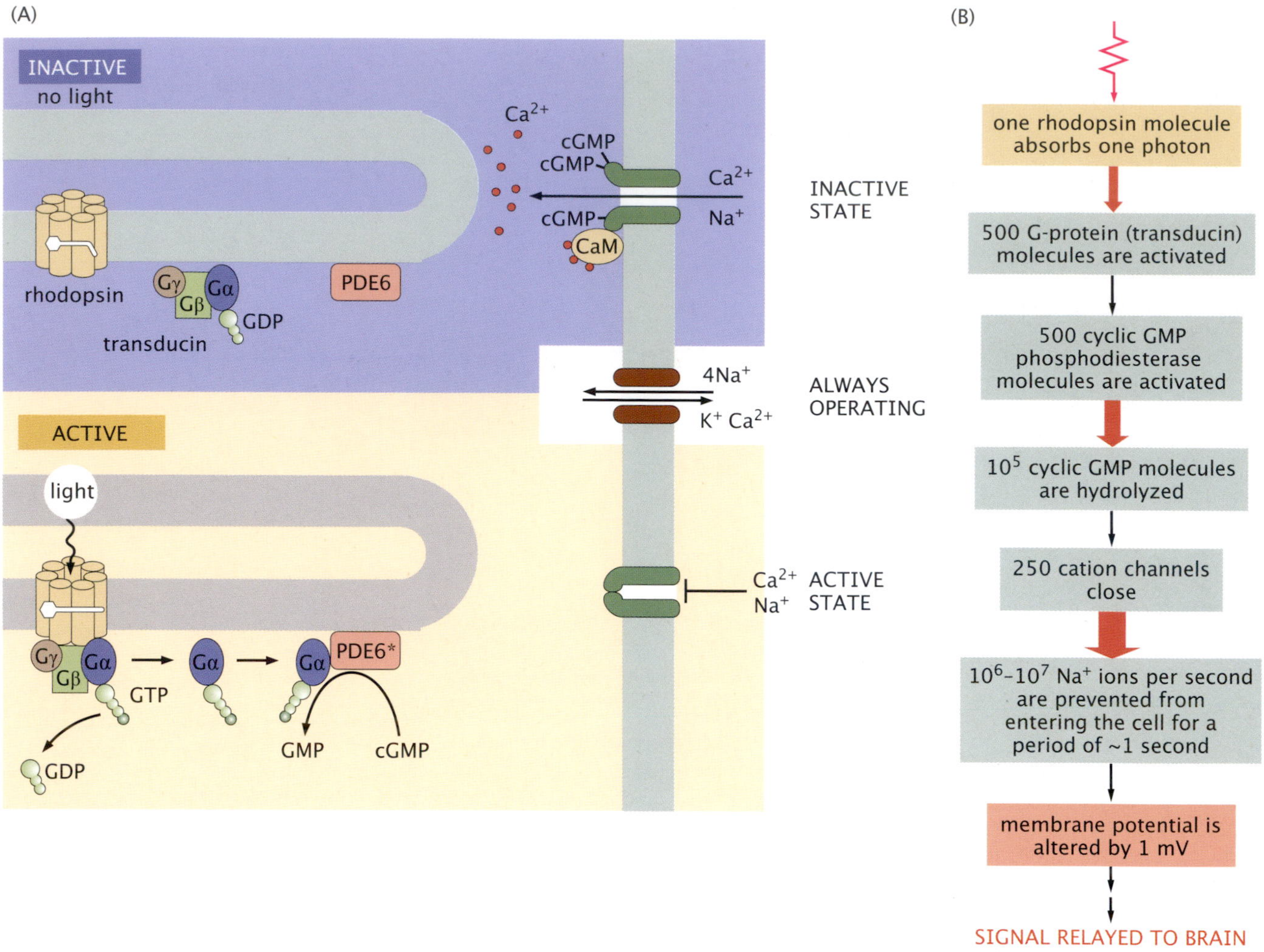

Figure 5.15

Cascade after rhodopsin activation. Activation of rhodopsin results in the hydrolysis of cGMP, causing cation channels to close. (A) In the dark (top), the cGMP phosphodiesterase PDE6 is inactive, and cGMP is able to accumulate inside the rod cell. cGMP binds to a ligand-gated ion channel (dark green) that is permeable to both sodium and calcium ions. Calcium is transported back out again by an exchanger (shown in brown) that uses the energy from allowing sodium and potassium ions to run down their electrochemical gradients to force calcium ions to be transported against their gradient. When a photon activates a rhodopsin protein (bottom), this triggers GTP-for-GDP exchange on transducin, and the activated subunit of transducin then activates PDE6, which cleaves cGMP. The ligand-gated channels close, and the transmembrane potential becomes more negative (hyperpolarized). (B) Signal amplification is achieved at several steps of the pathway, such that the energy of one photon eventually triggers a net charge change of about one million sodium ions. (A) adapted from Stockman et al. (2008). (B) adapted from Alberts et al. (2008); see Further Reading.

G protein paradigm described in Figure 5.10 (p. 179), with the one difference that the initial event is not the binding of a ligand but, rather, the absorption of a photon. The first of the series of signaling proteins that transmits information from the activated rhodopsin to the ion channels at the plasma membrane is called transducin, one of a family of very important signaling proteins in animals that, as introduced earlier, are generally known as "heterotrimeric G proteins," because they are typically composed of three distinct subunits, one of which binds GTP (or GDP). In most cases the $G\alpha$ subunit of the G protein is bound to GDP in the resting state and is also bound to the other two subunits, $G\beta$ and $G\gamma$. The membrane-associated G protein trimers are able to diffuse in the plane of the membrane. When a trimer comes into contact with an activated receptor (in this case, a rhodopsin protein with an all-trans retinal), the $G\alpha$ subunit releases the GDP in exchange for a GTP, and also unbinds from the $G\beta$ and $G\gamma$ subunits. The free, GTP-associated $G\alpha$ subunit can then interact with its downstream targets. For transducin activated by rhodopsin, the critical downstream target is the enzyme cyclic GMP phosphodiesterase (PDE). This enzyme cleaves one of the bonds in cGMP, converting it into GMP form.

Though we have not explored the behavior of rhodopsin here in any quantitative detail, the biological story makes it clear that signal transduction in vision is dependent upon a suite of allosteric proteins.

5.3.3   Light as a Ligand: Optogenetics

The rise of optogenetics over the last several decades is one of the most impressive technologies I have seen in my career. The beauty of this technique is that it makes it possible to address individual molecules, not by exposing them to messy externally controlled ligands but, rather, by applying a beam of focused light, making it possible to activate molecules when and where we want. As already seen in the context of rhodopsin and retinal in the previous section, these proteins as seen from the point of view of this book are allosteric proteins whose ligand is a photon. In this brief section, the idea is simply to revel in the science and technology of macromolecules that undergo conformational transitions in response to light as a ligand, leaving aside for a brief interlude the chapter's primary subject, the G proteins and GPCRs.

We never know from where the next big thing is going to come. We are currently living through the CRISPR craze (and it is exciting!), but it has its roots in unexpected quarters, having first emerged from obscure beginnings in considerations of bacterial immunity. Given the paramount role of probability in modern science, as witnessed in every chapter of this book, I never cease to be amused by the roots of the right rules of reasoning in the analysis of games of chance by luminaries such as Pascal and Fermat. And the rise of optogenetics has its roots in the humble and until now largely forgotten labors of botanist Andrei Sergeyevich Famintsyn, who was interested in the motility of single-celled organisms that exhibited behaviors that clearly involved "decision making".

As shown in Figure 5.16, Famintsyn's experiments consisted of exposing photosynthetic microbes to light and examining their motility. As seen in the figure, depending upon the intensity of the light, the *Chlamydomonas* cells

Figure 5.16

Phototaxis in *Chlamydomonas*. (A) Classic experiments of Famintsyn of dish phototaxis assay. At low light levels, the cells migrate as close to the light as they can. In high light levels, they migrate to that part of the dish that is optimal. (B) Dish phototaxis assay showing how over time, *Chlamydomonas* cells migrate in response to light. (A) adapted from Deisseroth and Hegemann (2017); see Further Reading. (B) adapted from Hegemann and Nagel (2013).

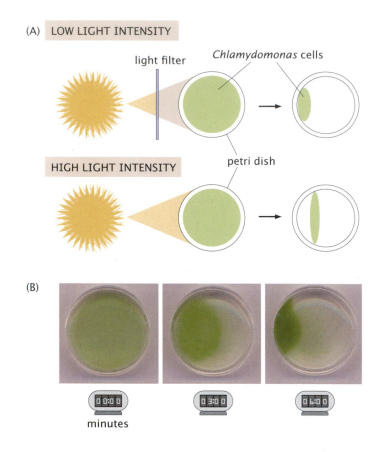

migrated differently. In low light levels, they migrated as far toward the light source as they could. However, at higher light intensity, they would move to some intermediate position. This experiment is reminiscent of the brilliant works that established the science of the chemotaxis phenomenon, as described in chapter 4 (p. 124). Here there were similar questions in play in trying to figure out the microscopic underpinnings of the phenomenon of phototaxis.

The analysis of phototaxis in the modern era of molecular biology led to the discovery of the underlying molecular mechanisms behind this process. First, at the level of the cell biology as shown in Figure 5.17, cells have a variety of ultra-structures devoted to determining the direction of the incident light and what to do with that detected light. Already in the nineteenth century, these single-celled organisms were observed to have a characteristic yellow spot, always at the same equatorial position in the cell body, dubbed the "eye spot." Using techniques similar to the classic experiments of Baylor, as shown in Figure 5.18, it was possible to capture individual *Chlamydomonas* cells and to measure light-induced photocurrents. These currents lead to voltage-sensitive channel gating in the flagellar membrane, which leads to a flux of calcium ions into the flagella. At this point, the *Chlamydomonas* cells abandon their characteristic breast stroke gait and replace it with a flagellar undulation which redirects their motion. To determine the genes responsible for these light-driven behaviors, candidate genes were expressed in *Xenopus laevis* oocytes and showed a light-gated conductance only when the gene of interest was present.

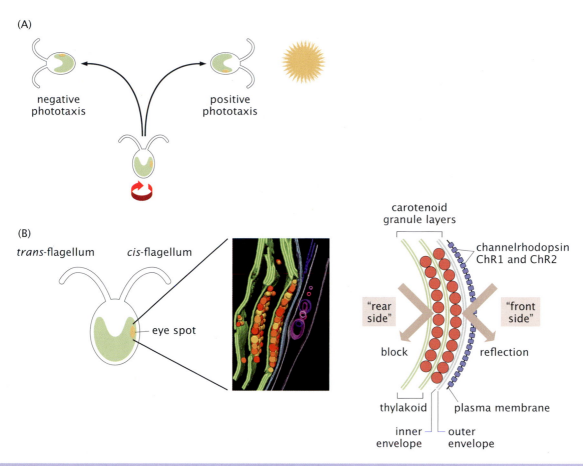

Figure 5.17

Anatomy of the *Chlamydomonas* eyespot. (A) Schematic of the classic model photosynthetic organism *Chlamydomonas reinhardtii* showing its chloroplast (green) and eyespot (yellow). (B) Eyespot at higher resolution. Vesicle layers (orange) with different indexes of refraction focus the light. The membrane is populated with channelrhodopsin-1 (ChR1), channelrhodopsin-2 (ChR2), and a voltage-gated calcium channel. Adapted from Ueki et al. (2016); cryo-em images from Engel et al. (2015).

What has emerged in the field of microbial opsins as a result of decades of work like that just described is summarized in Figure 5.19. Picking up once again on the main theme of the chapter, we now know there are a number of different opsins with the characteristic seven-transmembrane helixes already featured throughout the chapter. Much useful information on the fundamental mechanism of animal vision has been derived from studies of a salt-loving archaeon called *Halobacterium salinarum*. This archaeon has a brilliant purple color caused by large arrays in its membrane of a protein called bacteriorhodopsin that carries a single retinal pigment. In *Halobacterium*, bacteriorhodopsin uses the energy of light to generate a transmembrane proton gradient, which can be used to synthesize ATP. In other words, the archaeon is using light to generate biochemical energy in a manner generally analogous to photosynthesis, although electron transfer is not involved. Interestingly, the bacteriorhodopsin protein is not at all similar to the photosynthetic reaction center. Instead, it bears a striking structural and functional resemblance

Figure 5.18

Response to light as a ligand. (A) Experimental setup shows a single rod cell from the retina of a toad in a glass capillary and subjected to a beam of light. (B) Current traces as a function of time for photoreceptor subjected to light pulses in an experiment like that shown in part (A). (C) Experimental setup for measuring the light-induced currents in *Xenopus* oocytes due to bacteriorhodopsin as a function of light exposure. (D) Channel currents through channelrhodopsin-2 due to light exposure of *Xenopus* oocytes. (A and B) from Milo, R. and R. Phillips (2016), *Cell Biology by the Numbers*. Reproduced by permission of Taylor & Francis LLC, a division of Informa plc. (C) and (D) adapted from Hegemann and Nagel (2013), CC BY-NC 4.0.

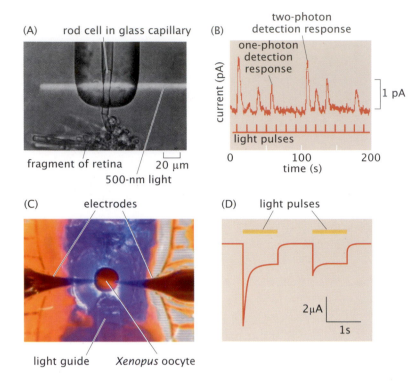

to the rhodopsins that underlie animal vision. The key to this process is the ability of the small molecule retinal to undergo a light-dependent structural isomerization. When the retinal is embedded in a protein, its light-dependent isomerization can drive a larger-scale conformational change in the protein, which can then have further biological consequences.

In *Halobacterium* bacteriorhodopsin, a molecule of retinal is covalently attached in the middle of a seven-transmembrane domain receptor. The retinal begins the photochemical cycle in the extended all-trans conformation. Absorption of a photon causes the retinal to switch to the 13-*cis* conformation. This causes strain within the protein's transmembrane helixes. In relieving the strain the protein undergoes a series of conformational changes that expose amino acid side-chain residues that can be protonated or deprotonated on either side of the membrane. The arrangement of a series of residues down the center of the seven-helix bundle facilitates the transport of a single proton from the inside of the cell to the outside. At the end of the cycle, the retinal returns to the lower-energy all-trans state. In the living archaeon the bacteriorhodopsin protein is expressed at a very high level in the membrane and arranged in densely packed two-dimensional arrays. The parallel action of the many individual pumps enables the bacterium to generate a significant transmembrane proton gradient from light energy alone.

As seen in Figure 5.19, *Halobacterium* also harbors a second opsin protein known as halorhodopsin, which is also an ion pump. In this case rather than pumping hydrogen ions, as does bacteriorhodopsin, halorhodopsin pumps chloride ions in the presence of green or yellow light. Halorhodopsin is another seven-transmembrane helix protein and, as in the case of bacteriorhodopsin,

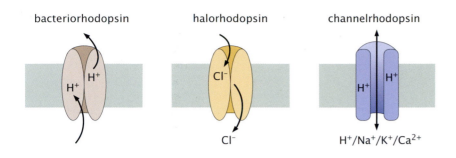

bacteriorhodopsin   halorhodopsin   channelrhodopsin

Figure 5.19

Opsin proteins in microbes. Bacteriorhodopsin is a light-driven pump of hydrogen ions. Halorhodopsin is a light-driven pump of chloride ions found in Archaea. Channelrhodopsins are light-gated ion channels.

opsin family	ion transported	mode of transport	peak activation wavelength (nm)
bacteriorhodopsin	H^+	pump	568
halorhodopsin	Cl^-	pump	580
channelrhodopsins (ChR1/ChR2/VChR1/VChR1)	$H^+>Na^+>K^+>Ca^{2+}$	channel	500/470/540/470

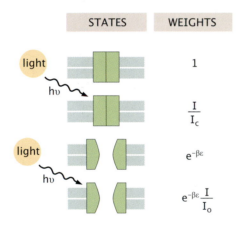

STATES	WEIGHTS
	1
	$\dfrac{I}{I_c}$
	$e^{-\beta\varepsilon}$
	$e^{-\beta\varepsilon}\dfrac{I}{I_o}$

Figure 5.20

States and weights for light as a ligand. The top two states correspond to the closed configuration and the bottom two states to the open configuration. Here we replace the familiar K_c and K_o with threshold light intensities I_c and I_o, respectively.

the conformational change induced in retinal due to photon absorption results in a conformational change in the protein itself, which leads to chloride ion import.

These microbial opsins offer us the chance to generalize our efforts put forth throughout the book to now consider the analysis of allosteric molecules that bind light instead of physical ligands. Figure 5.20 provides a states-and-weights analysis of the various states of our simplest model of such light-liganded molecules. In the scenario hypothesized here, the light intensity I at the relevant wavelength has taken the place of the ligand concentration c used elsewhere in the book. Similarly, the constants I_c and I_o serve the same role as the dissociation constants have done elsewhere. Effectively, I_o sets the intensity at which the channelrhodopsin will be open half the time. This analysis is phenomenological and amounts to reasoning by analogy with our earlier treatment of allostery. Using the states and weights shown in Figure 5.20, we can write the hypothesized probability that channelrhodopsin will be open as

$$p_{active}(I) = \frac{(1 + \frac{I}{I_o})}{(1 + \frac{I}{I_o}) + e^{-\beta\varepsilon}(1 + \frac{I}{I_c})}. \tag{5.1}$$

It would be very interesting to carry out rigorous experimental tests to see if our reasoning by analogy with conventional allosteric proteins is consistent with measurements.

The biology of the various opsin proteins described is itself amazing. But in characteristically inventive form, the science does not stop there. In the early 2000s, a number of different research groups saw the possibility of using these light-induced conformational changes as a tool to control and read out the state of neurons. Figure 5.21 gives an example of optical control of neuronal activity using both channelrhodopsin and halorhodopsin. Here we see that through the application of a tailored light history, the action potential can be sculpted to the experimentalist's will.

Finally, the use of light as a ligand has not been confined to neuroscience. As a tool in cell biology, light-controlled proteins are of great interest, because they make it possible to reprogram individual proteins. As seen in Figure 5.22, light has been used to induce protein associations (also shown in Figure 5.23, where the cross linking of molecular motors is induced by light), to induce gene expression, to cluster or sequester proteins, to localize proteins to particular regions such as the membrane, or to induce conformational changes.

Figure 5.23 shows how approaches using light as a ligand have been used to study the interactions between molecular motors and microtubules. A simplified model experimental system that consists of stabilized microtubules, kinesin motor proteins (dimers that can be linked by light), and ATP has been used to study aster formation. The simplified system is capable of *spontaneous* and programmed action as it generates complicated internal patterns of force and flow. The ability to control the motors with light shows the light protocol facilitates widely different patterns and dynamics of microtubules. With the application of a circular disc of light, the spontaneous formation of asters can be induced. Similarly, several discs of light can be used to induce the formation of several asters. From the point of view of this chapter, this is a sophisticated example of light as ligand, and from the perspective of the entire book, it fits the general theme of how conformational changes can completely change the function of proteins.

5.4   G Protein–Coupled Ion Channels

Our final story related to the GPCRs and G proteins brings us back to the ligand-gated channels we considered in chapter 3, (p. 77). Here we consider the G protein–gated inward rectifier K^+ channels (GIRK), a key part of our own repertoire of physiological responses. As anyone who jumps into the cold ocean knows, besides producing the sensation of difficulty catching our breath, counterintuitively perhaps, the action reduces our heart rate, as well. The GIRK channels that are the subject of this section are key players in this process, since after the vagus nerve is stimulated, acetylcholine acts on GPCRs in pacemaker cells, which leads in turn to activation of GIRK channels, increasing the time between pacemaker action potentials.

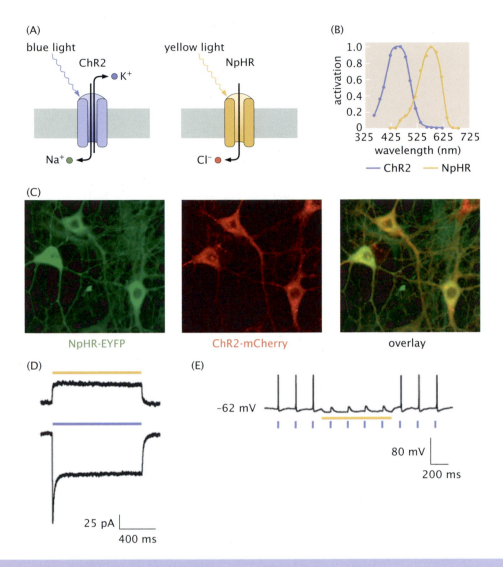

Figure 5.21

Optical control of neuronal activity. (A) Channelrhodopsin (left) is a light-activated cation channel, allowing transmembrane movements of potassium and sodium ions down their electrochemical gradient. Halorhodopsin (right) is a light-activated anion pump, primarily passing chloride ions. (B) The absorption spectra of channelrhodopsin and halorhodopsin are distinct. Channelrhodopsin is activated by blue light, while halorhodopsin is activated by lower-energy yellow light. (C) Both proteins can be coexpressed in cultured neurons. Here, the protein distribution is visualized using fluorescent tags. (D) These neurons can now be electrically stimulated by blue or yellow light. At a constant voltage, the exposure of the neurons to blue light (bottom) results in an inward current (shown here as downward), because the inward movement of positively charged sodium ions dominates. In contrast, the exposure of the neurons to yellow light (top) causes an inward flow of negatively charged chloride ions, and so results in a current of opposite sign. (E) Short pulses of blue light are capable of inducing action-potential firing in these cells. Each blue light pulse causes the cell to rapidly and dramatically depolarize from its resting potential of about −62 mV, and after each voltage spike the cell returns to its normal state. Exposure to yellow light completely blocks the action potentials triggered by the blue light pulses, as the influx of negatively charged chloride ions prevents depolarization. From Phillips, R., J. Kondev, J. Theriot, and H. Garcia (2013), *Physical Biology of the Cell*, 2nd ed. Reproduced by permission of Taylor & Francis LLC, a division of Informa plc. (C–E) adapted with permission from Springer Nature via Copyright Clearance Center: Zhang, F., et al. (2007), "Multimodal fast optical interrogation of neural circuitry," *Nature*.

Figure 5.22

Signaling and light as a ligand. Each row shows a different mechanism of light-induced activity. In the presence of light, proteins can be induced to associate, dissociate, or undergo a conformational change, with each one of these structural changes resulting in a change in signaling activity. Adapted from Tischer and Weiner (2014).

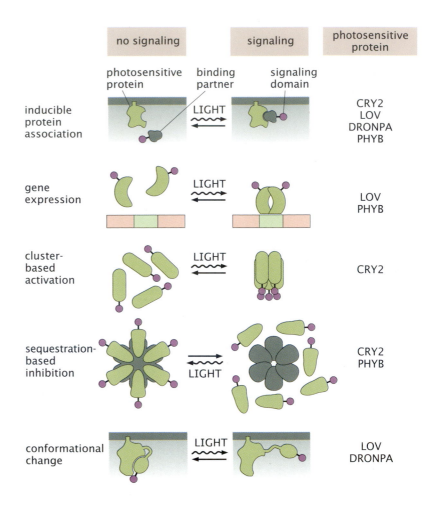

The story of these GIRK channels is summarized mechanistically in Figure 5.24. A neurotransmitter binds to a GPCR, as shown in the left part of the figure. This activates the GPCR, with the result that the G protein itself is activated. In its activated form, the G protein acts as a ligand that favors the open state of the GIRK channel. Four such activated G proteins bind and activate the channel. However, the behavior is richer and more complex. Not only do these channels respond to the activated $G_{\beta\gamma}$, but they also respond to the lipid PIP_2, and the process is amplified by the presence of Na^+. These results imply that, as we will study in chapter 9, these channels are subject to combinatorial control. To be more precise, the GIRK channel responds with AND logic to $G_{\beta\gamma}$ and PIP_2, requiring both to be present for channel gating to occur, with Na^+ serving to amplify opening. In this section, ideas from many different chapters converge.

As hinted at previously the structure of the GIRK channel itself is tetrameric, with four sites for binding both the $G_{\beta\gamma}$ subunits, PIP_2 and Na^+, as shown in Figure 5.25. The affinity of the $G_{\beta\gamma}$ subunits is different in the closed and open states, as indicated schematically in Figure 5.25(B). The structure as shown here provides the central hint to how we will formulate our MWC model for this problem. Indeed, rather than rehashing the states and weights, we instead

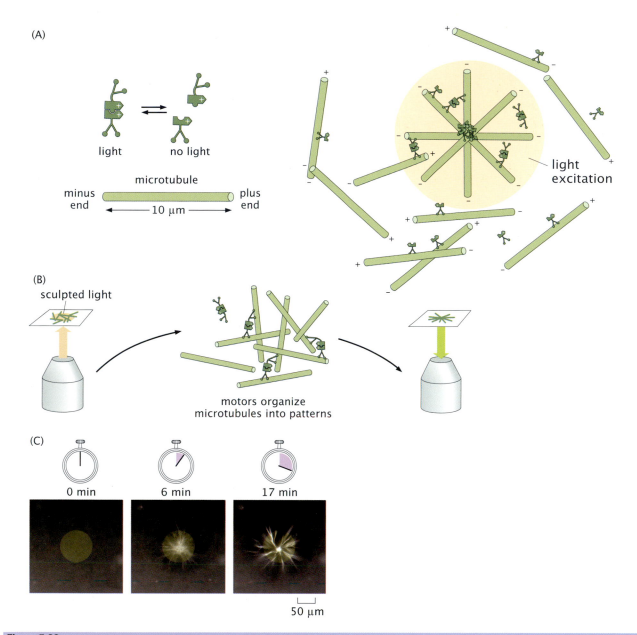

Figure 5.23

Light as a ligand and protein dimerization. (A) Light-induced dimerization of microtubule motors. Dimerization of the kinesin motors leads to crosslinking of the microtubules, which induces structural rearrangements of the microtubules into patterns such as asters. (B) Schematic of the time evolution of the system after exposure to sculpted light. (C) Experimental realization of the system showing the formation of an aster. Courtesy of Tyler Ross, Heun Jin Lee, Rachel Banks, and Matt Thomson.

Figure 5.24

Key molecular players in the function of the G protein–gated inward rectifier K$^+$ channel (GIRK). Binding of a neurotransmitter to the GPCR results in a signaling cascade. When activated, the membrane-anchored G$_{\beta\gamma}$ binds to GIRK, resulting in channel activation. Adapted from Wang et al. (2016) CC BY-NC 4.0.

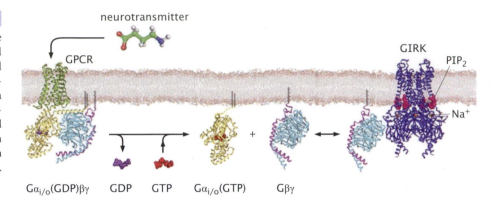

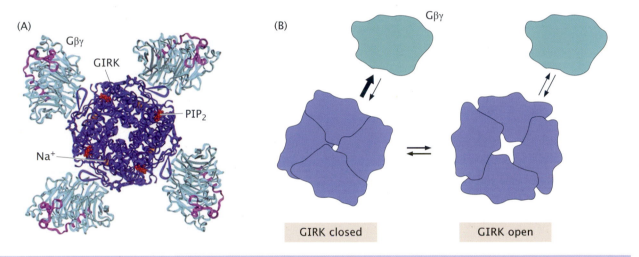

Figure 5.25

Top view of the GIRK channel. (A) Crystal structure of the GIRK ion channel showing the binding sites for G$_{\beta\gamma}$, PIP$_2$, and Na$^+$. (B) Binding of G$_{\beta\gamma}$ in the closed and open states of the channel. Adapted from Wang et al. (2016).

acknowledge that the three ligands can each bind independently to their corresponding binding sites. As usual, we can thus compute the activity of our putative MWC molecule as a function of the control parameters, in this case the concentrations of G$_{\beta\gamma}$, PIP$_2$, and Na$^+$. To construct the MWC activity function, we appeal to all the same ideas that have played out repeatedly throughout the book. In this case, we note that there are four binding sites for each of these ligands. Thus, the activity is given as

$$p_{active}(c_{G_{\beta\gamma}}, c_{PIP_2}, c_{Na^+}) = \tag{5.2}$$

$$\frac{(1 + \frac{c_{G_{\beta\gamma}}}{K_{G,O}})^4 (1 + \frac{c_{PIP_2}}{K_{P,O}})^4 (1 + \frac{c_{Na^+}}{K_{N,O}})^4}{(1 + \frac{c_{G_{\beta\gamma}}}{K_{G,O}})^4 (1 + \frac{c_{PIP_2}}{K_{P,O}})^4 (1 + \frac{c_{Na^+}}{K_{N,O}})^4 + e^{-\beta\varepsilon} (1 + \frac{c_{G_{\beta\gamma}}}{K_{G,C}})^4 (1 + \frac{c_{PIP_2}}{K_{P,C}})^4 (1 + \frac{c_{Na^+}}{K_{N,C}})^4}. \tag{5.3}$$

We have introduced the dissociation constants $K_{i,O}$ and $K_{i,C}$ for the open and closed states of the channel, with $i = G$ corresponding to binding of G$_{\beta\gamma}$, $i = P$

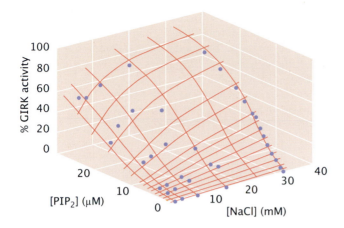

Figure 5.26

Gating of the GIRK channels depends upon multiple inputs. Surface plot shows normalized current as a function of both NaCl and PIP_2 concentrations. The model of equation 5.3 was used to compute how the channel activity depends upon the concentrations of PIP_2 and Na^+. Data from W Wang, KK Touhara, K Weir, BP Bean and R MacKinnon eLife, 5:e15751 (2016). Fits using the MWC model courtesy of Parijat Sil, Chandana Gopalakrishnappa, and Tal Einav.

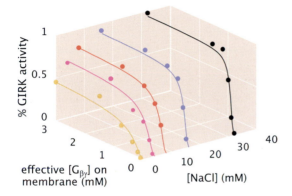

Figure 5.27

Comparison of MWC model of GIRK gating and experimental data. The model of equation 5.3 was used to compute how the channel activity depends upon the concentrations of $G_{\beta\gamma}$ and Na^+. Data from Wang et al. (2016). Fits using the MWC model courtesy of Parijat Sil, Chandana Gopalakrishnappa, and Tal Einav.

corresponding to binding of PIP_2, and $i = N$ corresponding to binding of Na^+. The difference in energy between the closed and open states of the channel is given by ε. This activity function provides a framework that we can use to interpret large swaths of experimental data.

To determine the response of the GIRK channel to the different control ligands, a series of elegant experiments were performed using an in vitro reconstitution of the channels and the $G_{\beta\gamma}$ subunits in supported lipid bilayers. By calibrating the amount of PIP_2 and $G_{\beta\gamma}$, it was then possible to measure the dose-response functions, as shown in Figures 5.26 and 5.27. To perform these experiments in a quantitative fashion that made it possible to obtain dose-response curves for this complicated example of combinatorial control, the concentration of $G_{\beta\gamma}$ was controlled using NTA lipids in the supported lipid bilayer in which the in vitro reconstitution was performed. These lipids are used to anchor the $G_{\beta\gamma}$ subunits, and hence by titrating the amount of NTA lipid, one is effectively tuning the amount of $G_{\beta\gamma}$. Note also that in the data reported in these figures, NaCl concentration is reported, which in turn tells us how much Na^+ is present.

In addition to the data, Figures 5.26 and 5.27 show the corresponding model fits using the MWC model characterized by equation 5.3. The results shown here illustrate that a set of parameter values can be found that are consistent with an MWC interpretation of the dose-response curves ($K_{G,O} = 8 \, \mu M$,

$K_{G,C} = 170 \ \mu$M, $K_{P,O} = 5.6 \ \mu$M, $K_{P,C} = 100 \ \mu$M, $K_{N,O} = 9.6$ mM, and $K_{N,C} = 30$ mM). The experiments went further by examining how the G_α subunit can deactivate the channel by binding again to the $G_{\beta\gamma}$ subunits and thus rendering them unable to interact with the GIRK channel.

This final example was offered as an incomplete analysis of a rich problem that draws from many of the key themes of the book: ligand-gated ion channels, GPCRs, and combinatorial control. The experiments described here provide an elegant and rigorous challenge to the predictive MWC framework.

5.5 Summary

It is a common refrain to hear that G protein–coupled receptors are the most abundant signaling molecules in higher eukaryotes and that a large fraction of drugs target them. In this chapter, we provided an in-depth description of some of the rich biological uses of such proteins and of the long intellectual journey culminating in our current understanding. The color of multicellular organisms is one of their greatest charms. In mice and mammoths, a key mutation in a GPCR, the melanocortin-1 receptor, leads to changes from a dark to a light phenotype. Several of the best-studied examples of GPCRs are offered by the adrenergic receptors and the opsins. We summarized the role of GPCRs and G proteins in both of these contexts, and included an analysis of the dose-response data that characterize their function. The chapter ended with a concrete quantitative example of G protein activation of a physiologically important potassium ion channel.

5.6 Further Reading and Viewing

Alberts, B., A. D. Johnson, J. Lewis, D. Morgan, M. Raff, K. Roberts, and P. Walter (2016) *Molecular Biology of the Cell*, 6th ed. New York: W. W. Norton. This is my go-to source on any new topic I want to learn about. It has also been our greatest source for examples in which we can mathematicize the sophisticated diagrammatic models of some interesting biological problem. The GPCRs are a perfect example.

Canals, M., P. M. Sexton, and A. Christopoulos (2011) "Allostery in GPCRs: 'MWC' revisited." *Trends Biochem. Sci.* 36:663. Christopoulos and coworkers have done a number of important studies on how the MWC model can be applied to GPCRs. This paper summarizes some of their results and thinking.

Deisseroth, K. and P. Hegemann (2017) "The form and function of channelrhodopsin." *Science* 357:eaan5544. This excellent article traces the history of the understanding of the molecules that have become today's technology of optogenetics.

Hoekstra, H. iBioSeminars. The genetic basis of evolutionary change in morphology, phenotypic adaptations, and behavior. More or less the entirety of my discussion of color is based on things learned from Hopi Hoekstra. Her papers are great, but this series of three iBioseminars is the place to start.

Hoekstra, H. E. (2010) "From Darwin to DNA: The Genetic Basis of Color Adaptation." In *In the Light of Evolution: Essays from the Laboratory and Field*, edited by J. B. Losos. Greenwood Village, CO: Roberts and Company. This excellent article gives an overview of Hoekstra's take on the story of coat color I attempted to retell here. This paper let's you hear it from the master herself.

Kaufman, D. W. (1974) "Adaptive coloration in *Peromyscus polionotus*: Experimental selection by owls." J. Mammal. 55 (2): 271–283. This classic paper describes the predation of owls in enclosures on mice of different colors.

Lefkowitz, R. iBioSeminars. As with the lectures of Hopi Hoekstra mentioned here, Lefkowitz's treatment of GPCRs in general and the β-adrenergic receptor in particular is exceedingly well done. Reading his Nobel Lecture is also an excellent source.

Nathans, J. iBioSeminars. Jeremy Nathans gives three beautiful lectures on vision. These lectures are attended by a number of excellent molecular vignettes that discussion vision and the signaling cascades that attend it.

Phillips, R., J. Kondev, J. Theriot, and H. G. Garcia (2013) *Physical Biology of the Cell*. New York: Garland Press. Some of the section on vision in this chapter is lifted directly from the treatment in chapter 18 of PBOC.

Ptashne, M., and A. Gann (2002) *Genes and Signals*. Cold Spring Harbor, NY: Cold Spring Harbor Laboratory Press.

Rodieck, R. W. (1998) *The First Steps in Seeing*. Sunderland, MA: Sinauer Associates. This is unequivocally one of the best science books I have seen in a long life of reading. Rodieck's storytelling, his appeal to simple order-of-magnitude estimates, and his exquisite diagrams all conspire to produce a book that at once reads like a novel, a detective story, and a biography of great ideas. Many of the figures in this chapter that concern vision are updates of Rodieck's clear thinking.

Roth, S., and F. J. Bruggeman (2014) "A conformation-equilibrium model captures ligand-ligand interactions and ligand-biased signaling by G-protein coupled receptors." *FEBS J.* 281:4659–4671. This paper gives an indication of how the states-and-weights approach advocated throughout the book can be used to examine the function of G protein–coupled receptors.

Tischer, D., and O. Weiner (2014) "Illuminating cell signalling with optogenetic tools." *Nat. Rev. Mol. Cell Biol.* 15 (8): 551–558. This paper gives a thoughtful and thorough analysis of the many uses of light as a ligand as a tool in cell biology.

5.7 REFERENCES

Barton, N. H., D. E. G. Briggs, J. A. Eisen, D. B. Goldstein, and N. H. Patel (2007) *Evolution*. Cold Spring Harbor, NY: Cold Spring Harbor Laboratory Press.

Baylor, D. A., T. D. Lamb, and K.-W. Yau (1979) "The membrane current of single rod outer segments." *J. Physiol.* 288:589–611.

Calvin, M., and A. A. Benson (1948) "The path of carbon in photosynthesis." *Science* 107 (2784): 476–480.

Engel, B. D., M. Schaffer, L. K. Cuellar, E. Villa, J. M. Plitzko, and W. Baumeister (2015) "Native architecture of the *Chlamydomonas* chloroplast revealed by in situ cryo-electron tomography." *eLife* 4:e04889.

Fawcett, W. (1966) *The Cell, Its Organelles and Inclusions: An Atlas of Fine Structure*. Philadelphia: W. B. Saunders.

Hegemann, P., and G. Nagel (2013) "From channelrhodopsins to optogenetics." *EMBO Mol. Med.* 5:173–176.

Hoekstra, H. E., R. J. Hirschmann, R. A. Bundey, P. A. Insel, and J. P. Crossland (2006) "A single amino acid mutation contributes to adaptive beach mouse color pattern." *Science* 313:101.

Lefkowitz, R. J. (2012) "A Brief History of G Protein Coupled Receptors." Nobel Lecture, December 8, Stockholm.

Manceau, M., V. S. Domingues, C. R. Linnen, E. B. Rosenblum, and H. E. Hoekstra (2010) "Convergence in pigmentation at multiple levels: Mutations, genes and function." *Phil. Trans. R. Soc.* B365:2439–2450.

Rieke, F., and D. A. Baylor (1998) "Single-photon detection by rod cells of the retina." *Rev. Mod. Phys.* 70:1027–1036.

Stockman, A., H. Jägle, M. Pirze, and L. T. Sharpe (2008) "The dependence of luminous efficiency on chromatic adaptation." *J. Vision* 8 (16):1.

Tischer, D., and O. Weiner (2014) "Illuminating cell signalling with optogenetic tools." *Nat. Rev. Mol. Cell Biol.* 15 (8): 551–558.

Ueki, N., T. Ide, S. Mochiji, Y. Kobayahi, R. Tokutsu, N. Ohnisihi, K. Yamaguchi, S. Shigenobu, K. Tanaka, J. Mingawa, T. Hisabori, M. Hirono, and K.-I. Wakabayashi (2016) "Eyespot-dependent determination of the phototactic sign in *Chlamydomonas reinhardtii*." *Proc. Natl. Acad. Sci.* 113:5299–5304.

Vignieri, S. N., J. G. Larson, and H. E. Hoekstra (2010) "The selective advantage of crypsis in mice." *Evolution* 64 (7): 2153–2158.

Wang, W., K. K. Touhara, K. Weir, B. P. Bean, and R. MacKinnon (2016) "Cooperative regulation by G proteins and Na^+ of neuronal GIRK2 K^+ channels." *eLife* 5:e15751.

Zhang, Z., L. P. Wang, and M. Brauner (2007) "Multimodal fast optical interrogation of neural circuitry." *Nature* 446:633.

DYNAMICS OF MWC MOLECULES: ENZYME ACTION AND ALLOSTERY

6

> The caterpillar does all the work but the butterfly gets all the publicity.
>
> —George Carlin

One of the main arguments of this book is that the idea of allostery has enormous reach for explaining broad swaths of biology. Whether we think of the binding curves of oxygen to hemoglobin, the gating properties of ligand-gated ion channels, or the action of membrane-bound receptors to phosphorylate cellular substrates, allostery turns out to be the right language again and again. In the early 1960s, enzyme activity and its control provided one of the central playing fields for the study of allostery. Experiments on several key enzymes such as aspartate transcarbamoylase, an enzyme performing an early step in the pyrimidine synthesis pathway, led to the recognition that the activity of these enzymes was different depending upon environmental circumstances such as whether purines or pyrimidines were present in the reaction. In this chapter, we examine the phenomenology of enzyme action and present a unified statistical mechanical view of their dynamics. We then use these ideas to examine processes such as "feedback inhibition," the phenomenon in which the product of an enzyme acts back on that enzyme to alter its activity. We will see how the allostery concept figured into the development of understanding of enzyme action. In addition, we will explore the mathematical implementation of the MWC concept in the context of enzyme reactions and see how it provides a common language for much familiar enzyme data but also gives a simple explanation for less intuitive enzyme behaviors, such as the inhibition of an enzyme by a substrate or the speeding up of an enzyme reaction by an inhibitor.

6.1 Enzyme Phenomenology

In his paper entitled "Bacterial growth: Constant obsession with dN/dt," Frederick Neidhardt (1999) tells us: "One of life's inevitable disappointments—one felt often by scientists and artists, but not only by them—comes from expecting others to share the particularities of one's own sense of awe and wonder. This truth came home to me recently when I picked up Michael Guillen's fine book *Five Equations That Changed the World* and discovered that my equation—the one that shaped my scientific career—was not considered one of the five."

The sense of awe and wonder that Neidhardt was speaking of comes from one of life's most astonishing feats, which can routinely be seen by any high-school student. Take a 5 mL plastic tube and fill it with bacterial growth media (LB media is best for thinking about this example). Then, use a pipette to seed this media with a single bacterial cell with its tiny femtoliter volume and picogram mass. Then put this growth tube into a 37° C incubator and let it shake for 12 hours. When you come back, the initially colorful but transparent solution will now have a milky and opaque appearance owing to the roughly 5 billion bacterial cells that now fill the tube. What has happened over those short 12 hours is that this one cell has grown and divided roughly every half hour, and then those daughter cells have repeated that process over and over again. Neidhardt's beloved equation that describes this process is

$$\frac{dN}{dt} = kN, \tag{6.1}$$

where $N(t)$ is the number of bacteria at time t with $N(0) = 1$. This equation tells us of the exponential growth of our collection of cells as long as sufficient resources are present to maintain that growth. The familiar solution to this equation is

$$N(t) = e^{kt}, \tag{6.2}$$

where we used the fact that at $t = 0$ there is only one cell present in the tube.

ESTIMATE

Estimate: The Census of an *E. coli* Cell

To get a sense of the magnitude of this feat of exponential growth, we review some simple orders of magnitude that help us appreciate what it takes to make a new cell, driven by a huge variety of different enzymes whose functions form the centerpiece of this chapter. First, a typical bacterial cell, several microns long, has a volume of roughly 1 μm^3, or 1 fL. If we make the reasonable order of magnitude suggestion that the density of these cells is comparable to that of water, then their mass is roughly $m = \rho V \approx 1 \text{ kg/L} \times 10^{-15} \text{ L} = 10^{-15} \text{ kg} = 1 \text{ pg}$. Given the picogram mass of a bacterial cell and the realization that roughly 70% of that mass is water, we are left with roughly 0.3 pg of macromolecular mass that needs to be synthesized over the 20–30 minutes between cell divisions. Approximately half of the dry mass (i.e., 0.15 pg) is tied up in proteins. Thus, we can estimate the number of proteins per cell as

$$\text{number of moles of protein} \approx \frac{0.15 \times 10^{-12} \text{ g}}{30,000 \text{ g/mol}} \approx \frac{1}{2} \times 10^{-17} \text{ mol}. \tag{6.3}$$

We can convert this value to number of proteins as

$$\text{number of proteins} \approx \frac{1}{2} \times 10^{-17} \text{ mol} \times 6 \times 10^{23} \text{ proteins/mol}$$

$$\approx 3 \times 10^6 \text{ proteins}. \tag{6.4}$$

Given that on average each of these few million proteins has roughly 300 amino acids, this tells us that to double the protein content of the cell requires

$$N_{\text{amino acids}} \approx \frac{300 \text{ amino acids}}{\text{protein}} \times \frac{3 \times 10^6 \text{ proteins}}{\text{cell}} \approx \frac{10^9 \text{amino acids}}{\text{cell}}.$$

(6.5)

The point of this straightforward estimate is to get a sense of the burden of natural resources needed to build a new cell; that is, on average a cell will need a total of roughly 50 million copies of each of the 20 distinct amino acids used to build proteins.

Not only must the protein content of a cell be doubled during each cell division, so, too, must all the other macromolecules such as the genome. Ultimately, the genome is one of the distinguishing features of living matter, and the replication of the genetic material is a signature process of the cell cycle. For a bacterium such as *E. coli*, the genome is made up of roughly 5×10^6 bp, meaning that to make a new genome requires $\approx 10^7$ nucleotides.

An important hint about the makeup of the genetic material articulated in the early 1950s is known as *Chargaff's rules*, which say there is a 1:1 ratio of pyrimidines and purines. A more specific way of stating this idea is that the number of As and Ts and the number of Gs and Cs are equal. The Watson-Crick structure for DNA gives a natural interpretation of this constraint on the basis of the pairing between bases. But how well obeyed is this constraint at the level of the precursor building blocks, namely, the purines and pyrimidines available in the bacterial cytoplasm? The answer to that question brings us to the role of allostery in the context of enzyme kinetics, since the cell's "decision" to tune the level of pyrimidine biosynthesis is dictated by allosteric signaling.

The enzyme that mediates the first step in the pyrimidine biosynthesis pathway is aspartate transcarbamoylase, several structural views of which are given in Figure 6.1. This protein is a 12-subunit complex made up of two trimers that are the catalytic subunits and three dimers of subunits that are responsible for regulating the activity of the enzyme. Interestingly, these regulatory subunits make it possible for the enzyme to tune its rate of pyrimidine production based upon the concentrations of purines and pyrimidines already present.

In some ways, we can think of this enzyme as the Chargaff enforcer that helps guarantee that the pool of nucleotides used to assemble nucleic acids is balanced. Specifically, accumulation of the molecule CTP, which is the end product of the pyrimidine pathway, slows down the activity of the enzyme. In contrast, the accumulation of the familiar ATP, the molecular outcome of the purine pathway, drives catalytic activity of the enzyme. An example of the kind of data that led to these insights is shown in Figure 6.2 where the enzyme activity is defined as the rate of product formation per enzyme molecule. Hence, the products of the purine and pyrimidine pathways serve as "governors" in much the same way as the mechanical contraption shown in Figure 1.14 (p. 15) tunes the uptake of fuel by an engine.

Figure 6.1

Structure of aspartate transcarbamoylase (ATCase). (A) Classic illustration of the allosteric transition from Irving Geis. (B) Modern impression of the structure of ATCase from David Goodsell. The catalytic subunits are shown in red, and the regulatory parts of the enzyme are shown in blue. The activity of the enzyme is regulated by CTP, a nucleotide that has a pyrimidine ring. (C) Structure of the enzyme in different allosteric states. The T state is inactive, and the R state is the active state competent for enzyme action. The key point of this part of the figure is to give the reader a structural impression of how large the conformational change is that accompanies this allosteric transition. (A) from Geis, I. (1988), "ATCase T-State," and (1988) "ATCase R-state." Used with permission from the Howard Hughes Medical Institute (www.hhmi.org). All rights reserved. (B) courtesy of David Goodsell. (C) adapted from Changeux (2013).

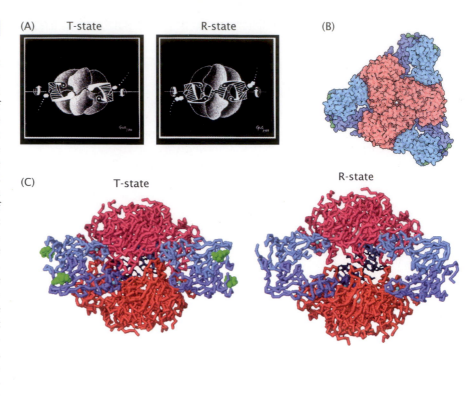

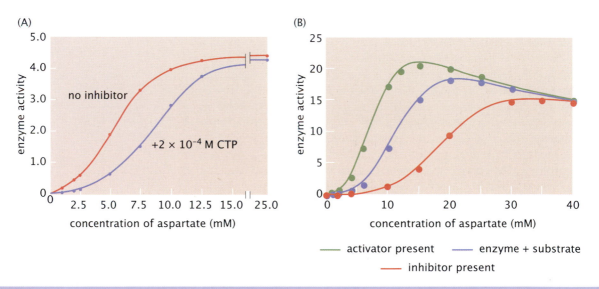

Figure 6.2

Enzyme activity of aspartate transcarbamoylase. (A) Enzyme activity is shown as a function of concentration of substrate (aspartate) with and without inhibitor present. The data illustrate that in the presence of CTP, the reaction rate is reduced. (B) Modern measurement of enzyme activity as a function of aspartate concentration in the presence of both activator (ATP - green) and inhibitor (CTP - red). Substrate inhibition is seen at high concentrations of aspartate for the pH used in this experiment. (A) Adapted from Gerhart and Pardee (1962); see Further Reading. (B) adapted from Kantrowicz (2012); see Further Reading.

One of the reasons the behavior of this enzyme was surprising is that it is inhibited by CTP, a molecule which is notably different from the substrates of the reaction itself. When CTP binds to the regulatory part of the enzyme, this reflects that the levels of pyrimidine in the cell are sufficiently high, and thus the pathway should be inhibited. When pyrimidine levels are low, this enzyme catalyzes a condensation reaction between aspartate and carbamoyl phosphate to form N-carbamoylaspartate and orthophosphate. What made the regulation of the enzyme puzzling is that one of the early ideas about such feedback inhibition was that the inhibitor would somehow fit into the same place (i.e., the active site) as the substrate itself. However, in cases like this, it was seen that the inhibitor and the substrates bear little if any structural resemblance. This observation left in its wake a puzzle as to how substrate and inhibitor interacted with the enzyme to tune the production rate of product. Of course, this entire book is one long soliloquy on that subject, showing again and again how allosteric regulators can bind at places other than the active site while still controlling the activity of the molecule in question. Before embarking on an analysis of how the allostery concept answers that challenge in the context of enzymes, we first revisit classical enzyme kinetics, as embodied in the Michaelis-Menten equations. The point of that exercise is to set the stage for the remainder of this chapter, which generalizes the statistical mechanical thinking that has dominated the book thus far to entertain a dynamical view of allostery.

6.2   Statistical Mechanics of Michaelis-Menten Enzymes

We begin by introducing the textbook Michaelis-Menten treatment of enzymes. This will serve both to introduce basic notation and to explain the states-and-weights methodology we will use for describing enzymes more generally. Many enzyme-catalyzed biochemical reactions are characterized by Michaelis-Menten kinetics. Such enzymes constitute a simple but important class for studying the relationship between the traditional chemical kinetics based on reaction rates with a physical view dictated by statistical mechanics. According to the Michaelis-Menten model, enzymes are single-state catalysts that bind a substrate and promote its conversion into a product. Although this scheme precludes allosteric interactions, many enzymes (e.g., acetylcholinesterase, triosephosphate isomerase, bisphosphoglycerate mutase, adenylate cyclase) are well described by Michaelis-Menten kinetics.

The key players in this reaction framework are a monomeric enzyme E which binds a substrate S (forming an enzyme-substrate complex ES) and converts it into a product P. This product is then removed from the system and cannot return to its original state as substrate. In the language of concentrations, this reaction can be written as

$$[E] + [S] \underset{k_{off}}{\overset{k_{on}}{\rightleftharpoons}} [ES] \overset{k_{cat}}{\rightarrow} [E] + [P], \tag{6.6}$$

where the rate of product formation is given by

$$\frac{d[P]}{dt} = k_{cat}[ES]. \tag{6.7}$$

Note that the second step in equation 6.6 has a reaction arrow pointing in only one direction, indicating that the reaction is irreversible: once substrate is converted into product, it cannot be converted back into substrate again. However, in writing the enzyme dynamics in this way, we commit what has been called the "original thermodynamic sin" (see Gunawardena in Further Reading). While the one-way reaction is probably an appropriate approximation for the in vitro conditions considered by Michaelis and Menten themselves, such strict irreversibility is likely wide of the mark in the cellular interior and ultimately is inconsistent with thermodynamics. The reader is invited to generalize the sinful version presented here to properly account for the conversion of substrate back into product.

To make analytical progress with this simple model of enzyme action, Briggs and Haldane (1925) assumed a time-scale separation in which the substrate and product concentrations ($[S]$ and $[P]$) changed slowly over time while the free and bound enzyme states ($[E]$ and $[ES]$) changed much more rapidly. This modification allows us to approximate this system over short time scales by assuming that the slow components (in this case $[S]$) remain constant and can be absorbed into the rates, resulting in the *effective* kinetic scheme

$$[E] \underset{k_{off}+k_{cat}}{\overset{k_{on}[S]}{\rightleftarrows}} [ES]. \tag{6.8}$$

Assuming that the system described by equation 6.8 reaches steady state (over the short time scale of this approximation) quickly enough that the substrate concentration does not appreciably diminish, this implies

$$[E][S]k_{on} = [ES]\left(k_{off} + k_{cat}\right), \tag{6.9}$$

which we can rewrite as

$$\frac{[ES]}{[E]} = \frac{[S]k_{on}}{k_{off} + k_{cat}} \equiv \frac{[S]}{K_M}, \tag{6.10}$$

where

$$K_M = \frac{k_{off} + k_{cat}}{k_{on}} \tag{6.11}$$

is called the Michaelis constant. K_M incorporates the binding and unbinding of ligand, as well as the conversion of substrate into product; in the limit $k_{cat} = 0$, K_M reduces to the familiar dissociation constant $K_D = k_{off}/k_{on}$. Thus, we can now work out a more revealing expression for the rate of product formation. From equations 6.7 and 6.10, we can write the rate of product formation as

$$\frac{d[P]}{dt} = k_{cat}\frac{[E][S]}{K_M}, \tag{6.12}$$

which we can simplify by noting that the maximum rate at which the reaction can go is $v_{max} = k_{cat}[E_{tot}]$, where $[E_{tot}] = [E] + [ES]$, since when this condition is satisfied every enzyme is part of an enzyme-substrate complex. Hence, we have

$$\frac{d[P]}{dt} = k_{cat}[E_{tot}]\frac{\frac{[E][S]}{K_M}}{[E_{tot}]} = v_{max}\frac{\frac{[E][S]}{K_M}}{[E] + \frac{[E][S]}{K_M}}, \tag{6.13}$$

(A)

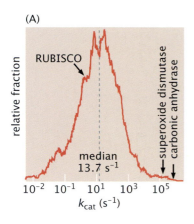

(B)

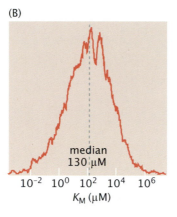

Figure 6.3

Parameters in the Michaelis-Menten framework. (A) Distribution of values of k_{cat}. (B) Distribution of values of K_M, the Michaelis constant.

which can be simplified to

$$\frac{d[P]}{dt} = v_{max} \frac{\frac{[S]}{K_M}}{1 + \frac{[S]}{K_M}}, \tag{6.14}$$

the famed and familiar form of Michaelis-Menten kinetics. Figure 2.7 (p. 49) illustrates how ubiquitous this functional form is across broad swaths of biology.

We are interested not only in developing statistical mechanical intuition for the problems considered in this book but also quantitative parametric intuition. The enzyme kinetics description we have considered thus far in equation 6.14 features several key enzyme-specific parameters. Figure 6.3 presents a compendium of measured values for the constants appearing in the Michaelis-Menten model of enzyme kinetics, namely, k_{cat} and K_M. In light of these data, we can adopt simple rules of thumb, such as that typical enzymatic k_{cat} values are $k_{cat} \approx 10\ \mathrm{s}^{-1}$, and typical $K_M \approx 100\ \mu M$.

Since we are trying to make contact with the statistical mechanical language of probabilities used throughout the book, it is convenient to write the probabilities of finding free enzyme and enzyme-substrate complex, namely, $p_E = [E]/[E_{tot}]$ and $p_{ES} = [ES]/[E_{tot}]$, respectively. The kinetic scheme shown in equation 6.8 maps the dynamics of the Michaelis-Menten enzyme onto an effective two-state system. We can visualize the microscopic states of the enzyme using a modified diagram of states and weights, shown in Fig. 6.4, that incorporates the nonequilibrium character of the underlying chemical kinetics. Unlike in the equilibrium examples described thus far in the book, for which states and weights come from the Boltzmann distribution, in the nonequilibrium setting of enzyme kinetics the states-and-weights table serves as a mnemonic for understanding the different states of the enzyme and their corresponding probabilities. Thus, we can now write the probabilities of the two states as

$$p_E = \frac{w_E}{Z_{tot}} = \frac{1}{1 + \frac{[S]}{K_M}}, \tag{6.15}$$

and

$$p_{ES} = \frac{w_{ES}}{Z_{tot}} = \frac{\frac{[S]}{K_M}}{1 + \frac{[S]}{K_M}}, \tag{6.16}$$

Figure 6.4

States, weights, and rates for the Michaelis-Menten enzyme. The probability of finding an enzyme (green) in either the free or bound state equals the weight of that state divided by the sum of all weights $(1 + [S]/K_M)$, where $[S]$ is the concentration of substrate (dark red), and $K_M = (k_{off} + k_{cat})/k_{on}$ is the Michaelis constant. At $[S] = K_M$, half the enzyme population exists in the free form, and half exists in the bound form.

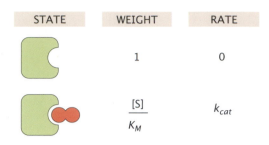

STATE	WEIGHT	RATE
	1	0
	$\dfrac{[S]}{K_M}$	k_{cat}

Figure 6.5

Properties of a Michaelis-Menten enzyme. (A) Probabilities of the free enzyme (p_E) and bound enzyme (p_{ES}) states as a function of substrate concentration. As the amount of substrate $[S]$ increases, more enzyme is found in the bound state rather than the free state. (B) The rate of product formation for a non-allosteric enzyme. The rate of product formation has the same functional form as the probability p_{ES} of the enzyme-substrate complex. Parameters were chosen to reflect "typical" enzyme kinetics values: $k_{cat} = 10^3$ s^{-1}, $K_M = 10^{-4}$ M, and $[E_{tot}] = 10^{-9}$ M.

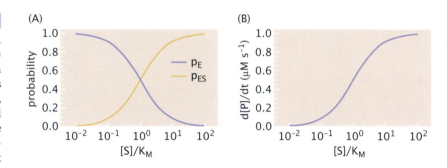

where

$$Z_{tot} = w_E + w_{ES} \qquad (6.17)$$

is the sum of all weights. Using these probabilities, we can reinterpret the equation describing the rate of product formation as

$$\frac{d[P]}{dt} = v_{max} p_{ES}. \qquad (6.18)$$

Figure 6.5 shows the probabilities of the different states and the rate of product formation as a function of the substrate concentration for an enzyme described by the kinetic scheme of Michaelis-Menten. The parameter k_{cat} scales $d[P]/dt$ vertically, while K_M rescales the substrate concentration $[S]$ by the midpoint in activity. Increasing K_M by a factor of 10 implies that 10 times as much substrate is needed to have the same rate of product formation; on the semilog plots in Figure 6.5 this corresponds to shifting all curves to the right by a factor of 10. The weights in Figure 6.4 allow us to understand Figure 6.5: when $[S] < K_M$, $w_E > w_{ES}$, so that an enzyme is more likely to be in the substrate-free state; when $[S] > K_M$, $w_E < w_{ES}$, and an enzyme is more likely to be found as an enzyme-substrate complex. Increasing K_M shifts the concentration at which the bound ES state begins to dominate over the free E state.

In equilibrium statistical mechanics, a ligand binding to a receptor would yield the exact same states and weights as in Figure 6.4 except that the Michaelis constant K_M would be replaced by the dissociation constant K_d. As a result, by using the Michaelis-Menten approximation we can apply the well-established tools and intuition of equilibrium statistical mechanics when analyzing the inherently nonequilibrium problem of enzyme kinetics. However, we note that the responses of enzymes such as aspartate transcarbamoylase (ATCase) exhibit a kinetics that is quite distinct from the Michaelis-Menten form. In particular, as with many of the activity curves we will see throughout the book, their distinguishing feature is the renowned "sigmoidal" behavior characterized mathematically by a change in sign of the second derivative of the binding curve, as shown in Figure 2.14 (p. 58). Note that the activity curve for a Michaelis-Menten enzyme shown in Figure 6.5 appears sigmoidal, but this false impression comes from the fact that the concentrations (*x*-axis) are plotted on a log scale. In the next several sections, we will show how to generalize this method of states and weights to MWC enzymes with competitive inhibitors, allosteric regulators, and multiple substrate binding sites.

6.3   Statistical Mechanics of MWC Enzymes

The whole point of the introduction to this chapter was to note the revolutionary shift in thinking that came about in the early 1960s as a result of enzyme processes such as feedback inhibition. How do we go beyond the Michaelis-Menten framework to include both the sigmoidal dependence of the enzyme upon the concentration of substrate and the presence of feedback mechanisms that alter the bare rates of the enzyme? The answer to those questions came from the realization that enzymes are not static entities. Rather, many enzymes are dynamic, constantly fluctuating between different conformational states. This notion was initially conceived by Monod-Wyman-Changeux (MWC) and Gerhart and Pardee to characterize enzymes such as the complex multisubunit enzyme aspartate transcarbamoylase (ATCase) introduced at the beginning of the chapter and shown in Figure 6.1. The case study of ATCase is instructive in revealing how an enzyme might exhibit the kinds of nuanced behavior that go beyond the Michaelis-Menten enzyme. Bolstered especially by recent insights from structural biology, not surprisingly, here, too, we will invoke our standard allosteric battle cry by arguing that the enzyme has two distinct conformational states with different enzyme rates in those two states.

What is especially impressive about the ATCase case study is that it has been thoroughly and systematically unpacked from a number of distinct and important points of view. First, structural biology has shown how both the negative and positive feedback properties of this enzyme are implemented at the level of atomic positions, as described in the article by Kantrowitz referenced in the Further Reading section at the end of the chapter. Second, a variety of functional assays with subunits and mutants have shown how the various pieces in the protein complex conspire to give rise to the full repertoire of responses. For example, if the catalytic subunits are isolated, their kinetics reduce to the familiar Michaelis-Menten form with kinetic parameters consistent with those already associated with only the active state of the molecule. In the MWC model, the ATCase is assumed to exist in two supramolecular states for which nomenclature from the literature on hemoglobin to unfold in the next chapter is sometimes used: a relaxed R state, which has high affinity for a substrate and in which the enzyme is active, and a tense T state, which has low affinity for a substrate (see Figure 6.1) and in which the enzyme has low activity. Although for ATCase, the transition between the T and R states is primarily induced by an external ligand, recent advances in single-molecule techniques have shown that many proteins intrinsically fluctuate between these different states even in the absence of ligand (as shown for ion channels in Figure 3.8 (p. 86)). These observations imply that the MWC model can be applied to a wide range of enzymes beyond those with multisubunit complexes.

We will designate an enzyme with two possible states (an active state E_A and an inactive state E_I) as an MWC enzyme. Note that although the R and T language is often used to describe these states, for the remainder of the chapter we will instead use the more transparent notation A and I to indicate active and inactive states, respectively. For generality, we will assume that even in the inactive state, the enzyme can still perform the enzymatic reaction, albeit at a lower catalytic rate. The kinetics of the simplest MWC enzyme is given by

$$[E_A]+[S] \underset{k_{off}^A}{\overset{k_{on}^A}{\rightleftharpoons}} [E_AS] \xrightarrow{k_{cat}^A} [E_A]+[P]$$

$$k_{trans}^A \Big\updownarrow k_{trans}^I \qquad k_{trans}^{AS} \Big\updownarrow k_{trans}^{IS}$$

$$[E_I]+[S] \underset{k_{off}^I}{\overset{k_{on}^I}{\rightleftharpoons}} [E_IS] \xrightarrow{k_{cat}^I} [E_I]+[P] \quad , \tag{6.19}$$

Note that as written, we once again commit Gunawardena's "original thermodynamic sin" (see Gunawardena in Further Reading) in which the product formation step is characterized by a one-way arrow signifying strict irreversibility. As in the case of a Michaelis-Menten enzyme, we assume that the time scale for the different enzyme species (i.e., active and inactive) to reach steady state is much shorter than the time scale over which the concentration of substrate and product will appreciably change. As a result, we can incorporate the slowly changing quantities $[S]$ and $[P]$ into the rates and use only the enzyme concentrations as nodes, resulting in the renormalized kinetic scheme

$$[E_A] \underset{k_{off}^A+k_{cat}^A}{\overset{k_{on}^A \, [S]}{\rightleftharpoons}} [E_AS]$$

$$k_{trans}^A \Big\updownarrow k_{trans}^I \qquad k_{trans}^{AS} \Big\updownarrow k_{trans}^{IS}$$

$$[E_I] \underset{k_{off}^I+k_{cat}^I}{\overset{k_{on}^I \, [S]}{\rightleftharpoons}} [E_IS] \quad . \tag{6.20}$$

We assume that the quasi-steady-state condition holds for all enzyme states, so that for the time period when our analysis is applicable, we have

$$\frac{d[E_AS]}{dt} = \frac{d[E_A]}{dt} = \frac{d[E_IS]}{dt} = \frac{d[E_I]}{dt} = 0. \tag{6.21}$$

To see how to put this two-state picture of the enzyme into mathematical practice, we begin by observing that the accumulation of product P in an enzyme reaction can be generically written as

$$\frac{d[P]}{dt} = k_{cat}^I p_{E_IS}[E_{tot}] + k_{cat}^A p_{E_AS}[E_{tot}], \tag{6.22}$$

where we have assumed that there is no degradation of the product and that the rate of production is given by the amount of enzyme ($p_i[E_{tot}]$) associated

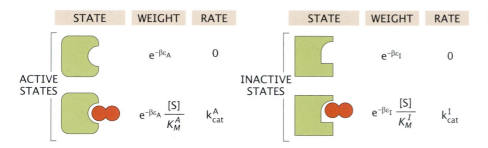

Figure 6.6

States and weights for an MWC enzyme. In this simple case, the enzyme has two conformational states and a single active site. Our first foray into the problem illustrates the key states and substrate binding *without* reference to the influence of inhibitor molecules. K_M^A and K_M^I denote the substrate concentrations at which half the active and half the inactive enzymes are bound, respectively. The energies ε_A and ε_I provide the free-energy scale for the ligand-free conformations, thus dictating their relative probabilities. Decreasing the energy ε_A of the active state would increase the probability of all the active enzyme conformations relative to the inactive conformations.

with being in either the I or the A state with substrate bound (p_{E_IS} and p_{E_AS}, respectively), times the rate of production (k_{cat}^I) when the enzyme is in each of those states. To complete the analysis, we need to know how the probability of the two states depends upon the concentration of substrate. To that end, we consider the states and weights as shown in Figure 6.6. The figure asserts that there are two different conformational states and that the rate of enzyme action is different in those two states. So far, we are ignoring the regulatory ligand and simply exploring how the conformations are altered by the substrate itself.

The states and weights of Figure 6.6 tell us that there is an energy cost ε_I for the enzyme to be in the inactive state and a corresponding energy cost ε_A for the system to be in the active state. Also, by analogy with the way we handled the states and weights of the Michaelis-Menten enzyme in Figure 6.4, we assign different K_Ms to the inactive (K_M^I) and active (K_M^A) states.

Thus, we can write the dynamical equation for the rate of product formation as

$$\frac{dP}{dt} = k_{cat}^I [E_{tot}] \frac{e^{-\beta\varepsilon_I}\frac{[S]}{K_M^I}}{e^{-\beta\varepsilon_I}(1+\frac{[S]}{K_M^I}) + e^{-\beta\varepsilon_A}(1+\frac{[S]}{K_M^A})}$$

$$+ k_{cat}^A [E_{tot}] \frac{e^{-\beta\varepsilon_A}\frac{[S]}{K_M^A}}{e^{-\beta\varepsilon_I}(1+\frac{[S]}{K_M^I}) + e^{-\beta\varepsilon_A}(1+\frac{[S]}{K_M^A})}, \tag{6.23}$$

where $[S]$ is the concentration of substrate, k_{cat}^I is the rate of product formation from the inactive state, and k_{cat}^A is the rate of product formation from the active state.

The probabilities of the different states shown in Figure 6.6 and the rate of product formation given by equation 6.23 are shown in Figure 6.7. Although we use the same parameters for the active state as in Figure 6.5, namely, $k_{cat}^A = 10^3$ s^{-1} and $K_M^A = 10^{-4}$ M, the p_{E_A} and p_{E_AS} curves in Figure 6.7(A) look markedly different from the p_E and p_{ES} curves in Figure 6.5(A), because there are four states in play rather than just two. When $[S] = 0$ the enzyme exists only in the unbound states E_A and E_I, whose relative probabilities are given by

$$\frac{p_{E_A}}{p_{E_I}} = e^{-\beta(\varepsilon_A - \varepsilon_I)}, \tag{6.24}$$

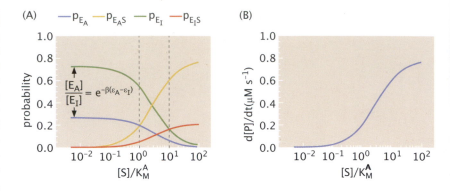

Figure 6.7

State probabilities and product formation rates. (A) Probabilities of each enzyme state. While the active state has the same catalytic rate k_{cat}^A and Michaelis constant K_M^A as the non-MWC enzyme in Figure 6.5(A), the inactive state significantly alters the forms of p_{E_A} and $p_{E_A S}$. (B) Assuming $k_{cat}^A \gg k_{cat}^I$, $d[P]/dt$ is dominated by the active enzyme-substrate complex, $p_{E_A S}$. Parameters used were $k_{cat}^A = 10^3 \text{ s}^{-1}$, $k_{cat}^I = 10 \text{ s}^{-1}$, $K_M^A = 10^{-4} \text{ M}$, $K_M^I = 10^{-3} \text{ M}$, $e^{-\beta(\varepsilon_A - \varepsilon_I)} = e^{-1}$, and $[E_{tot}] = 10^{-9} \text{ M}$.

indicating that most of the enzymes in the ensemble are in the unbound inactive state. When $[S] \to \infty$, the bound states $E_A S$ and $E_I S$ dominate with relative probabilities

$$\frac{p_{E_A S}}{p_{E_I S}} = e^{-\beta(\varepsilon_A - \varepsilon_I)} \frac{K_M^I}{K_M^A}, \tag{6.25}$$

indicating that most enzymes are in the bound active state. Note that the p_{E_A} and $p_{E_A S}$ curves intersect when $p_{E_A} = p_{E_A S}$ at $[S] = K_M^A$, while the two inactive state curves intersect at $[S] = K_M^I$, analogously to the Michaelis-Menten enzyme in Figure 6.5(A).

Using the framework just described, we can compute properties of the enzyme kinetics curve shown in Figure 6.7(B). One important property is the dynamic range of an enzyme, the difference between the maximum and minimum rate of product formation. In the absence of substrate ($[S] \to 0$) and a saturating concentration of substrate ($[S] \to \infty$), the rate of product formation equation 6.23 becomes

$$\lim_{[S] \to 0} \frac{d[P]}{dt} = 0 \tag{6.26}$$

$$\lim_{[S] \to \infty} \frac{d[P]}{dt} = [E_{tot}] \frac{k_{cat}^A \frac{e^{-\beta \varepsilon_A}}{K_M^A} + k_{cat}^I \frac{e^{-\beta \varepsilon_I}}{K_M^I}}{\frac{e^{-\beta \varepsilon_A}}{K_M^A} + \frac{e^{-\beta \varepsilon_I}}{K_M^I}}. \tag{6.27}$$

From these two expressions, we can write the dynamic range as

$$\text{dynamic range} = \left(\lim_{[S] \to \infty} \frac{d[P]}{dt} \right) - \left(\lim_{[S] \to 0} \frac{d[P]}{dt} \right)$$

$$= [E_{tot}] k_{cat}^A \left(1 - \frac{1 - \frac{k_{cat}^I}{k_{cat}^A}}{1 + e^{-\beta(\varepsilon_A - \varepsilon_I)} \frac{K_M^I}{K_M^A}} \right), \tag{6.28}$$

where every term in the fraction has been written as a ratio of the active and inactive state parameters. We find that the dynamic range increases as k_{cat}^I / k_{cat}^A, $e^{-\beta(\varepsilon_A - \varepsilon_I)}$, and K_M^I / K_M^A increase (assuming $k_{cat}^A > k_{cat}^I$).

Another important property is the concentration of substrate at which the rate of product formation lies halfway between its minimum and maximum value, which we will denote as $[EC_{50}]$, as we have done in previous chapters. It is straightforward to show that the definition

$$\lim_{[S] \to [EC_{50}]} \frac{d[P]}{dt} = \frac{1}{2} \left(\lim_{[S] \to \infty} \frac{d[P]}{dt} + \lim_{[S] \to 0} \frac{d[P]}{dt} \right) \tag{6.29}$$

is satisfied when

$$[EC_{50}] = K_M^A \frac{e^{-\beta(\varepsilon_A - \varepsilon_I)} + 1}{e^{-\beta(\varepsilon_A - \varepsilon_I)} + \frac{K_M^A}{K_M^I}}. \tag{6.30}$$

With increasing $e^{-\beta(\varepsilon_A - \varepsilon_I)}$, the value of $[EC_{50}]$ increases if $K_M^A > K_M^I$ and decreases otherwise. $[EC_{50}]$ always decreases as K_M^A / K_M^I increases. Lastly, we note that in the limit of a Michaelis-Menten enzyme, $\varepsilon_I \to \infty$, we recoup the familiar results

$$\text{dynamic range} = [E_{tot}] k_{cat}^A, \tag{6.31}$$

and

$$[EC_{50}] = K_M^A. \tag{6.32}$$

We now turn to the question of how enzyme activity is altered by the presence of allosteric effectors.

6.3.1 Modulating Enzyme Activity with Allosteric Effectors

The catalytic activity of many enzymes is controlled by molecules that bind to sites on the enzyme other than the active sites themselves. Indeed, it is these situations that really gave rise to the allostery concept in the first place (see Gerhart (2014) in Further Reading for an enlightening discussion of the early history of feedback inhibition). Binding of these regulatory ligands induces conformational changes that alter the substrate binding site and act as allosteric regulators by increasing or reducing $d[P]/dt$. Allosterically controlled enzymes are important regulatory nodes in metabolic pathways, as shown in Figure 1.7 (p. 10), and are often responsible for keeping cells in homeostasis. Some well-studied examples of allosteric control include glycogen phosphorylase, phosphofructokinase, glutamine synthetase, and aspartate transcarbamoylase (ATCase). In many cases these systems are characterized by Hill functions, but the Hill coefficients thus obtained can be difficult to interpret and do not provide much information about the organization or regulation of an enzyme, nor do they reflect the underlying state space of possible conformations and states of occupancy, with the role of effector molecules being captured particularly indirectly. In this section we will show how the statistical mechanical approach to enzymes described thus far in the chapter can be directly extended to consider the role of allosteric effectors.

So far, our analysis has centered only on how a sigmoidal activity curve might arise as a result of the enzyme having several different conformational states. But this non-Michaelis-Menten functional form is but only one of the interesting features of these regulated enzymes. A second key feature of their function is the existence of mechanisms for feedback such that the product of the particular

Figure 6.8

States, weights, and rates for an MWC enzyme. This enzyme is subject to control by the binding of an effector molecule (yellow) at concentration $[R]$ and acts on a substrate molecule (red) at concentration $[S]$. The four states on the left correspond to the "active" state of the enzyme, and the four states on the right to the "inactive" state of the enzyme. We make the simplifying assumption that the two active states with substrate bound have the same rate of product formation, and similarly for the two substrate-bound inactive states.

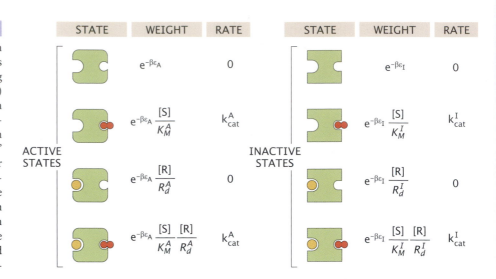

pathway of interest (or some other pathway) can tune the activity of the enzyme. We consider an MWC enzyme with one site for an allosteric regulator R and a different site for a substrate molecule S that will be converted into product. In Figure 6.8 we show the states and weights for an enzyme with distinct binding regions for the substrate and the regulatory effector. Note that we have a total of eight distinct states corresponding to the four states of occupancy (i.e., empty, bound with inhibitor, bound with substrate, and bound with both) of the two different conformational states.

For the states and weights of Figure 6.8, the weight of each state that contains substrate has a factor $[S]/K_M$, where K_M is a Michaelis constant. The weights corresponding to states with an allosteric regulator have a factor $[R]/R_d$, where R_d is a dissociation constant (because the regulator can only bind and unbind to the enzyme, whereas substrate can be turned into product). The regulator R contributes a factor $[R]/R_d^A$ when it binds to the active state, and a factor $[R]/R_d^I$ when it binds to the inactive state, where R_d^A and R_d^I are the dissociation constants between the regulator and the active and inactive states of the enzyme, respectively. An allosteric activator has a smaller dissociation constant $R_d^A < R_d^I$ for binding to the active-state enzyme, so that adding activator increases the relative weights of the active states. Since the active state transforms substrate at a much faster rate than the inactive state, $k_{cat}^A \gg k_{cat}^I$, adding activator increases the rate of product formation $d[P]/dt$. An allosteric inhibitor would switch the ordering of the dissociation constants to $R_d^A > R_d^I$.

Using the states and weights in Figure 6.8, we can compute the probability of each enzyme state. For example, the probabilities of the four states that form product are given by

$$p_{E_A S} = e^{-\beta \varepsilon_A} \frac{\frac{[S]}{K_M^A}}{Z_{tot}}, \tag{6.33}$$

$$p_{E_A SR} = e^{-\beta \varepsilon_A} \frac{\frac{[S]}{K_M^A} \frac{[R]}{R_d^A}}{Z_{tot}}, \tag{6.34}$$

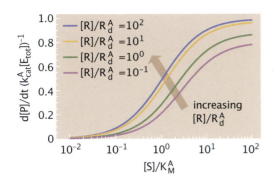

Figure 6.9

Effects of an allosteric regulator R on the rate of product formation. The regulator's greater affinity for the active enzyme state increases the fraction of the active conformations and hence $d[P]/dt$. Parameters used were $R_d^A/R_d^I = 10^{-2}$ M and the parameters from Figure 6.7.

$$p_{E_I S} = e^{-\beta \varepsilon_I} \frac{\frac{[S]}{K_M^I}}{Z_{tot}}, \tag{6.35}$$

and

$$p_{E_I SR} = e^{-\beta \varepsilon_I} \frac{\frac{[S]}{K_M^I} \frac{[R]}{R_d^I}}{Z_{tot}}, \tag{6.36}$$

where

$$Z_{tot} = e^{-\beta \varepsilon_A} \left(1 + \frac{[S]}{K_M^A}\right) \left(1 + \frac{[R]}{R_d^A}\right) + e^{-\beta \varepsilon_I} \left(1 + \frac{[S]}{K_M^I}\right) \left(1 + \frac{[R]}{R_d^I}\right) \tag{6.37}$$

is the sum of all weights in Figure 6.8.

The total enzyme concentration $[E_{tot}]$ is a conserved quantity which equals the sum of the concentrations of all the distinct enzyme states ($[E_A]$, $[E_A S]$, $[E_A R]$, $[E_A SR]$ and their inactive-state counterparts. Using the probabilities worked out above, we can write these concentrations as $[E_A S] = [E_{tot}]p_{E_A S}$, $[E_A SR] = [E_{tot}]p_{E_A SR}$, and so on, so that the rate of product formation is given by

$$\frac{d[P]}{dt} = k_{cat}^A \left([E_A S] + [E_A SR]\right) + k_{cat}^I \left([E_I S] + [E_I SR]\right)$$

$$= [E_{tot}] \frac{k_{cat}^A e^{-\beta \varepsilon_A} \frac{[S]}{K_M^A} \left(1 + \frac{[R]}{R_d^A}\right) + k_{cat}^I e^{-\beta \varepsilon_I} \frac{[S]}{K_M^I} \left(1 + \frac{[R]}{R_d^I}\right)}{e^{-\beta \varepsilon_A} \left(1 + \frac{[S]}{K_M^A}\right) \left(1 + \frac{[R]}{R_d^A}\right) + e^{-\beta \varepsilon_I} \left(1 + \frac{[S]}{K_M^I}\right) \left(1 + \frac{[R]}{R_d^I}\right)}. \tag{6.38}$$

The rate of product formation given in equation 6.38 for different $[R]$ values is shown in Figure 6.9. It is important to realize that by choosing the weights in Figure 6.8, we have selected a particular model for the allosteric regulator, namely, one in which the regulator binds equally well to an enzyme with or without substrate. There are many other possible models. For example, we could add an interaction energy between an allosteric regulator and a bound substrate. However, the simple model in Figure 6.8 already possesses the important feature that adding more allosteric activator yields a larger rate of product formation $d[P]/dt$, as shown in Figure 6.9.

We end this section by noting that we can think of the action of an allosteric regulator differently. In particular, we can reimagine the role of the effector as redefining the energies of the active and inactive states. To better understand this, consider the probability of an active-state enzyme-substrate complex (with or without a bound regulator). From equation 6.36, these probabilities are given by

$$
p_{E_AS} + p_{E_ASR} = \frac{e^{-\beta \varepsilon_A} \frac{[S]}{K_M^A} \left(1 + \frac{[R]}{R_d^A}\right)}{e^{-\beta \varepsilon_A} \left(1 + \frac{[S]}{K_M^A}\right) \left(1 + \frac{[R]}{R_d^A}\right) + e^{-\beta \varepsilon_I} \left(1 + \frac{[S]}{K_M^I}\right) \left(1 + \frac{[R]}{R_d^I}\right)}. \tag{6.39}
$$

We can now rewrite this equation using effective-energy parameters as

$$
p_{E_AS} + p_{E_ASR} \equiv \frac{e^{-\beta \tilde{\varepsilon}_A} \frac{[S]}{K_M^A}}{e^{-\beta \tilde{\varepsilon}_A} \left(1 + \frac{[S]}{K_M^A}\right) + e^{-\beta \tilde{\varepsilon}_I} \left(1 + \frac{[S]}{K_M^I}\right)}, \tag{6.40}
$$

where we have defined new effective-energy parameters

$$
\tilde{\varepsilon}_A = \varepsilon_A - \frac{1}{\beta} \log\left(1 + \frac{[R]}{R_d^A}\right) \tag{6.41}
$$

and

$$
\tilde{\varepsilon}_I = \varepsilon_I - \frac{1}{\beta} \log\left(1 + \frac{[R]}{R_d^I}\right). \tag{6.42}
$$

Note that by defining these *effective* energies, we are "integrating out" the effector degrees of freedom, exactly as described in section 2.6 (p. 63). We are saying that an MWC enzyme in the presence of a regulator at concentration $[R]$ is equivalent to an MWC enzyme with no regulator provided that we use the new energies $\tilde{\varepsilon}_A$ and $\tilde{\varepsilon}_I$ for the active and inactive states. An analogous statement holds for all the conformations of the enzyme, so that the effects of a regulator can be completely absorbed into the energies of the active and inactive states! In other words, adding an allosteric regulator allows us to tune the parameters ε_A and ε_I of an allosteric enzyme, and thus change its rate of product formation, in a quantifiable manner.

One application of this novel result is that we can easily compute the dynamic range of an enzyme as well as the concentration of substrate for half-maximal rate of product formation. Both these quantities follow from the analogous expressions for an MWC enzyme that we developed in equations 6.28 and 6.30. In the case with an effector, the substitution of the effective energies $\tilde{\varepsilon}_A$ and $\tilde{\varepsilon}_I$ leads to a dynamic range of the form

$$
\text{dynamic range} = [E_{tot}] k_{cat}^A \left(1 - \frac{1 - \frac{k_{cat}^I}{k_{cat}^A}}{1 + e^{-\beta(\varepsilon_A - \varepsilon_I)} \frac{1 + [R]/R_d^A}{1 + [R]/R_d^I} \frac{K_M^I}{K_M^A}}\right) \tag{6.43}
$$

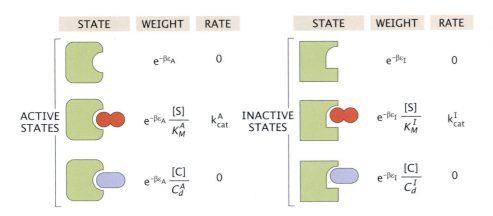

Figure 6.10

States and weights for an MWC enzyme in the presence of substrate (red) and a competitive inhibitor (blue). While the substrate S can be transformed into product, the inhibitor C can occupy the substrate binding site but the substrate cannot be enzymatically modified.

and an EC_{50} value of

$$EC_{50} = K_M^A \frac{e^{-\beta(\varepsilon_A - \varepsilon_I)} \frac{1 + [R]/R_d^A}{1 + [R]/R_d^I} + 1}{e^{-\beta(\varepsilon_A - \varepsilon_I)} \frac{1 + [R]/R_d^A}{1 + [R]/R_d^I} + \frac{K_M^A}{K_M^I}}. \tag{6.44}$$

As expected, the dynamic range of an enzyme increases with regulator concentration $[R]$ for an allosteric activator ($R_d^A < R_d^I$). Adding more activator will shift $[EC_{50}]$ to the left if $K_M^A < K_M^I$ (as shown in Figure 6.9) or to the right if $K_M^A > K_M^I$. The opposite effects hold for an allosteric inhibitor ($R_d^I < R_d^A$).

6.3.2   Competitive Inhibitors and Enzyme Action

Another level of control found in many enzymes is competitive inhibition. When a competitive inhibitor C binds to the same active site as the substrate S, the number of possible states is increased, because the state of competitive inhibitor occupancy decreases the probability that the enzyme's active state will bind with substrate. In this case, an enzyme can exist in the unbound state E, as an enzyme-substrate complex ES, or as an enzyme-competitor complex EC. As more inhibitor is added to a system, it crowds out the substrate from the enzyme's active site, which decreases product formation. Many cancer drugs (e.g., Lapatinib, Sorafenib, and Erlotinib) are competitive inhibitors for kinases involved in signaling pathways.

To construct the appropriate states and weights for this case, we build upon the states already shown in Figure 6.6 but now add two additional states of occupancy corresponding to competitor binding, as shown in Figure 6.10. Only the enzyme-substrate complexes in the active ($E_A S$) and inactive ($E_I S$) states lead to product formation. The probabilities of each of these states is given by dividing their statistical weights as shown in Figure 6.10 with the partition function (which includes the competitive inhibitor states) given by

$$Z_{tot} = e^{-\beta \varepsilon_A} \left(1 + \frac{[S]}{K_M^A} + \frac{[C]}{C_d^A} \right) + e^{-\beta \varepsilon_I} \left(1 + \frac{[S]}{K_M^I} + \frac{[C]}{C_d^I} \right). \tag{6.45}$$

Figure 6.11

Figure 6.11

Effects of a competitive inhibitor C on the rate of product formation. When $[C] \lesssim C_d^A, C_d^I$, the inhibitor cannot outcompete the substrate at high substrate concentrations, while the free form of enzyme dominates at low substrate concentrations. Therefore, increasing $[C]$ up to values of $\approx C_d^A$ or C_d^I has little effect on $d[P]/dt$. Once $[C] \gtrsim C_d^A, C_d^I$, the inhibitor can outcompete substrate at large concentrations, pushing the region where the enzyme-substrate complex dominates farther to the right. Parameters used were $C_d^A/C_d^I = 1$ and the parameters from Figure 6.7.

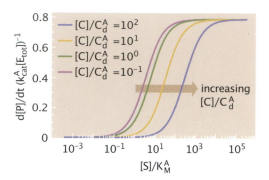

Following exactly the same logic already introduced several times throughout the chapter, we can write the rate of product formation as

$$\frac{d[P]}{dt} = k_{cat}^A [E_A S] + k_{cat}^I [E_I S]$$

$$= [E_{tot}] \frac{k_{cat}^A e^{-\beta \varepsilon_A} \frac{[S]}{K_M^A} + k_{cat}^I e^{-\beta \varepsilon_I} \frac{[S]}{K_M^I}}{e^{-\beta \varepsilon_A}\left(1 + \frac{[S]}{K_M^A} + \frac{[C]}{C_d^A}\right) + e^{-\beta \varepsilon_I}\left(1 + \frac{[S]}{K_M^I} + \frac{[C]}{C_d^I}\right)}. \tag{6.46}$$

We see that adding more competitive inhibitor decreases $d[P]/dt$ (since the denominator increases), as expected.

Figure 6.11 shows the rate of product formation for various inhibitor concentrations $[C]$. Adding more competitive inhibitor increases the probability of the inhibitor-bound states and thereby decreases the probability of binding by those states competent to form product. As with our analysis of allosteric regulators, we can absorb the effects of the competitive inhibitor C_d^A and C_d^I in equation 6.46 into the enzyme parameters ε_A and ε_I and K_M^A and K_M^I, resulting in

$$\frac{d[P]}{dt} = [E_{tot}] \frac{k_{cat}^A e^{-\beta \varepsilon_A}\left(1 + \frac{[C]}{C_d^A}\right) \frac{[S]}{K_M^A\left(1 + \frac{[C]}{C_d^A}\right)} + k_{cat}^I e^{-\beta \varepsilon_I}\left(1 + \frac{[C]}{C_d^I}\right) \frac{[S]}{K_M^I\left(1 + \frac{[C]}{C_d^I}\right)}}{\left(e^{-\beta \varepsilon_A}\left(1 + \frac{[C]}{C_d^A}\right)\left(1 + \frac{[S]}{K_M^A\left(1 + \frac{[C]}{C_d^A}\right)}\right)\right.}$$

$$\left. + e^{-\beta \varepsilon_I}\left(1 + \frac{[C]}{C_d^I}\right)\left(1 + \frac{[S]}{K_M^I\left(1 + \frac{[C]}{C_d^I}\right)}\right)\right)$$

$$\equiv [E_{tot}] \frac{k_{cat}^A e^{-\beta \tilde{\varepsilon}_A} \frac{[S]}{\tilde{K}_M^A} + k_{cat}^I e^{-\beta \tilde{\varepsilon}_I} \frac{[S]}{\tilde{K}_M^I}}{e^{-\beta \tilde{\varepsilon}_A}\left(1 + \frac{[S]}{\tilde{K}_M^A}\right) + e^{-\beta \tilde{\varepsilon}_I}\left(1 + \frac{[S]}{\tilde{K}_M^I}\right)}, \tag{6.47}$$

where we have defined the new energies and Michaelis constants,

$$\tilde{\varepsilon}_A = \varepsilon_A - \frac{1}{\beta}\log\left(1 + \frac{[C]}{C_d^A}\right), \tag{6.48}$$

$$\tilde{\varepsilon}_I = \varepsilon_I - \frac{1}{\beta}\log\left(1 + \frac{[C]}{C_d^I}\right), \tag{6.49}$$

$$\tilde{K}_M^A = K_M^A \left(1 + \frac{[C]}{C_d^A} \right), \tag{6.50}$$

$$\tilde{K}_M^I = K_M^I \left(1 + \frac{[C]}{C_d^I} \right). \tag{6.51}$$

Here we once again invoke the idea of integrating out degrees of freedom to create *effective* parameters, as described in section 2.6 (p. 63). Note that equation 6.47 has exactly the same form as the rate of product formation of an MWC enzyme without a competitive inhibitor, equation 6.23. In other words, a competitive inhibitor modulates both the effective energies and the Michaelis constants of the active and inactive states. Thus, an observed value of K_M may not represent a true Michaelis constant if an inhibitor is present. In the special case of a Michaelis-Menten enzyme ($e^{-\beta \varepsilon_I} \to \infty$), we recover the known result that a competitive inhibitor changes only the apparent Michaelis constant.

As shown for the allosteric regulator, the dynamic range and the concentration of substrate for half-maximal rate of product formation [EC_{50}] follow from the analogous expressions for an MWC enzyme given in equations 6.28 and 6.30 and using the parameters $\tilde{\varepsilon}_A$, $\tilde{\varepsilon}_I$, $\tilde{K}_M^A$, and $\tilde{K}_M^I$. That is, if we want to know the dynamic range and [EC_{50}] of an enzyme with a competitor, we can use the expressions for the enzyme without competitor given in equations 6.28 and 6.30, replacing the bare parameters with their effective values defined here. Hence, an allosteric enzyme with one active site in the presence of a competitive inhibitor has a dynamic range given by

$$\text{dynamic range} = [E_{tot}]k_{cat}^A \left(1 - \frac{1 - \frac{k_{cat}^I}{k_{cat}^A}}{1 + e^{-\beta(\varepsilon_A - \varepsilon_I)} \frac{K_M^I}{K_M^A}} \right) \tag{6.52}$$

and an EC_{50} value of

$$EC_{50} = K_M^A \frac{e^{-\beta(\varepsilon_A - \varepsilon_I)} \left(1 + \frac{[C]}{C_d^A} \right) + \left(1 + \frac{[C]}{C_d^I} \right)}{e^{-\beta(\varepsilon_A - \varepsilon_I)} + \frac{K_M^A}{K_M^I}}. \tag{6.53}$$

We note that in equation 6.52 the dynamic range will take its largest value when $k_{cat}^A = k_{cat}^I$, a condition that is approximately satisfied by many allosteric enzymes, raising questions about the possible evolutionary origins of such parameter values. Notice that in equation 6.52, the dynamic range of an MWC enzyme in the presence of a competitive inhibitor is exactly the same as equation 6.28, the dynamic range in the absence of an inhibitor. This makes sense, because in the absence of substrate ($[S] \to 0$) the rate of product formation must be zero, and at saturating substrate concentrations ($[S] \to \infty$) the substrate completely crowds out any inhibitor concentration. Instead of altering the rate of product formation at these two limits, the competitive inhibitor shifts the $\frac{d[P]}{dt}$ curve, and therefore EC_{50}, to the right as more inhibitor is added.

As we have seen, the effects of both an allosteric effector and a competitive inhibitor can be absorbed into the parameters of an MWC enzyme. This

Figure 6.12

States and weights of an MWC enzyme having two substrate binding sites.

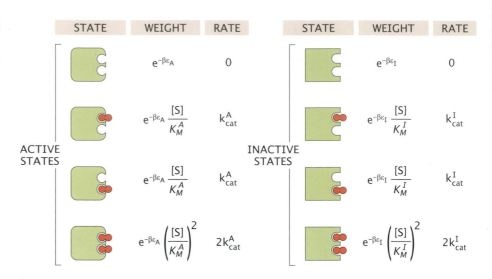

suggests that experimental data from enzymes that titrate these ligands can be collapsed into a one-parameter family of curves in which the single parameter is either the concentration of an allosteric effector or a competitive inhibitor.

6.3.3   Multiple Substrate Binding Sites

Many enzymes are composed of multiple subunits that contain substrate binding sites. Thus, an enzyme can ultimately have more than one substrate binding site, allosteric regulator site, or both. For example, the enzyme ATCase shown in Figure 6.1 (p. 204) has six substrate binding sites and six allosteric activator sites. We now extend the single-site model of an MWC enzyme introduced in Figure 6.6 (p. 211) to an MWC enzyme with two substrate binding sites, as shown in Figure 6.12. The reader is encouraged to generalize this result to the case of n binding sites. For the two-site case, assuming that both binding sites are independent, the states and weights of the system are shown in Figure 6.12. When the enzyme is doubly occupied, $E_A S^2$, we assume that it forms product twice as fast as a singly occupied enzyme $E_A S$. Note that in MWC models, cooperative interaction energies are not required to obtain cooperativity; cooperativity is inherently built into the fact that all binding sites switch concurrently from an active state to an inactive state. An explicit interaction energy, if desired, can always be added to the model.

Repeating the same procedure as in the previous sections, we compute the probability and concentration of each enzyme state. Because there are two singly bound enzyme states and one doubly bound enzyme state with twice the rate, the rate of product formation is given by

$$\frac{d[P]}{dt} = k_{cat}^A \left(2p_{E_A S}\right) + 2k_{cat}^A \left(p_{E_A S^2}\right) + k_{cat}^I \left(2p_{E_I S}\right) + 2k_{cat}^I \left(p_{E_I S^2}\right)$$

$$= 2[E_{tot}] \frac{k_{cat}^A e^{-\beta \varepsilon_A} \frac{[S]}{K_M^A}\left(1 + \frac{[S]}{K_M^A}\right) + k_{cat}^I e^{-\beta \varepsilon_I} \frac{[S]}{K_M^I}\left(1 + \frac{[S]}{K_M^I}\right)}{e^{-\beta \varepsilon_A}\left(1 + \frac{[S]}{K_M^A}\right)^2 + e^{-\beta \varepsilon_I}\left(1 + \frac{[S]}{K_M^I}\right)^2}. \quad (6.54)$$

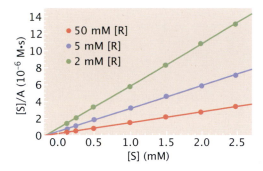

Figure 6.13

The effects of an allosteric effector on enzyme activity. Data points show experimentally measured activity for the enzyme α-amylase using substrate analog [S] (EPS) and allosteric activator [R] (NaCl). Best fit theoretical curves described by equation 6.57 are overlaid on the data. The best fit parameters are $e^{-\beta(\varepsilon_A - \varepsilon_I)} = 7.8 \times 10^{-4}$, $K_M^A = 0.6\,\text{mM}$, $K_M^I = 0.2\,\text{mM}$, $R_d^A = 0.03\,\text{mM}$, $R_d^I = 7.9\,\text{mM}$, $k_{cat}^A = 14\,\text{s}^{-1}$, and $k_{cat}^I = 0.01\,\text{s}^{-1}$. Data adapted from Feller et al. (1996).

6.3.4   What the Data Say

Thus far, this chapter has focused on the technical details of how to implement the MWC model in our thinking about enzymes. But as in previous chapters, it is of interest to explore how this thinking serves as a framework for considering real enzymes. Experimentally, the rate of product formation of an enzyme is often measured relative to the enzyme concentration and given as

$$A \equiv \frac{1}{[E_{tot}]} \frac{d[P]}{dt}, \qquad (6.55)$$

formally referred to as the activity. Enzymes are often characterized by their activity curves as substrate, inhibitor, and regulator concentrations are titrated. Such data not only determine important kinetic constants but can also characterize the nature of molecular players such as whether an inhibitor is competitive, uncompetitive, mixed, or noncompetitive. After investigating activity curves, we turn to a case study of the curious phenomenon of substrate inhibition, where saturating concentrations of substrate inhibit enzyme activity, and consider a minimal mechanism for substrate inhibition caused solely by allostery.

We begin with an analysis of α-amylase, one of the simplest allosteric enzymes, which has only a single catalytic site. This enzyme, also known as diastase, and sometimes said to be the first enzyme discovered and announced in a paper by Anselme Payen and Jean-Francois Persoz (1833), catalyzes the hydrolysis of large polysaccharides (e.g., starch and glycogen) into smaller carbohydrates so that they can be absorbed by gut transporters, which can absorb small sugars but not large polysaccharides. Amylase inhibitors are even used in patients with diabetes to reduce sugar absorption from foods. This enzyme is also responsible for giving some fruits their sweet taste. It is competitively inhibited by isoacarbose at the active site and is allosterically activated by chloride ions at a distinct allosteric site, though we will consider only the effect of the activator in our discussion. Readers are urged to explore the action of competitive inhibitors as an exercise. Figure 6.13 plots a measure of enzyme activity of α-amylase as a function of substrate concentration for different effector concentrations. How well are such data accounted for by the models worked out thus far in the chapter?

Recall that an enzyme with one active site and one allosteric site has activity given by equation 6.38, namely,

$$A = \frac{k_{cat}^A e^{-\beta \varepsilon_A} \frac{[S]}{K_M^A} \left(1 + \frac{[R]}{R_d^A}\right) + k_{cat}^I e^{-\beta \varepsilon_I} \frac{[S]}{K_M^I} \left(1 + \frac{[R]}{R_d^I}\right)}{e^{-\beta \varepsilon_A} \left(1 + \frac{[S]}{K_M^A}\right) \left(1 + \frac{[R]}{R_d^A}\right) + e^{-\beta \varepsilon_I} \left(1 + \frac{[S]}{K_M^I}\right) \left(1 + \frac{[R]}{R_d^I}\right)}. \tag{6.56}$$

Before the era of powerful computing, many clever tricks were invented for plotting enzyme activity, one of which is to plot substrate concentration divided by activity, $[S]/A$, as a function of substrate $[S]$. From the activity equation, we expect the $[S]/A$ curves in Figure 6.13 to be linear in $[S]$, as seen here,

$$\frac{[S]}{A} = \frac{e^{-\beta \varepsilon_A} \left(1 + \frac{[S]}{K_M^A}\right) \left(1 + \frac{[R]}{R_d^A}\right) + e^{-\beta \varepsilon_I} \left(1 + \frac{[S]}{K_M^I}\right) \left(1 + \frac{[R]}{R_d^I}\right)}{k_{cat}^A e^{-\beta \varepsilon_A} \frac{1}{K_M^A} \left(1 + \frac{[R]}{R_d^A}\right) + k_{cat}^I e^{-\beta \varepsilon_I} \frac{1}{K_M^I} \left(1 + \frac{[R]}{R_d^I}\right)}. \tag{6.57}$$

Figure 6.13 shows that the experimental data are well characterized by this functional form, so that the rate of product formation at any other substrate and allosteric activator concentration can be predicted by this model.

In the special case of a Michaelis-Menten enzyme ($\varepsilon_I \to \infty$), the preceding equation becomes

$$\frac{[S]}{A} = \frac{K_M^A + [S]}{k_{cat}^A}. \tag{6.58}$$

The x-intercept of all lines in such a plot intersect at the point $(-K_M^A, 0)$ which allows an easy determination of K_M^A. This is why plots of $[S]/A$ versus $[S]$, called *Hanes plots*, are often seen in enzyme kinetics data. Care must be taken, however, when extending this analysis to allosteric enzymes, for which the form of the x-intercept is more complicated.

6.4 Glycolysis and Allostery

For cells growing on glucose, the source of the energy used to synthesize ATP comes from metabolic breakdown of glucose. The pathway of these reactions is known as glycolysis and is illustrated schematically in Figure 1.6 (p. 8). For *E. coli* grown in the presence of oxygen, a single molecule of glucose can be metabolically broken down to form up to 30 molecules of ATP from ADP, because the pyruvate emerging from the glycolysis pathway can be used to fuel further energy-producing reactions. This process results in carbon dioxide as a waste product.

After the six-carbon glucose is taken up by the bacterial cell, it is broken down by the process of glycolysis to form two copies of the three-carbon molecule pyruvate. This sequential set of chemical transformations takes place through 10 distinct chemical steps, as shown at the molecular level in Figure 1.6. The pyruvate that emerges from this pathway can then be used to synthesize a variety of amino acids or fatty acids. As glucose is broken down to form pyruvate, some of the chemical energy stored in its covalent bonds is used to synthesize ATP and NADH. These high-energy carrier molecules can

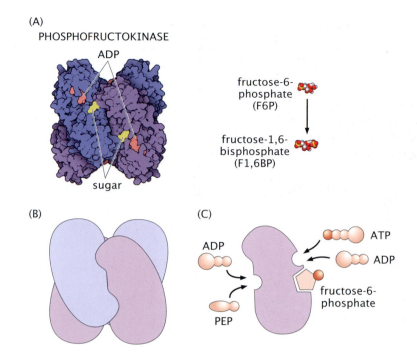

(A) PHOSPHOFRUCTOKINASE

Figure 6.14

Abstraction of phosphofructokinase structure. (A) A space filling view of the enzyme and its substrate. (B) Cartoon representation of the tetrameric enzyme. (C) Schematic of one of the monomeric units making up the tetrameric structure showing binding pockets for several of its binding partners. Fructose-6-phosphate and ATP bind in the active-site region of the enzyme, and ADP (and PEP) bind elsewhere on the molecule and alter its activity, though at sufficiently high concentrations, ADP inhibits the reaction by competing with ATP. The γ-phosphate group of ATP which transfers to F6P is shaded darker. Figures courtesy of David Goodsell.

then donate their energy to drive forward biosynthetic reactions that are not intrinsically energetically favorable. Interestingly, the reactions of the glycolysis pathway give us an opportunity to invoke the ideas about allostery developed thus far in the book.

6.4.1 The Case of Phosphofructokinase

One of the key nodes in the glycolysis pathway of Figure 1.6 is the enzyme phosophofructokinase, whose role is to add a phosphate group to fructose 6-phosphate during glycolysis. A central question that provides a backdrop to the analysis to follow is, how can glycolysis maintain a high ATP:ADP ratio in the presence of impressive 100-fold changes in the ATP consumption rate?

Our abstraction of this complex enzyme is shown in Figure 6.14, where we begin with the three-dimensional space-filling structure as represented by the elegant paintings of David Goodsell. In the preface, I alerted the reader to the fact that many of our treatments of biological problems would be caricatures. Figure 6.14 is such a caricature, because it paints an incomplete picture of many of the known regulatory interactions of phosphofructokinase with the universe of small molecules also present within a cell and its environment. In part (B) of the figure, we replace the molecular view with a streamlined view using icons. As seen in Figure 1.7 (p. 10), this enzyme interacts with a panoply of other molecules. We amplify on that earlier figure by showing the binding of phosphofructokinase with some of its small-molecule partners, in Figure 6.14(C).

How does the enzyme rate for bacterial phosphofructokinase depend upon the concentration of its substrate fructose-6-phosphate? Figure 6.15 shows activity curves for the enzyme, including a quantitative opportunity for us to see

Figure 6.15

Activity curves for phosphofructokinase. Enzyme rate as a function of the concentration of the substrate fructose-6-phosphate is shown for a number of concentrations of the ADP (red) and PEP (blue). ADP concentration was varied from 0 to 2 mM. The curves represent fits to the data using a Hill function. Adapted from Blangy, Buc, and Monod (1968) (data) and Olsman and Goentoro (2016) (curves).

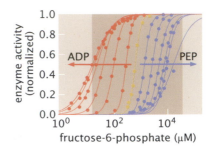

Figure 6.16

Computing the partition function for bacterial phosphofructokinase. There are three sites, each of which can be occupied independently. As a result, the partition function for the active state is given by the product of the partition functions for each site. The partition function for the inactive state is given by the product of the partition functions for each site when in the inactive state. The total partition function is obtained by adding those two partial partition functions together.

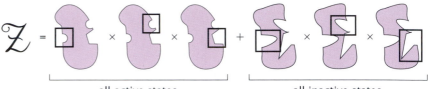

all active states   all inactive states

how the activity of the enzyme is modulated by the presence of small-molecule regulators like those shown in Figure 6.14. Specifically, in the data of Figure 6.15 we see ADP in its role as a competitive inhibitor. However, as indicated in Figure 6.14(C), at low concentrations, ADP is an allosteric effector, increasing the rate of enzyme action. These data hark back to the mantra of physical biology, namely, quantitative data demands quantitative models. Thus, we now turn to the approaches highlighted throughout the chapter to explore the dynamics of this enzyme.

As we have done repeatedly throughout the book, we can use the states-and-weights concept to work out a model of phosphofructokinase activity. For the particular case shown in Figure 6.14, we consider only four of the enzyme's broad array of binding partners, namely, ATP, ADP, phosphoenolpyruvate (PEP), and fructose-6-phosphate (F6P). Already, this presents a daunting proliferation of states. There are three binding sites in both the inactive and active forms of phosphofructokinase. Each of these sites can either be empty or occupied with one of the ligands that binds at that site. As seen in Figures 6.14 and 6.16, site 1 has three states of occupancy, site 2 has three states of occupancy, and site 3 has two states of occupancy. Thus, for one subunit in the active state alone, there are a daunting $3 \times 3 \times 2 = 18$ distinct states. Since there are four subunits per molecule, each of which can exist in both an active and an inactive conformation, there are a huge number of distinct conformational states of different occupancy. One of the beautiful points about the MWC framework is that each site, behaves independently, implying that the partition function for the active and inactive states can be written as a product of the individual partition functions for each site, as shown in Figure 6.16.

To compute the partition function for the active site, we proceed as shown in Figures 6.17 and 6.18. These two figures show how to account for the full complexity of the three sites per monomer, as well as to account for the four separate monomers. Using the states, weights, and rates, we can now determine

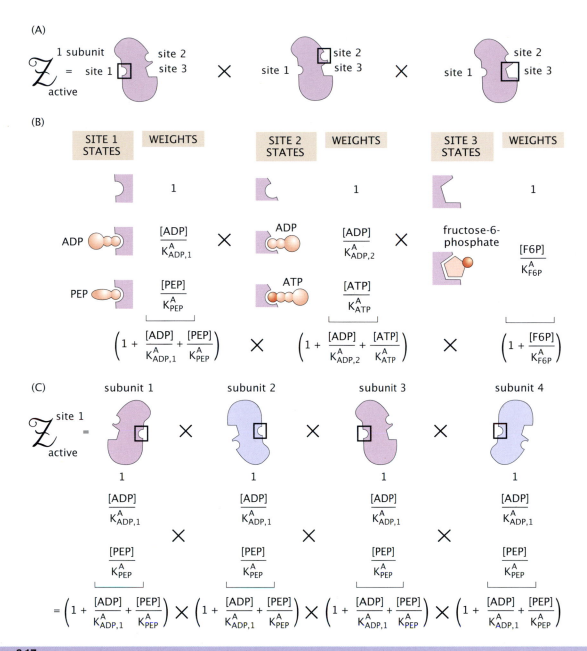

Figure 6.17

States and weights for phosphofructokinase in the active state. (A) Constructing the partition function for one subunit of phospho-fructokinase as a product over the states of the different sites. (B) States and weights for each binding site on one monomeric subunit of the enzyme when in the active state. (C) Constructing the entire active state partition function for site 1.

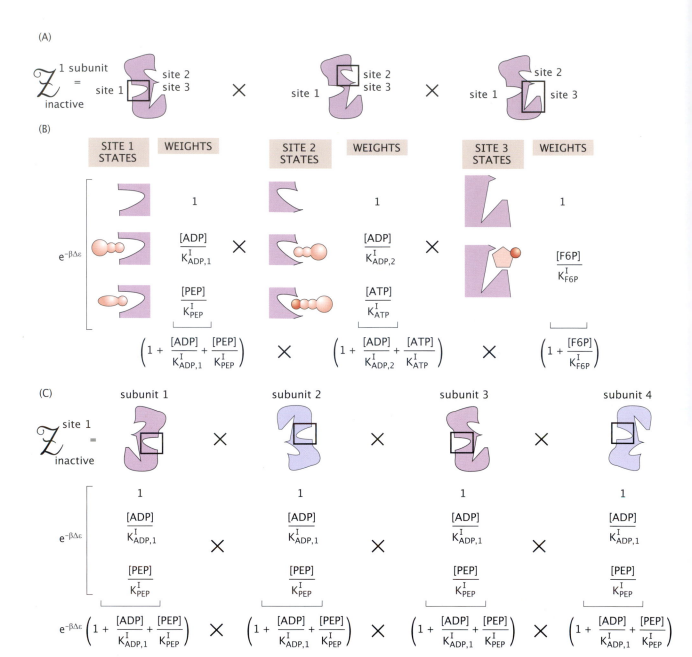

Figure 6.18

States and weights for phosphofructokinase in the inactive state. (A) Constructing the partition function for one subunit of phosphofructokinase as a product over the states of the different sites. (B) States and weights for each binding site on one monomeric subunit of the enzyme when in the inactive state. (C) Constructing the entire inactive state partition function for site 1.

the rate at which the enzyme produces product. First, we write the partition function by summing over all states. Specifically, we have $Z_{tot} = Z_A + Z_I$, with

$$Z_A = (1 + \frac{[F6P]}{K_{F6P}^A})^4(1 + \frac{[ATP]}{K_{ATP}^A} + \frac{[ADP]}{K_{ADP}^A})^4(1 + \frac{[ADP]}{K_{ADP,1}^A} + \frac{[PEP]}{K_{PEP}^A})^4 \quad (6.59)$$

and

$$Z_I = e^{-\beta\varepsilon}(1 + \frac{[F6P]}{K_{F6P}^I})^4(1 + \frac{[ATP]}{K_{ATP}^I} + \frac{[ADP]}{K_{ADP}^I})^4(1 + \frac{[ADP]}{K_{ADP,1}^I} + \frac{[PEP]}{K_{PEP}^I})^4.$$

$$(6.60)$$

We have used the notation $K_{ADP,1}^A$ and $K_{ADP,1}^I$ to signify that the binding of ADP to site 1 where it competes with PEP is different from the K_ds for binding to site 2 where ADP competes with ATP.

To obtain the enzyme rate, we now imagine that the inactive state is still able to produce product at a rate k_{cat}^I, and thus all those states for which there is substrate bound will produce product at that rate, weighted by the appropriate statistical weight. Assembling these terms results in a contribution to the overall rate of the form

$$k_{eff}^I = \frac{\left(4e^{-\beta\Delta\varepsilon}k_{cat}^I\frac{[F6P]}{K_{F6P}^I}\frac{[ATP]}{K_{ATP}^I}(1 + \frac{[F6P]}{K_{F6P}^I})^3(1 + \frac{[ATP]}{K_{ATP}^I} + \frac{[ADP]}{K_{ADP}^I})^3 (1 + \frac{[ADP]}{K_{ADP}^I} + \frac{[PEP]}{K_{PEP}^I})^4\right)}{Z_{tot}}. \quad (6.61)$$

Similarly, when the enzyme is in its active state, it can once again produce product, now at a rate k_{cat}^A leading to the definition

$$k_{eff}^A = \frac{4k_{cat}^A\frac{[F6P]}{K_{F6P}^A}\frac{[ATP]}{K_{ATP}^A}(1 + \frac{[F6P]}{K_{F6P}^A})^3(1 + \frac{[ATP]}{K_{ATP}^A} + \frac{[ADP]}{K_{ADP}^A})^3(1 + \frac{[ADP]}{K_{ADP}^A} + \frac{[PEP]}{K_{PEP}^A})^4}{Z_{tot}}.$$

$$(6.62)$$

The prefactor of 4 refers to the full tetrameric molecule, though some enzymologists prefer writing activity per monomeric unit. With these two rates defined, we can now write the total rate as

$$k_{cat}^{tot} = k_{eff}^I + k_{eff}^A. \quad (6.63)$$

The theoretical analysis just described in terms of states and weights can be compared with classic measurements on the activity of the bacterial phosphofructokinase as it phosphorylates its substrate fructose-6-phosphate, as shown in Figure 6.19. These measurements illustrate the rich and subtle interplay between the substrate itself and various other molecules which serve to regulate the enzymatic activity. For example, as seen in Figure 6.19(A), at low concentrations of ADP, this molecule serves as an activator of the enzyme activity. By way of contrast, at sufficiently high concentrations of ADP, the ADP will start to bind to the active site itself, competing with ATP, and thus to interfere with the ability of the enzyme to phosphorylate fructose-6-phosphate. Figure 6.19(B) shows a second regulatory intervention, this time in the form of phosphoenolpyruvate (PEP), which is a downstream intermediate. When its concentration gets too

Figure 6.19

Comparison of measurements on phosphofructokinase activity with theoretical treatment using the MWC model. All measurements were made in the presence of 100 mM ATP. (A) Enzyme *activity* as a function of substrate concentration (fructose-6-phosphate) as a function of ADP concentration. (B) Enzyme activity as a function of substrate concentration (fructose-6-phosphate) as a function of phosphoenolpyruvate concentration (PEP). Adapted from Blangy Buc, and Monod (1968).

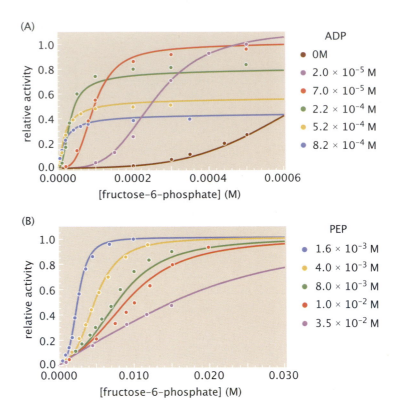

high, this means there is sufficient ATP, and hence the glycolysis pathway can be throttled back. Hence, PEP inhibit, phosphofructokinase.

Figure 6.19 shows data and features the results of using the MWC model to compute the enzyme activity. The various curves shown throughout the figure result from one global fit of the various parameters featured in equation 6.61 and shown in Figure 6.20, demonstrating a quite reasonable accord between the expectations from the theory and the observations, including the subtle influence of ADP both as activator (low concentrations) and inhibitor (high concentrations).

6.5   Summary

The study of enzyme structure and function is one of the crown jewels of biochemistry. I am particularly enamored of the early days of the field that pitted Pasteur against the Buchners as they tried to figure out the role of enzymes in living organisms and whether they constituted just another example of chemical reactions. The next critical episode in the narrative for our purposes was the years of effort that went into the study of enzyme kinetics and culminating in models such as that of Michaelis and Menten. But as the field advanced, it was clear that there were layers of regulation on enzyme behavior that defied these simplest models and called for the kinds of allostery ideas described in this chapter. Much of what we know about allostery got its start in the study of enzymes. This chapter gave us a chance to showcase some of these

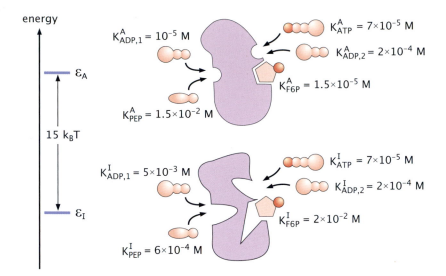

Figure 6.20

Parameters used in the MWC model of phosphofructokinase enzyme activity. Courtesy of Tal Einav and Denis Titov.

historic studies, especially that of aspartate transcarbamoylase. After reviewing the all-important Michaelis-Menten enzyme kinetics, we turned to various incarnations of the MWC enzyme concept, including cases involving allosteric effectors and competitive inhibitors. The glycolysis pathway and the famed enzyme phosphofructokinase served as a dominant case study. We ended by noting that, as warned in the preface, the book provides caricatures of many fields and that the field of enzyme kinetics has advanced much further than our discussion makes evident.

6.6   Further Reading

Barnett, J. A., and F. W. Lichtenthaler (2001) "A history of research on yeasts 3: Emil Fischer, Eduard Buchner, and their contemporaries, 1880–1900." *Yeast* 18:363–388. This is one of a whole series of excellent articles by Barnett. This one gives us a sense of the challenges faced in putting together a coherent picture of enzyme action.

Changeux, J-P. (2013) "50 years of allosteric interactions: The twists and turns of the models." *Nat. Rev. Mol. Cell Bio.* 14:1. This article gives an excellent overview of the rise of allostery and early thinking on enzymes.

Einav, T., L. Mazutis, and R. Phillips (2016) "Statistical mechanics of allosteric enzymes." *J. Phys. Chem. B* 120(26):6021–6037. Much of the work presented in this chapter is described in more details i this article.

Gerhart, J. (2014) "From feedback inhibition to allostery: the enduring example of aspartate transcarbamoylase." *FEBS J.* 281:612–620. This excellent article gives an insider's view of the rise of the allostery concept from one of its discoverers.

Gerhart, J. C., and A. B. Pardee (1962) "The enzymology of control by feedback inhibition." *J. Biol. Chem.* 247:891. The original paper from Gerhart and Pardee and still worth reading.

Gunawardena, J. (2014) "Time-scale separation: Michaelis and Menten's old idea, still bearing fruit." *FEBS J.* 281:473–488. This article provides important insights into the conceptual underpinnings of ideas such as Michaelis-Menten kinetics and separation of time scales more generally. It provides a deep appraisal of the many subtleties skirted in the present chapter. Further, it provides a unifying graph-theoretic interpretation of many of the developments made in the present chapter.

Hilser V. J., J. A. Anderson, and H. N. Motlagh (2015) "Allostery vs. allokairy." *Proc. Natl. Acad. Sci.* 112:11430–11431. This short article is a commentary that introduces another paper, but I find it to be a brilliantly executed introduction to a dynamical phenomenon christened "allokairy" (meaning other time/event). The key point is what some have called hysteretic enzymes, in which apparent allosteric behavior is attributed instead to a mismatch of time scales between substrate binding and the relaxation of the enzyme to the inactive state.

Kantrowitz, E. R. (2011) "Allostery and cooperativity in *Escherichia coli* aspartate transcarbamoylase." *Arch. Biochem. Biophys.* 519:81–90. Here a leader in the field gives us a tour of the paradigmatic and historic allosteric enzyme.

Segel, I. H. (1975) *Enzyme Kinetics.* New York: Wiley. Segel's book gives a nearly encyclopedic view of enzymes that goes well beyond the treatment given here.

6.7 REFERENCES

Blangy, D., H. Buc, and J. Monod (1968) "Kinetics of the allosteric interactions of phosphofructokinase from *Escherichia coli.*" *J. Mol. Biol.* 31:13–35.

Briggs, G. E32., and J.B.S. Haldane (1925) "A note on the kinetics of enzyme action." *Biochem J.* 19 (2): 338–339.

Changeux, J. P. (2013) "50 years of allosteric interactions: the twists and turns of the models." *Nat. Rev. Mol. Cell Biol.* 14:819–829.

Feller, G., O. le Bussy, C. Houssier, and C. Gerday (1996) "Structural and functional aspects of chloride binding to *Alteromonas haloplanctis* α-amylase." *J. Biol. Chem.* 271:23836–23841.

Neidhardt, F. C. (1999) "Bacterial growth: Constant obsession with dN/dt. *J. Bacteriol.* 181 (24): 7405–7408.

Olsman, N., and L. Goentoro (2016) "Allosteric proteins as logarithmic sensors." *Proc. Natl. Acad. Sci.* 113 (30): E4423–4430.

Payen, A., and J.-F. Persoz (1833) "Mémoire sur la diastase, les principaux produits de ses réactions et leurs applications aux arts industriels." *Annal. Chim. Phys.*, 2e série, 53:73–92.

HEMOGLOBIN, NATURE'S HONORARY ENZYME

7

Breathless, we flung us on a windy hill,
Laughed in the sun, and kissed the lovely grass.

—Rupert Brooke

In this chapter, we ask whether the allosteric framework presented thus far might actually hold further secrets about the unity of molecular mechanisms in different biological contexts. To that end, the chapter begins with a discussion of the long biochemical reach of the study of hemoglobin. We will argue that hemoglobin has many lessons to teach about allostery and its statistical mechanical incarnation. Over more than a century of effort devoted to understanding the structure and function of this important molecule, the kinds of allosteric models presented here were found to be a very useful language with which to describe the physiology of respiration as hidden within its binding curves. After an introduction to the phenomenology of hemoglobin and respiration, we will describe the statistical mechanics of this famed protein's behavior.

7.1 Hemoglobin Claims Its Place in Science

There is a great beauty in the way science uses the specific to illustrate the general. There are a host of examples in which a generation of scientists will fixate on some particular "model system" with the consequence that once this topic has been plumbed, it is realized that the lessons learned there can be used as the basis for investigations into other seemingly unrelated problems. The historic achievements in the study of hemoglobin are one great example of this observation. Indeed, the history of hemoglobin research reads like a who's who of the last 150 years of celebrated scientists from many domains of science. One of the ways in which hemoglobin has been central is through its influence on the way we think about binding in "two-state" systems, and the remainder of this chapter illustrates that concretely.

Figure 7.1

Respiration, red blood cells, and hemoglobin. Oxygen is taken into the lungs and passes by diffusion between the alveoli and the blood capillaries, where it is taken up by red blood cells for transport to tissues throughout the body. The red blood cells are full of oxygen-carrying hemoglobin molecules.

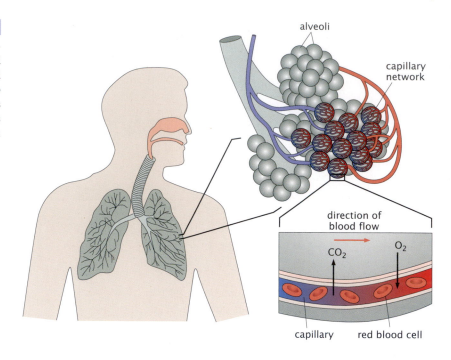

7.1.1   Hemoglobin and Respiration

We begin with a brief introduction to respiration. As shown in Figure 7.1, with every breath our lungs are filled with several liters of air. The oxygen in each such breath is at a sufficiently high partial pressure (or concentration) that there is a high probability the tetrameric hemoglobin molecules in our red blood cells will be occupied with four molecules of O_2 each. As those blood cells leave the lungs and travel throughout the circulatory system to our distant tissues, they are faced with a reduced oxygen partial pressure, resulting in the liberation of oxygen from the nearly saturated hemoglobin molecules. These oxygen molecules then diffuse to those cells that are within roughly 100 μm of the capillary, where they are absorbed and used for the many purposes of cellular physiology.

One of the ways we can understand this process is through the binding curves that show the degree of saturation of hemoglobin with oxygen. Examples of such binding curves from the original work of Christian Bohr (1904) are shown in Figure 7.2. The concentration of oxygen is plotted on the x-axis, while the y-axis tells us the percentage of hemoglobin binding sites occupied by oxygen. The different curves reveal the effect that bears Bohr's name and show the dependence of oxygen binding upon the concentration of CO_2. Explaining the mechanistic underpinnings of these curves provided a fundamental quantitative challenge to biochemists and biologists alike and led to a number of famous models such as the Pauling and Adair models and variants of the MWC model that are our main focus here.

The molecular basis of the binding curves shown in Figure 7.2 is revealed in Figure 7.3. As seen in the figure and as is evident to anyone receiving the results of her/his complete blood count (CBC) test (see Figure 7.4), each of our red blood cells harbors hundreds of millions of hemoglobin molecules.

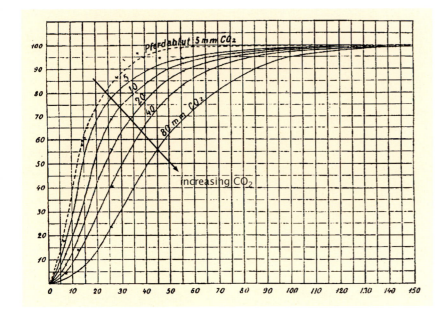

Figure 7.2

Binding curves for hemoglobin and the Bohr effect. Data from the original paper by Christian Bohr illustrating the effect that bears his name. Each curve reports the fractional binding of hemoglobin (y-axis) as a function of oxygen concentration (x-axis) for differing concentrations of CO_2.

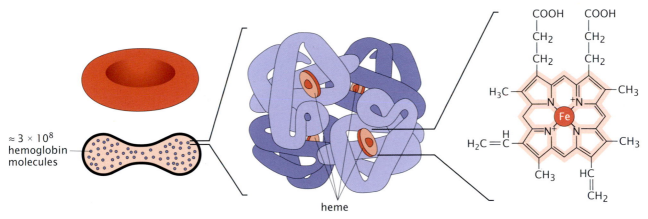

$\approx 3 \times 10^8$ hemoglobin molecules

heme

Figure 7.3

Red blood cells and hemoglobin. Each red blood cell harbors hundreds of millions of hemoglobin tetramers. Oxygen binds to the four heme sites within each hemoglobin molecule.

Estimate: Red Blood Cells and Hemoglobin

The result of a complete blood count test reveals the presence of on the order of 5×10^6 red blood cells per microliter of blood (see Figure 7.4). Given the total of 5×10^6 μL (i.e., ≈ 5 L) of blood in the adult human body leads to an estimate of $\approx 25 \times 10^{12}$ red blood cells in each of us, making them the most abundant cell type we have. Given also that there is roughly 15 g of hemoglobin for every deciliter of blood (the so-called mean corpuscular hemoglobin (MCH), also shown in Figure 7.4), we can estimate the number of hemoglobin molecules per cell. Once again using the estimate

ESTIMATE

that there are roughly 5 L of blood in a typical adult human means that each of us contains in total some 750 g of hemoglobin. Given that the mass of a hemoglobin molecule is roughly 64,000 Da, we can compute the number of hemoglobins as

$$\text{amount of hemoglobin} = \frac{750 \text{ g of hemoglobin}}{64,000 \text{ g/mol}} \approx 0.018 \text{ mol}, \qquad (7.1)$$

for the number of moles of hemoglobin, which we can then readily convert into number of molecules as

$$\text{\# of Hb molecules} \approx 0.018 \text{ mol} \times 6 \times 10^{23} \text{molecules/mol}$$

$$\approx 7 \times 10^{21} \text{ molecules} \qquad (7.2)$$

in the blood of the average adult. Finally, by merging all these observations from an everyday blood test we are led to the conclusion that in each red blood cell there are of the order of 3×10^8 hemoglobin molecules.

Table 4.1: Typical values from a CBC. Adapted from Maxwell (2002).

Test	Value
Red blood cell count (RBC)	Men: $\approx(4.3\text{–}5.7) \times 10^6$ cells/μL
	Women: $\approx(3.8\text{–}5.1) \times 10^6$ cells/μL
Hematocrit (HCT)	Men: $\approx$39–49%
	Women: $\approx$35–45%
Hemoglobin (HGB)	Men: $\approx$13.5–17.5 g/dL
	Women: $\approx$12.0–16.0 g/dL
Mean corpuscular hemoglobin (MCH)	$\approx$26–34 pg/cell
MCH concentration (MCHC)	$\approx$31–37%
Mean corpuscular volume (MCV)	$\approx$80–100 fL
White blood cell count (WBC)	$\approx(4.5\text{–}11) \times 10^3$ cells/μL
Differential (% of WBC):	
Neutrophils	$\approx$57–67
Lymphocytes	$\approx$23–33
Monocytes	$\approx$3–7
Eosinophils	$\approx$1–3
Basophils	$\approx$0–1
Platelets	$\approx(150\text{–}450) \times 10^3$ cells/μL

Figure 7.4

Results of a typical complete blood count test. These tests report cellular counts for red and white blood cells, as well as molecular counts for hemoglobin.

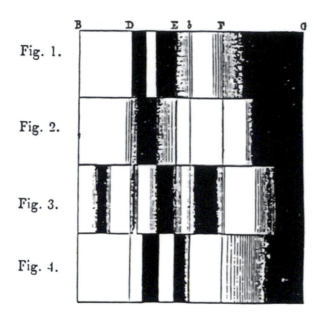

Figure 7.5

Stokes and the spectrum of hemoglobin. Stokes's paper (1863–1864) has a caption for these images: 1. Aqueous extract of ox blood clot; 2. same in alkaline solution (reduced condition); 3. in acidic solution of blood (decomposed coloring matter); 4. haematin in alkaline solution. Reprinted by permission from Springer Nature via Copyright Clearance Center: Holmes, F. L. (1995), "Crystals and carriers: The chemical and physiological identification of hemoglobin," in *No Truth Except in the Details*, eds. A. J. Kox and D. M. Siegel.

7.1.2   A Historical Interlude on the Colouring Matter

How did we come to an understanding of binding curves like those shown in Figure 7.2? To many of us, Stokes is a name either tied to a seemingly obscure theorem in mathematics or vaguely associated with spectroscopy. Though a professor of mathematics, Stokes interested himself in the study of the absorption characteristics of blood, resulting in one of his most famous papers, which he opens with the words: "Some time ago my attention was called to a paper by Professor Hoppe, in which he has pointed out the remarkable spectrum produced by the absorption of light by a very dilute solution of blood, and applied the observation to elucidate the chemical nature of the colouring matter. I had no sooner looked at the spectrum, than the extreme sharpness and beauty of the absorption-bands of blood excited a lively interest in my mind, and I proceeded to try the effect of various reagents." (1863–1864) The results of these experiments are shown in Figure 7.5, where we see that Stokes interested himself in the *changes* in the absorption spectrum of hemoglobin in the presence of different chemical conditions. Out of these efforts Stokes was able to advance a hypothesis (which turned out to be largely correct) of the role of hemoglobin as a carrier of oxygen in the blood. The spectroscopy of hemoglobin is now an everyday part of the modern medical world, so much so that each of us can do spectroscopy measurements ourselves simply by ordering a pulse oximeter. These devices exploit the different absorption properties of deoxyhemoglobin and oxyhemoglobin, providing a compelling example of the nearly unimaginable ways in which fundamental science is parlayed into powerful technologies.

There are other everyday examples besides the pulse oximeter in which hemoglobin and its effects can be seen with the naked eye. Under the right circumstances, if the reader was to break open a living tick, he or she would be struck by the appearance of intricate faceted shapes corresponding to crystals of hemoglobin like those shown in Figure 7.6. These small crystals hold in themselves another of the great stories of modern biology, namely, the way in

Figure 7.6

Crystals of hemoglobin. Otto Funke's drawings of hemoglobin crystals: 1. from normal human blood; 2. from blood from the heart of a young cat; 3. from venous blood of a guinea pig; 4. from venous blood of a squirrel; 5. from blood from the heart of a fish; 6. from normal human splenic blood. Reprinted by permission from Springer Nature via Copyright Clearance Center: Holmes, F. L. (1995), "Crystals and carriers: The chemical and physiological identification of hemoglobin," in *No Truth Except in the Details*, eds. A. J. Kox and D. M. Siegel.

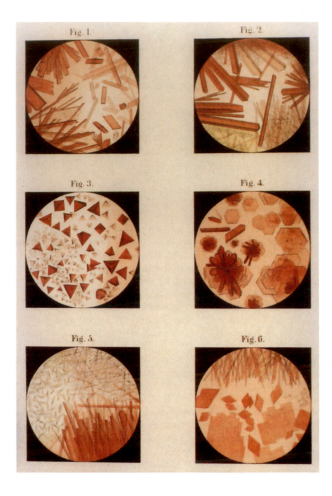

which the decades-long struggle to crack the structure of hemoglobin served as an impetus for the entire field of structural biology, and played into the parallel studies on the binding properties of this all-important molecule. For our purposes, the primary insight we will need to carry away is that the hemoglobin molecule has four distinct binding sites, as shown in Figure 7.3. The binding sites correspond to heme groups, each of which has a coordinated iron atom where oxygen can bind.

7.1.3   Hemoglobin as a "Document of Evolutionary History"

The story of hemoglobin and how it binds oxygen and other molecules such as carbon monoxide and carbon dioxide is long and fascinating and is tied to many important and key themes from our current understanding of the macromolecules of life, including both allostery and cooperativity. This topic will serve as the core substance of this chapter. But the insights did not stop with the statistical mechanics of binding. The very idea that molecules might serve as documents of evolutionary history also played out in part through considerations of the relatedness of hemoglobin from different organisms. In modern

PRIMATE HEMOGLOBINS

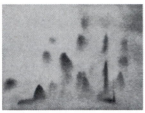

human

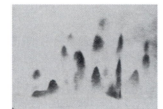

chimpanzee

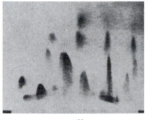

gorilla

orangutan

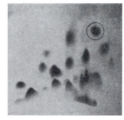

rhesus monkey

Figure 7.7

Molecules as documents of evolutionary history. Hemoglobin was used before the era of protein and DNA sequencing to determine the evolutionary relatedness of different organisms. This figure shows the use of electrophoresis and chromatography on paper to analyze the relatedness of the different molecules. If the molecular fragments of hemoglobin are different, they will migrate slightly differently, leading to a different pattern. Adapted from Zuckerkandl, Jones, and Pauling (1960). Courtesy of the Estates of Richard T. Jones and Linus Pauling.

terms, the documents of evolutionary history are revealed by exploring "protein taxonomy" in the form of amino acid and nucleotide sequences, yielding what Francis Crick (1958) referred to as "the most delicate expression possible of the phenotype of an organism." But in the late 1950s and early 1960s, when these ideas were first seriously in play, there were no sequence comparisons to be made because the technologies for sequencing both proteins and nucleic acids were only in their infancy. Instead, researchers such as Zuckerkandl and Pauling, and Sarich and Wilson, found different ways of examining molecular similarity.

What is so compelling about early work on evolution from the standpoint of hemoglobin is not the results themselves, which have been superseded by those emerging from the sequencing revolution. Rather, what was so impressive was the approach and ingenuity, which remain an inspiration to this day. As shown in Figure 7.7, in the days before the ability to routinely sequence proteins or DNA, it was still possible to carry out "protein taxonomy" to explore the evolutionary relatedness of different organisms. As seen in the figure, trypsin-digested hemoglobins from different species lead to slightly different fragments that will migrate differently when subjected to electrophoresis and chromatography. As a result, by comparing the patterns of these fragments, it is possible in turn to learn about the evolutionary relationships between the organisms that harbor those hemoglobins. What these pioneers had figured out how to do is genomics without sequences.

A more modern vision of thinking on evolutionary relatedness of different organisms as revealed by their hemoglobins is given in Figure 7.8. In this case, the amino acid sequences of the α-chain of hemoglobin humans, chimps, gorillas, cows, horses, donkeys, rabbits, and carp are compared. A satisfying first insight is the recognition that the sequences of human and chimp hemoglobin α-chains are strictly identical. We can organize our thinking about the relatedness of these different hemoglobins in the tree diagram of Figure 7.9.

Figure 7.8

Molecules as documents of evolutionary history redux. Sequence comparisons of hemoglobin. Comparison of hemoglobin sequences from different vertebrates. From top to bottom, the sequences from the α-chain of hemoglobin are shown from humans (accession number 57013850, 142 amino acids long), chimpanzees (110835747, 142 aa), gorillas (122407, 141 aa), cattle (*Bos taurus*, 13634094, 142 aa), horses (*Equus caballus*, 122411, 142 aa), donkeys (*Equus asinus*, 62901528, 142 aa), rabbits (*Oryctolagus cuniculus*, 229379, 141 aa), and carp (*Cyprinus carpio*, 122392, 143 aa). The alignment was performed using ClustalW. Hydrophobic residues are indicated in red, and polar residues in blue.

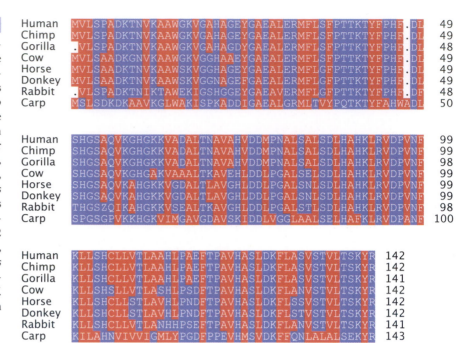

Human	MVLSPADKTNVKAAWGKVGAHAGEYGAEALERMFLSFPTTKTYFPHF.DL	49
Chimp	MVLSPADKTNVKAAWGKVGAHAGEYGAEALERMFLSFPTTKTYFPHF.DL	49
Gorilla	.VLSPADKTNVKAAWGKVGAHDGDYGAEALERMFLSFPTTKTYFPHF.DL	48
Cow	MVLSAADKGNVKAAWGKVGGHAAEYGAEALERMFLSFPTTKTYFPHF.DL	49
Horse	MVLSAADKTNVKAAWSKVGGHAGEYGAEALERMFLGFPTTKTYFPHF.DL	49
Donkey	MVLSAADKTNVKAAWSKVGGNAGEFGAEALERMFLGFPTTKTYFPHF.DL	49
Rabbit	.VLSPADKTNIKTAWEKIGSHGGEYGAEAVERMFLGFPTTKTYFPHF.DF	48
Carp	MSLSDKDKAAVKGLWAKISPKADDIGAEALGRMLTVYPQTKTYFAHWADL	50

Human	SHGSAQVKGHGKKVADALTNAVAHVDDMPNALSALSDLHAHKLRVDPVNF	99
Chimp	SHGSAQVKGHGKKVADALTNAVAHVDDMPNALSALSDLHAHKLRVDPVNF	99
Gorilla	SHGSAQVKGHGKKVADALTNAVAHVDDMPNALSALSDLHAHKLRVDPVNF	98
Cow	SHGSAQVKGHGAKVAAALTKAVEHLDDLPGALSELSDLHAHKLRVDPVNF	99
Horse	SHGSAQVKAHGKKVGDALTLDDLPGALSNLSDLHAHKLRVDPVNF	99
Donkey	SHGSAQVKAHGKKVGDALTLAVGHLDDLPGALSNLSDLHAHKLRVDPVNF	99
Rabbit	THGSZQIKAHGKKVSEALTKAVGHLDDLPGALSTLSDLHAHKLRVDPVNF	98
Carp	SPGSGPVKKHGKVIMGAVGDAVSKIDDLVGGLAALSELHAFKLRVDPANF	100

Human	KLLSHCLLVTLAAHLPAEFTPAVHASLDKFLASVSTVLTSKYR	142
Chimp	KLLSHCLLVTLAAHLPAEFTPAVHASLDKFLASVSTVLTSKYR	142
Gorilla	KLLSHCLLVTLAAHLPAEFTPAVHASLDKFLASVSTVLTSKYR	141
Cow	KLLSHSLLVTLASHLPSDFTPAVHASLDKFLANVSTVLTSKYR	142
Horse	KLLSHCLLSTLAVHLPNDFTPAVHASLDKFLSSVSTVLTSKYR	142
Donkey	KLLSHCLLSTLAVHLPNDFTPAVHASLFLSTVSTVLTSKYR	142
Rabbit	KLLSHCLLVTLANHHPSEFTPAVHASLDKFLANVSTVLTSKYR	141
Carp	KILAHNVIVVIGMLYPGDFPPEVHMSVDKFFQNLALALSEKYR	143

Figure 7.9

Evolutionary relatedness of organisms based on their hemoglobin sequences. The tree diagram shows the degree of similarity among each of the sequences. Very similar sequences are connected by short branches; increasing branch length indicates decreasing sequence similarity.

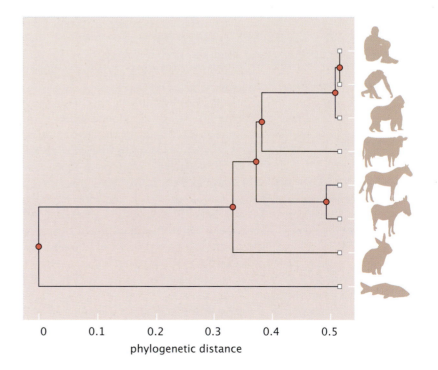

phylogenetic distance

Our aims in the remainder of this chapter will generally be more modest than these grand evolutionary questions and will focus instead on how the allostery concept can be used to understand the intriguing binding properties of this historic molecule.

7.2   States and Weights and Binding Curves

The central question we will examine in this section is how to calculate the fractional occupancy of the hemoglobin molecule as a function of the oxygen partial pressure (or concentration). As noted earlier in the chapter in conjunction with Figure 7.1, the "function" of hemoglobin is to carry oxygen in our red blood cells to our tissues. The way we characterize that function mathematically is a curve that relates the fractional occupancy and the oxygen concentration, the so-called binding curve (see Figure 7.2, p. 233). The binding curve has several clear limits. First, in the limit where there is almost no oxygen present the occupancy will be zero. That is, the binding sites within the hemoglobin molecule will be unoccupied. On the other hand, when the concentration of oxygen is very high (i.e., at "saturating" concentrations), all four sites within a given molecule of hemoglobin will be occupied. We can do better by noting that at low concentrations, the occupancy scales linearly with the oxygen concentration, a result that can be proved simply from the calculations to follow.

Generations of measurements have confirmed the intuitive picture just described. A classic example of such binding curves was shown in Figure 7.2. These data are especially compelling, since they illustrate not only how the occupancy of hemoglobin depends upon the oxygen concentration but how secondary ligands can alter the primary ligand binding. A more recent example of such binding curves for hemoglobins from a variety of different organisms and from human hemoglobin at different pHs is shown in Figure 7.10. As seen in the figure, the standard in the field is to denote the fractional occupancy as

$$Y = \frac{\langle N \rangle}{4}, \tag{7.3}$$

where $\langle N \rangle$ denotes the average number of oxygen molecules bound. Hence, Y itself runs from 0 to 1. Though our primary ambition is not to "fit" experimental data like those shown in the figure, such measurements serve as a motivation for the class of observations and data that any model must respond to.

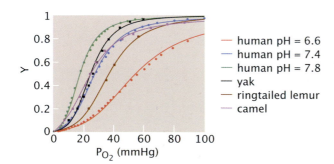

Figure 7.10

Oxygen-binding curves for different hemoglobins and at different pHs. Adapted from Milo et al. (2007).

Figure 7.11

States and weights of the hemoglobin molecule in the MWC model. The first row shows the various states of occupancy of the hemoglobin molecule in the T, state and the second row shows the various states of occupancy in the R state. The prefactors tell us the number of ways of realizing each such state, and μ is the chemical potential of oxygen.

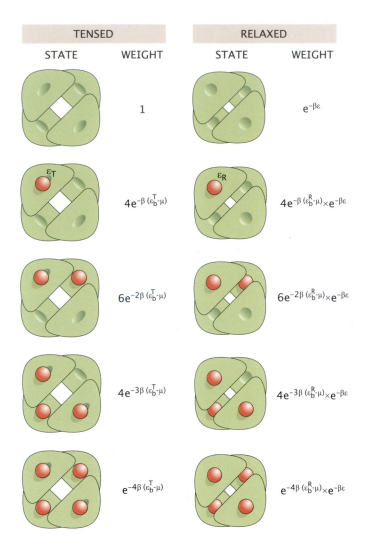

One of the first things a thoughtful reader might wonder is how the two-state modeling paradigm can be applied to the study of a complicated molecule such as hemoglobin with its four binding sites for oxygen molecules. The essence of the model is that, as with the molecules already described in previous chapters, hemoglobin is supposed to exist in two distinct conformational states, labeled tense (T) and relaxed (R), as shown in Figure 7.11. For historical reasons, this nomenclature was used instead of the more transparent inactive and active that we have favored throughout the book. As usual, it is illuminating to depict the various allowed states and their corresponding statistical weights, as shown in the figure. There are a total of 10 distinct states of conformation and occupancy. To see this, note that for the T state as shown in the figure, the hemoglobin can have 0, 1, 2, 3, or 4 oxygens bound to it, and similarly, for the R state, the hemoglobin can have 0, 1, 2, 3, or 4 oxygens bound to it, resulting in the 10 distinct states mentioned. Note also that the statistical weights have prefactors. For example, there are four ways that the R state can

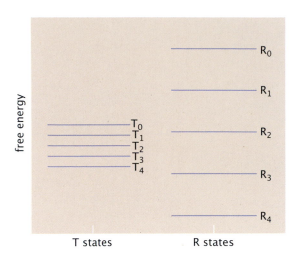

Figure 7.12

Energies of different hemoglobin states at a specific fixed concentration of oxygen. The binding of oxygen incurs a much larger advantage for the R state rather than the T state. Adapted from Milo et al. (2007).

have a single oxygen bound, since that oxygen molecule can be bound to each of the four hemes. Similarly, there are six ways that the four binding sites can be occupied by two oxygen molecules—four choices for where we put the first oxygen and three choices for where we put the second oxygen, for a total of 12 distinct binding configurations, but we then must divide by 2, since swapping the order in which we occupy the first and second sites leads to the same configuration.

In the absence of ligand, the T state of the protein is favored over the R state. We represent this unfavorable energy cost to access the R state with the energy ε. However, the interesting twist is that the oxygen-binding reaction has a higher affinity for the R state. Thus, with increasing oxygen concentration, the balance will be tipped toward the R state, despite the cost, ε, of accessing that state. We label the binding energies ε_b^T and ε_b^R, which signify the energy gained upon binding of oxygen to hemoglobin when it is in the T and R states, respectively. If we describe the binding to these two states in terms of the dissociation constants K_T and K_R, then they are rank-ordered in strength as $K_R < K_T$.

As seen in the states and weights, there is a hierarchy of different energies associated with the hemoglobin molecule in all its different states of occupancy with O_2. This hierarchy of energies can be captured pictorially, as seen in Figure 7.12, which shows the different energies at one particular O_2 concentration. As seen in the figure, the state R_0 signifying the R state with no oxygen bound has the highest energy. Similarly, T_0 signifies the T state with no oxygen bound and has a more favorable energy than R_0 in the absence of ligand, meaning that it would be the most probable state at low O_2 concentrations.

As described, Figure 7.12 is incomplete because it shows the free-energy differences between all the states for one particular choice of O_2 concentration. The free energies of the different states can be seen in the exponentials of Figure 7.11. For example, the free energy of the tense state with two O_2 molecules bound is given by

$$\text{free energy}(T_2) = 2(\varepsilon_b^T - \mu). \tag{7.4}$$

Figure 7.13

Energies of different hemoglobin states as a function of oxygen concentration. The energies used here are $\varepsilon = 2 \; k_B T$, which tells us that the R state is less favored than the T state, $\varepsilon_b^T = 0$ and $\varepsilon_b^R = -10 \; k_B T$, which indicates that binding to the R state is much more advantageous. The solid lines correspond to the T state, and the dotted lines correspond to the R state. At low concentrations, the 0-bound T state is most favored. For large concentrations of oxygen, the 4-bound R state has the lowest free energy.

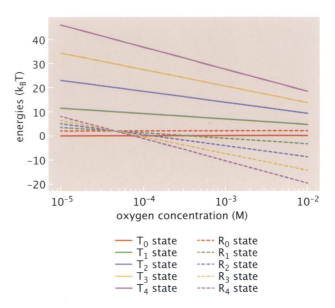

We can rewrite this in a way that makes the dependence on O_2 concentration explicit by invoking $\mu = \mu_0 + k_B T \ln [O_2]/c_0$ for the chemical potential with c_0 as a standard-state concentration, often taken as 1 M. In this case, we have

$$\text{free energy}(T_2) = 2(\varepsilon_b^T - \mu_0 - k_B T \ln [O_2]/c_0). \tag{7.5}$$

Recall from our discussion of chapter 2 (p. 35) that the chemical potential is a shorthand way of computing the change in the entropy of the reservoir when we take one ligand from it. Figure 7.13 shows how the free energy of all 10 states in the MWC model of hemoglobin varies as the oxygen concentration is increased, in each case using the energies revealed in the states and weights of Figure 7.11.

To find the occupancy (that is, $\langle N \rangle$) of the hemoglobin, we need to follow the central injunction of statistical mechanics and to use the statistical weights of each of the competing states, as we have done repeatedly throughout the book. As usual, when we compute the probability of some set of states, we must always divide the relevant statistical weights by the sum over *all* statistical weights to construct the partition function. In the case of hemoglobin, shown in Figure 7.11, the partition function is given by

$$\mathcal{Z} = \underbrace{e^{-\beta\varepsilon}(1 + 4e^{-\beta(\varepsilon_b^R - \mu)} + 6e^{-\beta(2\varepsilon_b^R - 2\mu)} + 4e^{-\beta(3\varepsilon_b^R - 3\mu)} + e^{-\beta(4\varepsilon_b^R - 4\mu)})}_{R \text{ terms}}$$

$$+ \underbrace{1 + 4e^{-\beta(\varepsilon_b^T - \mu)} + 6e^{-\beta(2\varepsilon_b^T - 2\mu)} + 4e^{-\beta(3\varepsilon_b^T - 3\mu)} + e^{-\beta(4\varepsilon_b^T - 4\mu)}}_{T \text{ terms}}.$$

$$\tag{7.6}$$

Note that the combinatorial factors as described earlier result from the number of ways of arranging the oxygens on the hemoglobin molecule. We can write this expression more transparently as

$$\mathcal{Z} = e^{-\beta\varepsilon}(1 + e^{-\beta(\varepsilon_b^R - \mu)})^4 + (1 + e^{-\beta(\varepsilon_b^T - \mu)})^4. \tag{7.7}$$

We have repeatedly translated fluently between the statistical mechanical language of energies and the thermodynamic language of dissociation constants. If we use the fact that for ideal solutions we can write the chemical potential as

$$\mu = \mu_0 + k_B T \ln \frac{[O_2]}{c_0}, \tag{7.8}$$

then we can rewrite our sum over all statistical weights in the more compact and intuitive form,

$$\mathcal{Z} = e^{-\beta\varepsilon}(1 + \frac{[O_2]}{K_R})^4 + (1 + \frac{[O_2]}{K_T})^4, \tag{7.9}$$

where we have also introduced the notation $K_T = c_0 e^{\beta \Delta \varepsilon_b^T}$, $K_R = c_0 e^{\beta \Delta \varepsilon_b^R}$, $\Delta \varepsilon_b^T = \varepsilon_b^T - \mu_0$, and $\Delta \varepsilon_b^R = \varepsilon_b^R - \mu_0$.

To find the average occupancy, we can use our knowledge of the individual statistical weights and the partition function derived earlier. Specifically, we have

$$\langle N \rangle = \frac{1}{\mathcal{Z}}(4 \times 1(e^{-\beta\varepsilon}\frac{[O_2]}{K_R} + \frac{[O_2]}{K_T}) + 6 \times 2(e^{-\beta\varepsilon}(\frac{[O_2]}{K_R})^2 + (\frac{[O_2]}{K_T})^2)$$
$$+ 4 \times 3(e^{-\beta\varepsilon}(\frac{[O_2]}{K_R})^3 + (\frac{[O_2]}{K_T})^3) + 1 \times 4(e^{-\beta\varepsilon}(\frac{[O_2]}{K_R})^4 + (\frac{[O_2]}{K_T})^4). \tag{7.10}$$

This cumbersome expression notes that there are four states of single occupancy, six states of double occupancy, four states of triple occupancy, and a single state of quadruple occupancy. However, we can streamline the expression by noting the shared factors, resulting in

$$\langle N \rangle = 4\frac{e^{-\beta\varepsilon}\frac{[O_2]}{K_R}(1 + \frac{[O_2]}{K_R})^3 + \frac{[O_2]}{K_T}(1 + \frac{[O_2]}{K_T})^3}{e^{-\beta\varepsilon}(1 + \frac{[O_2]}{K_R})^4 + (1 + \frac{[O_2]}{K_T})^4}. \tag{7.11}$$

Our previous analysis of the average occupancy required us to enumerate all the different states of occupancy by hand. As the molecular situations we encounter become more complicated, with not only the primary ligand such as oxygen present, but also other ligands such as carbon monoxide or carbon dioxide present, these by-hand analyses become more cumbersome. Fortunately, there is a simpler protocol for obtaining the average occupancy. It is here that the chemical potential idea introduced in section 2.3.1 really comes into its own.

We begin by examining the meaning of average occupancy, which for hemoglobin is

$$\langle N \rangle = \sum_{i=0}^{4} i p(i), \tag{7.12}$$

where $p(i)$ refers to the probability that hemoglobin has i O_2 molecules bound to it. Using our knowledge of the partition function, we can write this expression more explicitly as

$$\langle N \rangle = \sum_{i=0}^{4} \frac{i}{\mathcal{Z}} \frac{4!}{i!(4-i)!}(e^{-\beta\varepsilon}e^{-\beta i(\varepsilon_b^R - \mu)} + e^{-\beta i(\varepsilon_b^T - \mu)}), \tag{7.13}$$

where the factor involving factorials tells us how many ways there are of having i O_2 molecules bound to the four binding sites in hemoglobin.

We can make progress with this expression by noting that if we take the derivative of the exponential terms with respect to μ, we bring down a factor of i as follows,

$$\frac{\partial}{\partial \mu} e^{-\beta i (\varepsilon_b^R - \mu)} = i\beta e^{-\beta i (\varepsilon_b^R - \mu)}. \tag{7.14}$$

The reason this is useful is that it now allows us to compute the occupancy sum simply by taking a derivative with respect to the partition function, namely,

$$\langle N \rangle = \frac{1}{\beta \mathcal{Z}} \frac{\partial}{\partial \mu} [e^{-\beta \varepsilon} (1 + e^{-\beta(\varepsilon_b^R - \mu)})^4 + (1 + e^{-\beta(\varepsilon_b^T - \mu)})^4]. \tag{7.15}$$

The result of evaluating the required derivative is

$$\langle N \rangle = \frac{4}{\beta \mathcal{Z}} [e^{-\beta \varepsilon} \beta e^{-\beta(\varepsilon_b^R - \mu)} (1 + e^{-\beta(\varepsilon_b^R - \mu)})^3 + \beta e^{-\beta(\varepsilon_b^T - \mu)} (1 + e^{-\beta(\varepsilon_b^T - \mu)})^3]. \tag{7.16}$$

7.3   Y oh Y

Before the age of computers, studies of enzyme kinetics and binding processes used a variety of ways of plotting dose-response data, such as Scatchard plots and Hill plots, that made the connections to key parameters in the underlying models more transparent. In this section, we take a trip down memory lane to review the nature of the Hill plots that were a key part of people's thinking over the years as the allosteric nature of hemoglobin was explored.

One of the ways we can explore the relation between the formulation used here and that in the original work of Monod, Wyman, and Changeux is to calculate the dissociation constant for the reaction

$$R_0 \rightleftharpoons T_0. \tag{7.17}$$

We imagine the system in the absence of ligands switching intermittently between these two states, with the relative occupancies determined by the difference in energy of the two states in much the same way that Figure 3.8 (p. 86) revealed the spontaneous transitioning between closed and open states of a ligand-gated ion channel. We know that $N_R + N_T = N_{tot}$, where N_{tot} is the total number of molecules in question.

Using our statistical mechanics framework for a two-state system (i.e., considering only the two states R_0 and T_0, since there is no O_2, and therefore the states with oxygen occupancy are irrelevant), we can compute the relative occupancy of the system in the two states, as shown in the two unoccupied states of Figure 7.11. In particular, we have

$$R_0 = \frac{N_R}{N_{tot}} = p_R = \frac{e^{-\beta \varepsilon}}{1 + e^{-\beta \varepsilon}}, \tag{7.18}$$

and

$$T_0 = \frac{N_T}{N_{tot}} = p_T = \frac{1}{1 + e^{-\beta \varepsilon}}, \tag{7.19}$$

where we have related the macroscopic concentrations to our microscopic model by exploiting the idea that the energy of the R state is higher than that of the T state by an amount ε. In the language favored in many treatments of the MWC model, we employ the definition

$$L = \frac{[T_0]}{[R_0]}. \tag{7.20}$$

This means that we can use our expressions for T and R to compute this ratio, with the result $L = e^{\beta \varepsilon}$.

If we take the binding curve of equation 7.11 and simplify it through the definitions $c = K_R/K_T$ and $x = [O_2]/K_R$, we obtain an alternative expression for the occupancy,

$$Y = \frac{\langle N \rangle}{4} = \frac{x(1+x)^3 + Lcx(1+cx)^3}{(1+x)^4 + L(1+cx)^4}. \tag{7.21}$$

One of the preferred ways of plotting the oxygen-binding curves is the Hill plot, as shown in Figure 7.14. In this approach, the occupancy curves for hemoglobin are presented as a plot of $\ln(Y/(1-Y))$ versus $\ln pO_2$. To see why such plots are useful, we recall the nature of Hill functions as a way of describing cooperative binding (see sec. 2.5, p. 57). The fractional occupancy in the language of Hill functions is given by

$$Y = \frac{x^n}{1+x^n}. \tag{7.22}$$

Thus, we can compute $Y/(1-Y)$ using

$$1 - Y = 1 - \frac{x^n}{1+x^n} = \frac{1}{1+x^n}, \tag{7.23}$$

with the result that

$$\frac{Y}{1-Y} = x^n, \tag{7.24}$$

and hence

$$\ln \frac{Y}{1-Y} = n \ln x. \tag{7.25}$$

When written in this form, the sigmoidal Hill function offers up its Hill coefficient as the slope of the plot of $\ln(Y/(1-Y))$ versus $\ln pO_2$.

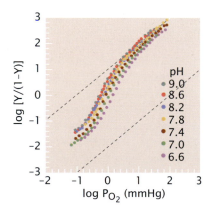

Figure 7.14

Oxygen-binding curves for hemoglobin. The different curves correspond to different pH values. Y is the fractional occupancy of the hemoglobin, with a value of 0 in the absence of ligand and a value of 1 when all four sites are occupied. Adapted from Yonetani et al. (2002).

With these preliminaries in hand, we now consider the application of these ideas to hemoglobin. Using our earlier definition of Y, we can compute $1 - Y$ as

$$1 - Y = \frac{(1+x)^3 + L(1+cx)^3}{(1+x)^4 + L(1+cx)^4}. \tag{7.26}$$

Using this result, we are now poised to compute $Y/(1 - Y)$, with the result that

$$\frac{Y}{1-Y} = \frac{x(1+x)^3 + Lcx(1+cx)^3}{(1+x)^3 + L(1+cx)^3}. \tag{7.27}$$

This result allows us to plot the theoretical response of the MWC model to the data shown in Figure 7.14.

We can take this analysis further by trying to understand the asymptotes revealed in Figure 7.14, where we see that for both small $[O_2]$ and large $[O_2]$, with the idea of small and large determined by the parameters K_T and K_R, that the binding curves adopt a particularly simple form. For the case in which $[O_2] \ll K_T, K_R$, our expression becomes

$$\frac{Y}{1-Y} = \frac{1+Lc}{1+L}x = \alpha_0 x. \tag{7.28}$$

Similarly, for the case in which $[O_2] \gg K_T, K_R$, our expression becomes

$$\frac{Y}{1-Y} = \frac{1+Lc^4}{1+Lc^3}x = \alpha_\infty x. \tag{7.29}$$

To develop a little more intuition for our results, we note that $L \gg 1$ and in general, $c \ll 1$, implying that $\alpha_\infty > \alpha_0$ and hence that the low-concentration asymptote will fall below the high-concentration asymptote, as we expect and as we see in Figure 7.14.

7.4   Hemoglobin and Effectors: The Bohr Effect and Beyond

Our discussion of binding in hemoglobin is incomplete without examining some of the many physiological subtleties such as how pH or the presence of effector or competitor molecules alter the binding curves we revealed in Figure 7.14, and how the MWC framework needs to be amended to account for these nuances.

The Bohr effect refers specifically to classic data like those of the Danish physiologist Christian Bohr, shown in Figures 7.2 (p. 233). Figure 7.10 (p. 239) gives a more modern view of the Bohr effect by showing how the oxygen-binding curves for human hemoglobin are shifted as a function of pH. This effect has profound physiological implications, since lowering pH means that more oxygen will be delivered to oxygen-deprived tissues. Interestingly, the feedback mechanism that drives this effect results from the generation of carbon dioxide in tissues such as muscle during exercise. From a mathematical perspective, the Bohr effect (see Figure 7.15) can be characterized through

$$\text{strength of Bohr effect} = \frac{\partial \log p_{50}}{\partial pH}, \tag{7.30}$$

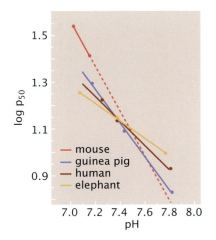

Figure 7.15

Bohr effect in hemoglobin for different mammals. The midpoint for oxygen occupancy as a function of pH is different for different animals. Adapted from Riggs (1960).

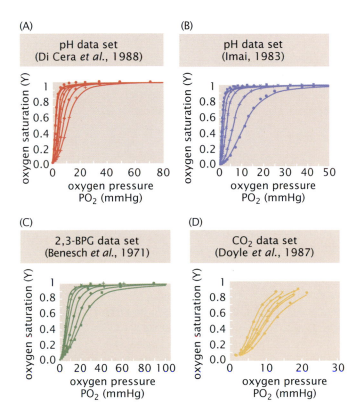

Figure 7.16

Global and self-consistent analysis of the binding properties of hemoglobin from the perspective of the MWC model. (A)–(D) Oxygen saturation curves under different conditions. Adapted from Rapp, and Yifrach (2017).

where p_{50} refers to the midpoint for oxygen saturation. As seen in Figure 7.16, other effectors such as BPG and CO_2 lead to shifts in the saturation curves, as well.

These effects have been of great interest from a microscopic and structural perspective, where the question has been to figure out how changes in effector concentration interact with the individual residues making up the hemoglobin molecule and thereby shift the binding curve. But instead we interest ourselves in a different question, namely, is there a simple unifying way of viewing such data from the perspective of the MWC framework? Here we see the way statistical mechanics can be tailored to account for these effects.

In keeping with the way we have treated allosteric effectors throughout the book, we begin by examining the states and weights, as shown in Figure 7.17. Here we imagine that, as appears to be the case for both CO_2 and H^+ ions, the effectors can bind all the different subunits. In this case, we write the partition function for this problem as

$$\mathcal{Z} = (1 + \frac{[O_2]}{K_R})^4 (1 + \frac{[effector]}{K_R^{effector}})^4 + e^{-\beta\varepsilon}(1 + \frac{[O_2]}{K_T})^4 (1 + \frac{[effector]}{K_T^{effector}})^4, \quad (7.31)$$

where we have now introduced two additional dissocation constants $K_R^{effector}$ and $K_T^{effector}$ to describe the binding of the effector to hemoglobin in the R and T states, respectively. Note that we can rewrite this result in the alternative form

$$\mathcal{Z} = e^{-\beta\varepsilon_R^{eff}} (1 + \frac{[O_2]}{K_R})^4 + e^{-\beta\varepsilon_T^{eff}} (1 + \frac{[O_2]}{K_T})^4, \quad (7.32)$$

where now all reference to the presence of the effector has been subsumed into the effective parameters ε_T^{eff} and ε_R^{eff}. By inspection, and also in light of the discussion of section 2.6 (p. 63), we can write the difference between these effective energies as

$$\Delta\varepsilon_{eff} = \varepsilon + k_B T \ln \frac{(1 + \frac{[effector]}{K_R^{effector}})^4}{(1 + \frac{[effector]}{K_T^{effector}})^4}. \quad (7.33)$$

This beautiful result tells us that the presence of effectors can be thought of as shifting the equilibrium between the T and the R states in the absence of oxygen, and nothing more. As a result of this analysis, we see that there is a family of occupancy curves, each of which differs only through the constant $L = [T_0]/[R_0]$, as seen in Figure 7.16.

A second famous and interesting effector molecule is 2,3-bisphosphoglycerate (also known as BPG). BPG is an intermediate in the glycolysis pathway and binds to the T state of hemoglobin, favoring the deoxygenated form of the molecule, thus rendering the hemoglobin molecule a better delivery agent in working tissues. BPG does not bind in a way that competes for the same sites as oxygen binds to. Instead, BPG binds into a hollow region between the subunits. To represent this situation in statistical mechanical language, Figure 7.18 shows our vision for the states and weights of this problem. We see that for every state of oxygen occupancy in the T state, we can have BPG either bound or not. Further, since BPG is not thought to bind to the R state of hemoglobin, we do not include that possibility in the model, in distinction with our usual approach of allowing all states of binding in both the T and R states.

Given the states and weights of Figure 7.18, we are prepared to write the partition function for this problem as

$$\mathcal{Z} = e^{-\beta\varepsilon}(1 + \frac{[O_2]}{K_R})^4 + (1 + \frac{[O_2]}{K_T})^4 (1 + \frac{[BPG]}{K_T^{BPG}}). \quad (7.34)$$

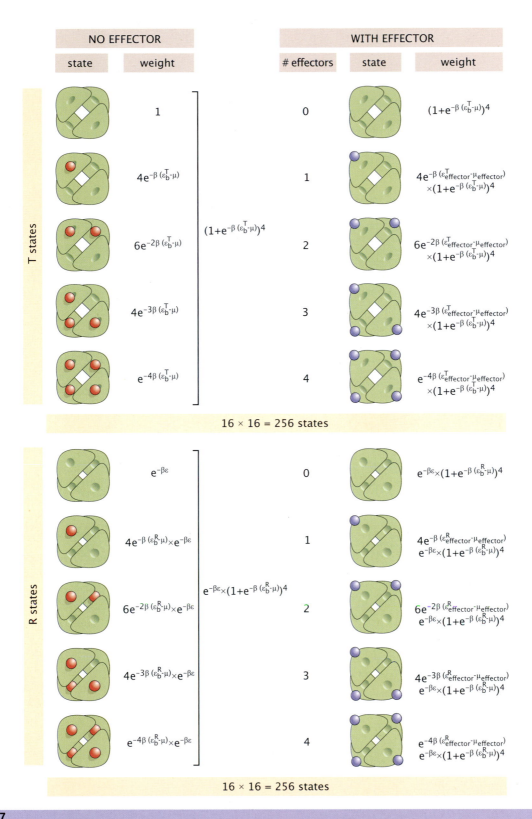

Figure 7.17

States and weights for hemoglobin with effectors. There are 256 individual states of occupancy for the T state and 256 individual states of occupancy for the R state. Since it is not possible to include all these states, the schematic shows a shorthand way of representing them.

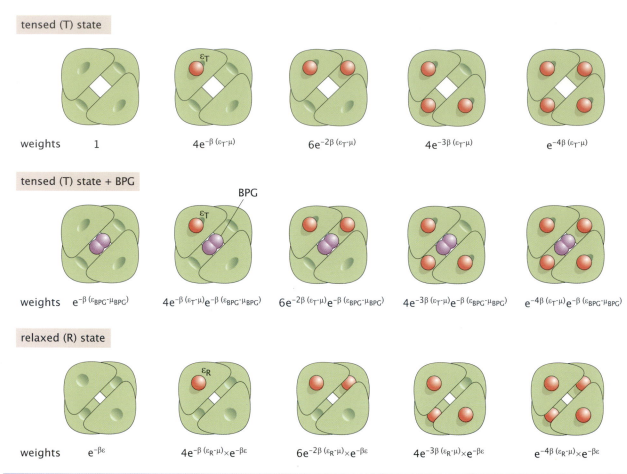

tensed (T) state

weights   1 $4e^{-\beta(\varepsilon_T-\mu)}$ $6e^{-2\beta(\varepsilon_T-\mu)}$ $4e^{-3\beta(\varepsilon_T-\mu)}$ $e^{-4\beta(\varepsilon_T-\mu)}$

tensed (T) state + BPG

weights $e^{-\beta(\varepsilon_{BPG}-\mu_{BPG})}$ $4e^{-\beta(\varepsilon_T-\mu)}e^{-\beta(\varepsilon_{BPG}-\mu_{BPG})}$ $6e^{-2\beta(\varepsilon_T-\mu)}e^{-\beta(\varepsilon_{BPG}-\mu_{BPG})}$ $4e^{-3\beta(\varepsilon_T-\mu)}e^{-\beta(\varepsilon_{BPG}-\mu_{BPG})}$ $e^{-4\beta(\varepsilon_T-\mu)}e^{-\beta(\varepsilon_{BPG}-\mu_{BPG})}$

relaxed (R) state

weights $e^{-\beta\varepsilon}$ $4e^{-\beta(\varepsilon_R-\mu)}\times e^{-\beta\varepsilon}$ $6e^{-2\beta(\varepsilon_R-\mu)}\times e^{-\beta\varepsilon}$ $4e^{-3\beta(\varepsilon_R-\mu)}\times e^{-\beta\varepsilon}$ $e^{-4\beta(\varepsilon_R-\mu)}\times e^{-\beta\varepsilon}$

Figure 7.18

States and weights for hemoglobin with BPG. BPG is assumed to bind only the T state (equivalent to saying $K_{BPG}^R \to \infty$).

Note that we can rewrite this result in the alternative form

$$\mathcal{Z} = e^{-\beta\varepsilon}(1+\frac{[O_2]}{K_R})^4 + e^{-\beta\varepsilon_T^{eff}}(1+\frac{[O_2]}{K_T})^4, \tag{7.35}$$

where now we have subsumed all reference to the presence of BPG into the effective parameter ε_T^{eff}. By inspection, and also in light of the discussion of section 2.6 (p. 63), we can write the difference between the effective energies as

$$\Delta\varepsilon_{eff} = \varepsilon_T - \varepsilon + k_BT\ln(1+\frac{[BPG]}{K_T^{BPG}}). \tag{7.36}$$

This is a very profound result, because what it shows us in turn is that the null model for the role of an allosteric effector such as BPG strictly leads to a renormalization of the parameter L, which tells us about the relative stability of the T and the R states in the absence of ligand. To be specific, the value of L in the

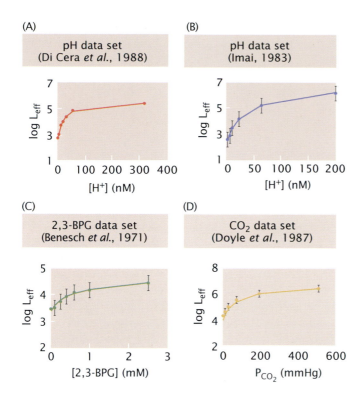

Figure 7.19

The effective equilibrium constant L_{eff} in the presence of effectors. The four curves correspond to the data shown in Figure 7.16 with the same color code used here as there. Adapted Rapp and Yifrach 2017.

presence of BPG, which we christen L_{eff}, is given by

$$L_{eff} = L(1 + \frac{[BPG]}{K_T^{BPG}}).$$ (7.37)

As a result of the analysis of the last few pages, we are now in a position to appeal to the MWC model to account for pure oxygen-binding data, as well as binding at different pHs and in the presence of CO_2 and BPG. Recently, an impressive effort was undertaken to see whether the entirety of the binding data for hemoglobin with oxygen and an array of effectors could be described using one self-consistent parameter set. The results of that study are shown in Figure 7.16, which shows oxygen-binding data at different pHs and in the presence of BPG and in the presence of CO_2. The central message of the binding data shown in Figure 7.16(A) is that a global fit to the various data sets can be realized with the only parameter changing from one binding curve to the next being L, the equilibrium constant between the T and R states in the absence of oxygen, which effectively measures the leakiness of the hemoglobin molecule. A more subtle aspect of the data shown in Figure 7.19 is the monotonic increase in L_{eff} as a function of the effector concentrations.

Philosophically, one of my biggest critiques of current biological research is insufficient respect for the high importance of rigorous work like that undertaken here, especially with the objective of self-consistency, similar to the labors undertaken by the likes of Lagrange and Laplace as they explored the stability of the solar system through detailed studies on the motions of Saturn and Jupiter and comets, detailed work aimed not at unearthing surprises but, rather, at making sure we actually know what we think we know, an important undertaking for the general mission of understanding the limits and validity of the MWC model.

7.5   Physiological versus Evolutionary Adaptation: High Fliers and Deep Divers

A theme that has emerged from characterizing allosteric molecules in terms of their biophysical parameters such as leakiness, EC_{50}, and dynamic range is the question of how these biophysical parameters are tuned both physiologically and evolutionarily. The previous section made it clear that in hemoglobin's role in physiology, its affinity for oxygen is adapted in many ways, so that oxygen is delivered more efficiently where it is needed. In chapter 4, we made our first foray into the all-important biological topic of adaptation. There, we noted that the chemotaxis receptors used posttranslational modifications as a way to change their EC_{50} for different ambient levels of chemoattractant. For hemoglobin, we have seen the way in which changes in pH and the presence of effectors such as CO_2 and BPG can alter the balance between the T and the R states, thus controlling the propensity of oxygen to bind to hemoglobin. In this section, I report on a series of captivating studies that were made of the ways in which hemoglobin (and myoglobin) adapts not only physiologically but evolutionarily as well.

The starting point for the discussion is the idea that certain key biophysical parameters are the targets of adaptation. Figure 7.20 shows two such parameters that occupy center stage in our analysis, namely, the oxygen partial pressure at which hemoglobin is half saturated (labeled here as p_{50} but throughout the book usually referred to as EC_{50}) and the effective Hill coefficient, n, which tells us about the sharpness of response of hemoglobin as the partial pressure of oxygen is increased. Figure 7.20(A) shows how both these parameters vary under different physiological conditions (i.e., different pH values). The data in Figure 7.20(B) take an evolutionary view of hemoglobin by comparing the saturation midpoint and effective Hill coefficients in the hemoglobins of a number of different species.

The trends exhibited in Figures 7.20(A) and (B) were interpreted to imply that physiological and evolutionary adaptation are to some extent independent and orthogonal. Specifically, the conclusion here is that the effective Hill coefficient does not change much for human hemoglobin at different pH values, but, rather (as we already know from the Bohr effect), the p_{50} changes as the pH is varied. By way of contrast, as seen in Figure 7.20(B), comparison of hemoglobins from different animals reveals that the p_{50} is relatively constant across species, and it is the effective Hill coefficient that seems to be the target of evolutionary adaptation. Personally, I find this study to be quite important for indicating the direction work on molecules and evolution should be heading more broadly. Although sequence and structure may be a convenient way to report on evolutionary change, the biophysical parameters that characterize the phenotype resulting from a given molecule may provide a more direct conduit to the fitness.

We now turn to several real-world biological examples I have found particularly compelling that illustrate the ideas of physiological and evolutionary

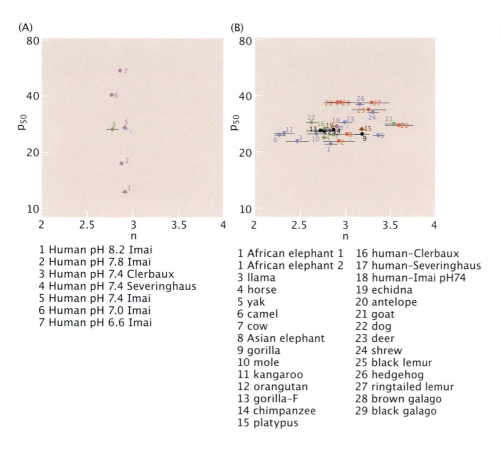

Figure 7.20

MWC and evolution of hemoglobin. (A) Plot of the saturation midpoint and the effective Hill coefficient for human hemoglobin at different pH values. (B) Plot of the saturation midpoint and the effective Hill coefficient for hemoglobins from different species. Adapted from Milo et al. (2007).

1 Human pH 8.2 Imai
2 Human pH 7.8 Imai
3 Human pH 7.4 Clerbaux
4 Human pH 7.4 Severinghaus
5 Human pH 7.4 Imai
6 Human pH 7.0 Imai
7 Human pH 6.6 Imai

1 African elephant 1
1 African elephant 2
3 llama
4 horse
5 yak
6 camel
7 cow
8 Asian elephant
9 gorilla
10 mole
11 kangaroo
12 orangutan
13 gorilla–F
14 chimpanzee
15 platypus

16 human–Clerbaux
17 human–Severinghaus
18 human–Imai pH74
19 echidna
20 antelope
21 goat
22 dog
23 deer
24 shrew
25 black lemur
26 hedgehog
27 ringtailed lemur
28 brown galago
29 black galago

adaptation. One case study features high-flying migratory birds—the bar-headed geese, which must traverse the Himalayas each year on their way to and from breeding grounds north of the dizzying heights of those mountains. The second case study focuses on elite divers such as cetaceans, some of which can stay underwater for several hours.

Figure 7.21 shows the migratory patterns of the bar-headed geese. These birds spend their summers in a region comprehending Mongolia to the Tibetan Plateau and then make their way south to India in the autumn, where they wait out the winter. As seen in the map, satellite telemetry has been used to measure the position and altitude of these birds, revealing that at times they fly higher than 4000 m, with their taxed hearts reaching 400 beats per minute. A number of avian traits are consistent with these demanding flight patterns, which, as seen in Figure 7.21(B), require steep climbs against headwinds.

As noted earlier in the chapter, from a physiological, medical, and evolutionary perspective, the sequencing revolution has rewritten our understanding of hemoglobin. The study of the high-altitude performance of the bar-headed geese is no exception. Figure 7.22 shows a few key mutations in the α- and β-globin genes for the bar-headed goose and its lowland relatives. As can be seen in the diagram, several of these mutations in the α^A and α^D globins are harbored only by the high-flying geese, and early "synthetic biology" work with engineered hemoglobins possessing these mutations showed enhanced oxygen-binding affinity.

Figure 7.21

Migration routes and corresponding altitudes for bar-headed geese. (A) Map showing the migration routes of the bar-headed geese as determined by satellite tracking of individual geese (separate colors for each goose). (B) Profile of altitude of birds during the migration. Adapted from Hawkes, L. A., et al (2013), "The paradox of extreme high-altitude migration in bar-headed geese Anser indicus," *Proc. R. Soc. B.* 280, fig. 1. Reproduced with permission of the Royal Society via Copyright Clearance Center.

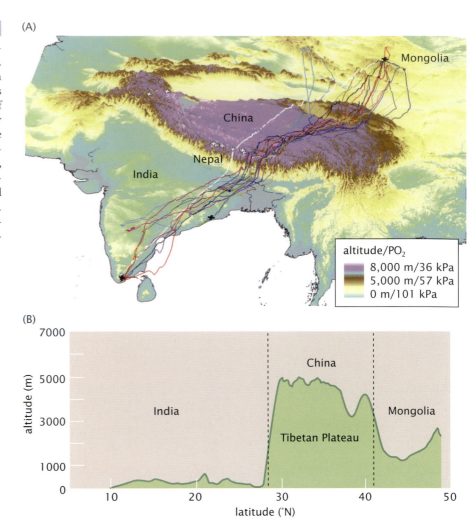

The consequences of these mutations can be explored by measuring both the equilibrium binding and kinetics of oxygen binding to the hemoglobins of these different species. Figure 7.23 shows comparative binding curves for hemoglobins from these different species of geese based on blood samples from the birds. The key physiological observation is that the high-flying bar-headed geese achieve a higher saturation at lower oxygen concentrations than do lowland geese. Interestingly, the work presented in Figure 7.23 can be described entirely satisfactorily using the MWC framework discussed throughout the chapter, with the key insight at the level of the biophysics of the bar-headed geese hemoglobin being that K_d for the R state is reduced, thus favoring oxygen binding at lower oxygen partial pressures. This example shows the linkage between the genotypes explored evolutionarily and the biophysical phenotypes achieved physiologically.

At the opposite end of the exploration of the third dimension (third dimension: $-z$ as opposed to the $+z$ explored by high-flying geese!) by animals, we find the world's elite divers, animals who can dive deep below the ocean's

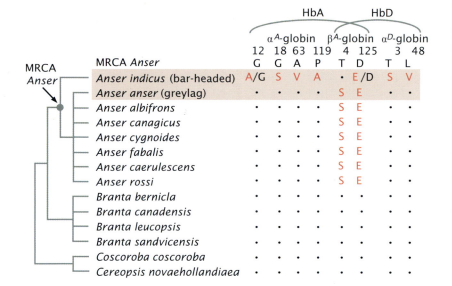

	αA-globin				βA-globin		αD-globin	
MRCA *Anser*	12	18	63	119	4	125	3	48
	G	G	A	P	T	D	T	L
Anser indicus (bar-headed)	A/G	S	V	A	·	E/D	S	V
Anser anser (greylag)	·	·	·	·	S	E	·	·
Anser albifrons	·	·	·	·	S	E	·	·
Anser canagicus	·	·	·	·	S	E	·	·
Anser cygnoides	·	·	·	·	S	E	·	·
Anser fabalis	·	·	·	·	S	E	·	·
Anser caerulescens	·	·	·	·	S	E	·	·
Anser rossi	·	·	·	·	S	E	·	·
Branta bernicla	·	·	·	·	·	·	·	·
Branta canadensis	·	·	·	·	·	·	·	·
Branta leucopsis	·	·	·	·	·	·	·	·
Branta sandvicensis	·	·	·	·	·	·	·	·
Coscoroba coscoroba	·	·	·	·	·	·	·	·
Cereopsis novaehollandiaea	·	·	·	·	·	·	·	·

HbA spans αA-globin and βA-globin. HbD spans βA-globin (125) and αD-globin.

Figure 7.22

Key mutations in the amino acid sequence of hemoglobin of high-flying geese. The brown highlighted region compares eight amino acids in the high-flying bar-headed goose (*Anser indicus*) and the lowland greylag goose (*Anser anser*) to 12 other waterfowl species, as well as the most recent common ancestor (MRCA) of the *Anser* genus.. Several unique amino acid substitutions are found in the hemoglobin sequence of the high-flying bar-headed goose in the α subunits of both the A and D isoforms. Adapted from Jendroszek et al. (2018).

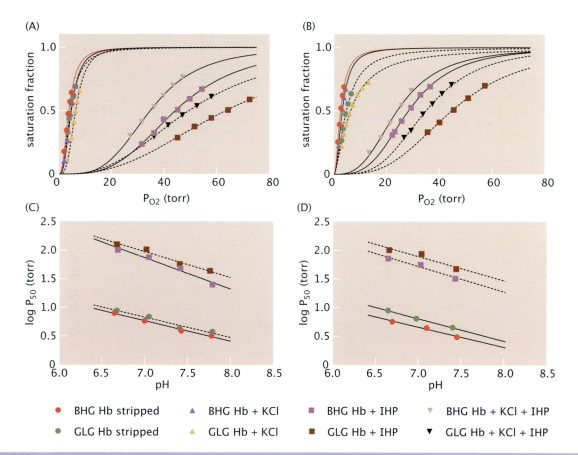

Figure 7.23

Binding properties of high-altitude goose hemoglobin. (A) Equilibrium binding curves for purified HbA under different conditions. (B) Equilibrium binding curves for purified HbD under different conditions. (C) and (D) Bohr effect for HbA and HbD, respectively. Adapted from Jendroszek et al. (2018).

surface and stay underwater for tens of minutes or more. Once again, oxygen binding has been implicated in this physiological feat, with myoglobin concentrations in these animals much higher than they are in humans, for example, as seen in Figure 7.24. Recall that our muscles and those of other mammals such as whales are full of myoglobin, a monomeric protein that has an oxygen-binding heme group like those found in the tetrameric structure of hemoglobin. Figure 7.24 shows that animals with aquatic lifestyles tend to have myoglobins that have more surface charges. The structures of these myoglobins are shown in Figure 7.25, with the extra charges highlighted in blue for the whale hemoglobin.

ESTIMATE

Estimate: Protein Mass per Wet Mass

The units featured on the y-axis of Figure 7.24 are meaningless unless we have some way to compare them to something we know and understand. Said differently, if we have 100 mg of some protein for every gram of wet mass, is that a lot or a little? To answer that question, we invoke some simple rules of thumb about the nature of cytoplasmic matter. We begin by noting that every 1 μm^3 (or equivalently, 1 fL) of cytoplasmic material contains roughly 10^6 proteins. The mass of 1 μm^3 of water is 1 pg (or, better, 1 fkg). To estimate the mass of those 10^6 proteins, we invoke another rule of thumb, which is that the typical protein has a mass of 30,000 Da (with the reminder that 1 Da $\approx 1.6 \times 10^{-27}$ kg). Hence, the mass of these proteins is

$$\text{mass of protein in 1 } \mu\text{m}^3 \approx 10^6 \times 30,000 \text{ Da} \times \frac{1.6 \times 10^{-27} \text{ kg}}{1 \text{ Da}}$$

$$\approx 5 \times 10^{-17} \text{ kg.} \tag{7.38}$$

Now we can compare this protein mass to the mass of water in 1 μm^3 with the result that

$$\frac{m_{protein}}{m_{water}} \approx \frac{5 \times 10^{-17} \text{ kg}}{10^{-15} \text{ kg}} = \frac{50 \text{ mg protein}}{1 \text{ g water}}. \tag{7.39}$$

This comparison now permits us to look at the y-axis units of Figure 7.24 with fresh eyes. In particular, we conclude that if the amount of myoglobin is 100 mg/g, this is a higher protein density than in typical cytoplasmic material, so clearly, there is lots of myoglobin in the tissues of elite divers. One hypothesis for the role of the extra charges on these myoglobins, seen in Figure 7.25, is that they inhibit aggregation of the protein.

A fascinating outcome of the study of the oxygen-carrying capacity of elite divers was an analysis of the evolutionary history of mammals that live in the ocean. Several versions of the outcome of that analysis are shown in Figures 7.26 and 7.27. Figure 7.26 plots the surface charge on myoglobin for a number of different organisms and presents hypothetical evolutionary timelines for how that surface charge differed for their ancestors. Figure 7.27 takes a different approach to these same evolutionary and physiological questions by providing

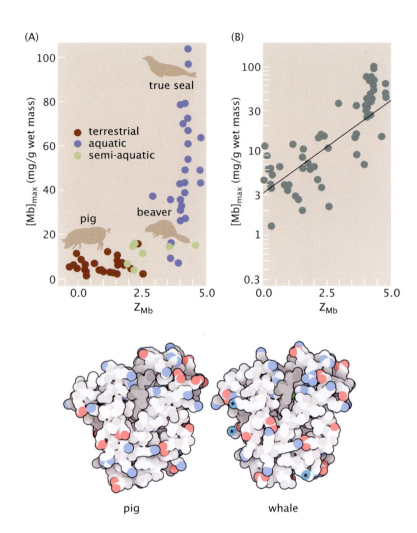

Figure 7.24

Myoglobin concentration and surface charge. (A) Amount of myoglobin per gram of wet mass as a function of myoglobin charge. (B) The same data as in (A) plotted on a log scale on the y-axis. Adapted from Mirceta, S., A. V. Signore, J. M. Burns, et al. (2013), "Evolution of mammalian diving capacity traced by myoglobin net surface charge," *Science* 340. Reprinted with permission from AAAS.

Figure 7.25

Myoglobin structures for pigs and cetaceans. The extra charges found in the whale are labeled with asterisks. Figures courtesy of David Goodsell.

diving times as a function of body mass for both extant and extinct organisms. As a window on trying to understand the details of what these figures are trying to tell us, let's recall the history of the whale.

By literary and scientific consensus, two of the greatest books ever written in English, appeared nearly contemporaneously in the 1850s, with both of them waxing poetic about whales and wondering about their mysterious origins. Chapter 104 ("The Fossil Whale") of Melville's 1851 classic *Moby Dick* cajoles us to consider the evolutionary history of the whale thus: "Having already described him in most of his present habatory and anatomical peculiarities, it now remains to magnify him in an archeological, fossiliferous, and antediluvian point of view. Applied to any other creature than the Leviathan—to an ant or a flea—such portly terms might justly be deemed unwarrantably grandiloquent. But when Leviathan is the text, the case is altered." In the remainder of the chapter, Melville gives us a scholarly account of the knowledge at his time of the cetacean fossil record, focusing on some of the most fascinating finds of all, discovered by slaves on a plantation in Alabama.

Melville was not the only one thinking deeply about the Leviathan. In chapter 6 of Darwin's 1859 masterpiece, *On the Origin of Species*, he, too, puzzled

Figure 7.26

Myoglobin surface charge. (A) Maximal myoglobin concentration and surface charge. (B) Inferred maximal myoglobin concentration and surface charge through the mammalian phylogeny over evolutionary time. Adapted from Mirceta, S., A. V. Signore, J. M. Burns, et al. (2013), "Evolution of mammalian diving capacity traced by myoglobin net surface charge," *Science* 340. Reprinted with permission from AAAS.

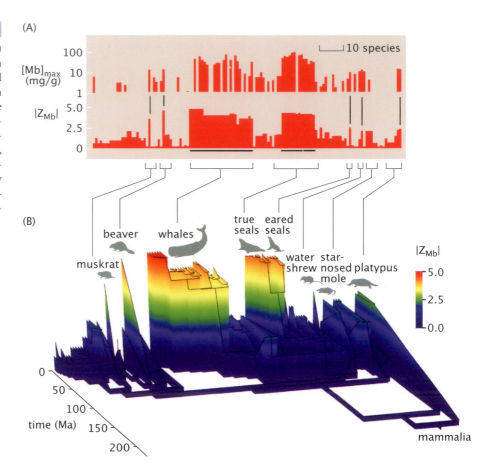

Figure 7.27

Diving times versus body mass and myoglobin concentration (red lines). Adapted from Mirceta, S., A. V. Signore, J. M. Burns, et al. (2013), "Evolution of mammalian diving capacity traced by myoglobin net surface charge," *Science* 340. Reprinted with permission from AAAS.

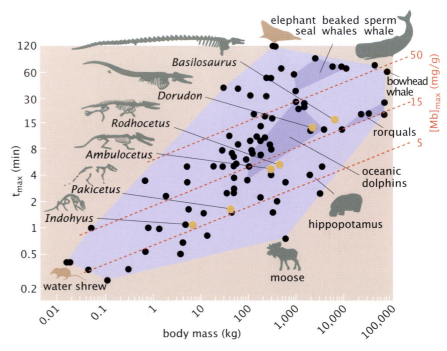

over the whale. That chapter, entitled "Difficulties on Theory" and set forth by the master himself to openly acknowledge those aspects of his theory that most troubled him, takes up the case of the whale: "In North America the black bear was seen by Hearne swimming for hours with widely open mouth, thus catching, like a whale, insects in the water. Even in so extreme a case as this, if the supply of insects were constant, and if better adapted competitors did not already exist in the country, I can see no difficulty in a race of bears being rendered, by natural selection, more and more aquatic in their structure and habits, with larger and larger mouths, till a creature was produced as monstrous as a whale." But the problem for Darwin, described in his chapter "On the Imperfection of the Geological Record" was how sparse our record was of organisms such as the whale, causing him to opine, "I look at the natural geological record, as a history of the world imperfectly kept, and written in a changing dialect; of this history we possess the last volume alone, relating only to two or three countries. Of this volume, only here and there a short chapter has been preserved; and of each page, only here and there a few lines. Each word of the slowly-changing language, in which the history is supposed to be written, being more or less different in the interrupted succession of chapters, may represent the apparently abruptly changed forms of life, entombed in our consecutive, but widely separated formations."

Interestingly, as hinted at in Figure 7.27, in the time since Darwin's attempts to make sense of the scant fossil record for cetaceans, the field has exploded with a long string of fossil discoveries such as *Indohyus*, *Pakicetus*, *Ambulocetus*, *Rhodocetus*, *Dorudon*, *Basilosaurus*, and many more, now making whale evolution one of the most compelling examples of Darwin's and Alfred Russel Wallace's theory of evolution, no longer befitting a chapter such as "Difficulties on Theory." Rather, the evolutionary story of the leviathan is suited for a chapter on success stories in evolution, from fossils to molecules, with myoglobin and hemoglobin providing one of the great windows onto evolutionary and physiological adaptation. In an amazing twist of history, DNA science came to the rescue, making it possible to advance and test evolutionary hypotheses about the myoglobins by comparing the myoglobins of extant species and thus making guesses about the myoglobins of their common ancestors, as seen dramatically in Figure 7.27.

7.6 Hemoglobin and Competitors: Carbon Monoxide Fights Oxygen

Carbon monoxide is a deadly gas that binds hemoglobin roughly 240 times as tightly as oxygen does (this means that CO has 1/240th the dissociation constant of O_2, or $240K_{CO} = K_{O_2}$, where $K_{O_2} = 26$ mmHg). We are interested in examining the competition that is set up when CO and O_2 are present simultaneously. But first, we appeal to experiments that were done on binding of carbon monoxide to hemoglobin in the absence of oxygen. Analogous to studies measuring the occupancy of hemoglobin by oxygen, experiments were performed to measure the occupancy of hemoglobin by carbon monoxide. An example of the outcome of such measurements is shown in Figure 7.28. Once again, the

Figure 7.28

Binding curves for carbon monoxide to hemoglobin. The binding properties were measured using light and hence the curve labeled "dark" is the saturation in the absence of light. Adapted from Brunori et al. (1972).

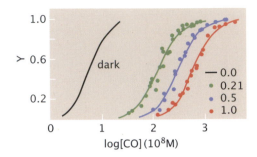

binding curves serve as a challenge to the statistical mechanics of allostery as a conceptual framework.

To compute the occupancy, we imitate the steps already taken in the context of oxygen occupancy by writing the states and weights, as shown in Figure 7.29. Thus, we can write the partition function directly as

$$\mathcal{Z} = e^{-\beta\varepsilon}(1 + \frac{[CO]}{K_R^{CO}})^4 + (1 + \frac{[CO]}{K_T^{CO}})^4, \qquad (7.40)$$

where we have introduced new dissociation constants that reflect the affinity of carbon monoxide for hemoglobin, with K_R^{CO} characterizing binding of CO to the R state, and K_T^{CO} characterizing binding of CO to the T state. Again, imitating in every detail the steps already described in section 7.2 (which really amounts to replacing $[O_2]$ in those equations with $[CO]$ here), we find the occupancy as a function of carbon monoxide concentration as

$$\langle N \rangle = 4 \frac{e^{-\beta\varepsilon}\frac{[CO]}{K_R^{CO}}(1 + \frac{[CO]}{K_R^{CO}})^3 + \frac{[CO]}{K_T^{CO}}(1 + \frac{[CO]}{K_T^{CO}})^3}{e^{-\beta\varepsilon}(1 + \frac{[CO]}{K_R^{CO}})^4 + (1 + \frac{[CO]}{K_T^{CO}})^4}. \qquad (7.41)$$

This occupancy function can be used to systematically fit data like those shown in Figure 7.28.

We now turn to the competition between oxygen and carbon monoxide. As usual, we start by formulating our thinking in the form of the various states we imagine our molecule and its partner ligands can adopt. To avoid a proliferation of too many states, which obscures the central message, we begin by considering a dimeric form of hemoglobin ("dimoglobin") with only two heme groups, as depicted in Figure 7.30, and then performing the generalization to four sites of occupancy by reasoning by analogy. The states depicted in the figure show that each of the sites can either be empty, occupied by O_2, or occupied by CO, and this is true for both the T and R states.

Using the states and weights presented in the figure, we can write the partition function in the usual MWC format as

$$Z = e^{-\beta\varepsilon}(1 + \frac{[O_2]}{K_R^{O_2}} + \frac{[CO]}{K_R^{CO}})^2 + (1 + \frac{[O_2]}{K_T^{O_2}} + \frac{[CO]}{K_T^{CO}})^2. \qquad (7.42)$$

We can adopt a useful notational simplification by defining $x_R = [O_2]/K_R^{O_2}$, $y_R = [CO]/K_R^{CO}$, $x_T = [O_2]/K_T^{O_2}$, and $y_T = [CO]/K_T^{CO}$. With these simplifications, we can now examine the occupancy of the dimoglobin molecule by both

tensed (T) state	weight	relaxed (R) state	weight

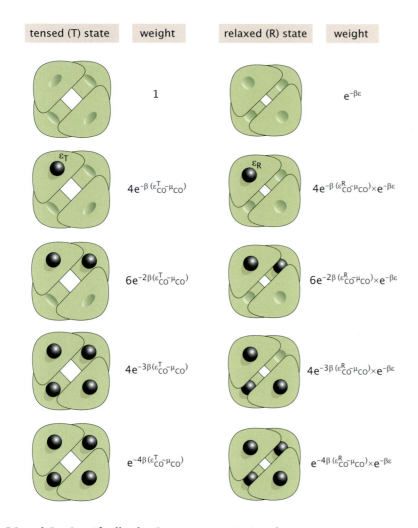

$$1$$

$$e^{-\beta\varepsilon}$$

$$4e^{-\beta(\varepsilon_{CO}^{T}-\mu_{CO})}$$

$$4e^{-\beta(\varepsilon_{CO}^{R}-\mu_{CO})}\times e^{-\beta\varepsilon}$$

$$6e^{-2\beta(\varepsilon_{CO}^{T}-\mu_{CO})}$$

$$6e^{-2\beta(\varepsilon_{CO}^{R}-\mu_{CO})}\times e^{-\beta\varepsilon}$$

$$4e^{-3\beta(\varepsilon_{CO}^{T}-\mu_{CO})}$$

$$4e^{-3\beta(\varepsilon_{CO}^{R}-\mu_{CO})}\times e^{-\beta\varepsilon}$$

$$e^{-4\beta(\varepsilon_{CO}^{T}-\mu_{CO})}$$

$$e^{-4\beta(\varepsilon_{CO}^{R}-\mu_{CO})}\times e^{-\beta\varepsilon}$$

Figure 7.29

States and weights for binding of carbon monoxide to the hemoglobin molecule in the MWC model. The first column shows the various states of occupancy of the hemoglobin molecule in the T state, and the second column shows the various states of occupancy in the R state. The prefactors tell us the number of ways of realizing each such state.

CO and O_2. Specifically, the O_2 occupancy is given by

$$\langle n_{O_2}\rangle = \frac{1}{Z}(2e^{-\beta\varepsilon}x_R + 2e^{-\beta\varepsilon}x_R^2 + 2e^{-\beta\varepsilon}x_Ry_R + 2x_T + 2x_T^2 + 2x_Ty_T), \quad (7.43)$$

and the CO occupancy is similarly given by

$$\langle n_{CO}\rangle = \frac{1}{Z}(2e^{-\beta\varepsilon}y_R + 2e^{-\beta\varepsilon}y_R^2 + 2e^{-\beta\varepsilon}x_Ry_R + 2y_T + 2y_T^2 + 2x_Ty_T). \quad (7.44)$$

The impact of CO on oxygen binding as revealed by this model is shown in Figure 7.31 where we see that with increasing CO concentration, the O_2 binding to dimoglobin is curtailed. This example of dimeric hemoglobin sets the stage for thinking about the case of tetrameric hemoglobin with a competition between oxygen and carbon monoxide.

The power of the approach taken earlier in the chapter using chemical potentials to write the states and weights now becomes especially evident. We are not going to be able to draw states-and-weights diagrams because the number of states is prohibitive. Each binding site has three states of occupancy (empty,

Figure 7.30

States and weights for "dimoglobin" accounting for competition between O_2 and CO for the same heme binding sites. The left column shows the various states of occupancy of the R state, and the right column shows the various states of occupancy of the T state.

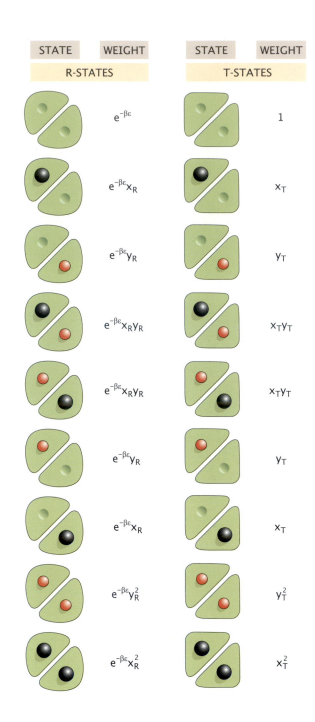

STATE	WEIGHT	STATE	WEIGHT
R-STATES		T-STATES	
	$e^{-\beta\varepsilon}$		1
	$e^{-\beta\varepsilon}x_R$		x_T
	$e^{-\beta\varepsilon}y_R$		y_T
	$e^{-\beta\varepsilon}x_R y_R$		$x_T y_T$
	$e^{-\beta\varepsilon}x_R y_R$		$x_T y_T$
	$e^{-\beta\varepsilon}y_R$		y_T
	$e^{-\beta\varepsilon}x_R$		x_T
	$e^{-\beta\varepsilon}y_R^2$		y_T^2
	$e^{-\beta\varepsilon}x_R^2$		x_T^2

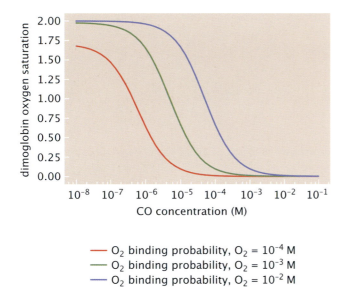

Figure 7.31

Fractional occupancy of dimoglobin by O_2 and CO. As CO concentration goes up, the probability of O_2 binding goes down. Parameters used for this toy model are $\varepsilon_b^R = 0$, $\varepsilon_b^T = 2\ k_B T$, $K_R^{O_2} = 10^{-3}$ M, $K_T^{O_2} = 10^{-5}$ M, $K_R^{CO} = (1/200)K_R^{O_2}$, and $K_T^{CO} = (1/200)K_T^{O_2}$.

— O_2 binding probability, $O_2 = 10^{-4}$ M
— O_2 binding probability, $O_2 = 10^{-3}$ M
— O_2 binding probability, $O_2 = 10^{-2}$ M

bound by CO, bound by O_2). Hence, there are a total of $3^4 = 81$ states for the R state and the same number for the T state, resulting in a total of 162 states. The alternative to obtaining the result by enumeration of states and weights is to exploit our knowledge of the total partition function given by

$$\mathcal{Z} = [e^{-\beta\varepsilon}(1 + e^{-\beta(\varepsilon_{O_2}^R - \mu_{O_2})} + e^{-\beta(\varepsilon_{CO}^R - \mu_{CO})})^4$$

$$+ (1 + e^{-\beta(\varepsilon_{O_2}^T - \mu_{O_2})} + e^{-\beta(\varepsilon_{CO}^T - \mu_{CO})})^4]. \qquad (7.45)$$

Given this partition function, we can in turn determine the average occupancy by taking the derivative with respect to the O_2 chemical potential as

$$\langle N \rangle = \frac{1}{\beta \mathcal{Z}} \frac{\partial}{\partial \mu_{O_2}} [e^{-\beta\varepsilon}(1 + e^{-\beta(\varepsilon_{O_2}^R - \mu_{O_2})} + e^{-\beta(\varepsilon_{CO}^R - \mu_{CO})})^4$$

$$+ (1 + e^{-\beta(\varepsilon_{O_2}^T - \mu_{O_2})} + e^{-\beta(\varepsilon_{CO}^T - \mu_{CO})})^4]. \qquad (7.46)$$

The result of evaluating the required derivative is

$$\langle N \rangle = \frac{4}{\mathcal{Z}} [e^{-\beta\varepsilon} e^{-\beta(\varepsilon_{O_2}^R - \mu_{O_2})}(1 + e^{-\beta(\varepsilon_{O_2}^R - \mu_{O_2})} + e^{-\beta(\varepsilon_{CO}^R - \mu_{CO})})^3$$

$$+ e^{-\beta(\varepsilon_{O_2}^T - \mu_{O_2})}(1 + e^{-\beta(\varepsilon_{O_2}^T - \mu_{O_2})} + e^{-\beta(\varepsilon_{CO}^T - \mu_{CO})})^3], \qquad (7.47)$$

which we can then simplify to the lovely and simple result

$$\langle N \rangle = 4\frac{e^{-\beta\varepsilon}\frac{[O_2]}{K_R^{O_2}}(1 + \frac{[O_2]}{K_R^{O_2}} + \frac{[CO]}{K_R^{CO}})^3 + \frac{[O_2]}{K_T^{O_2}}(1 + \frac{[O_2]}{K_T^{O_2}} + \frac{[CO]}{K_T^{CO}})^3}{e^{-\beta\varepsilon}(1 + \frac{[O_2]}{K_R^{O_2}} + \frac{[CO]}{K_R^{CO}})^4 + (1 + \frac{[O_2]}{K_T^{O_2}} + \frac{[CO]}{K_T^{CO}})^4}. \qquad (7.48)$$

This result is a clear statement of what the allosteric framework has to offer for the case of competitive binding between oxygen and the poison carbon monoxide. As seen in Figure 7.31, the physiological consequences of the binding of CO to hemoglobin are catastrophic, since the presence of low concentrations of carbon monoxide displaces oxygen from hemoglobin, thus depriving tissues of their required oxygen.

7.7 Pushing the MWC Framework Harder: Hemoglobin Kinetics

As seen throughout the book and this chapter in particular, the MWC model has been very successful in providing a quantitative conceptual framework for thinking about dose-response curves. A natural and powerful way to go beyond the dose-response curves is to interrogate this framework from the kinetic perspective, asking now whether the time evolution of these systems unfolds as would be predicted by a kinetic implementation of the MWC model.

To get a sense of the complexities of going to a kinetic description and the associated proliferation of parameters, Figure 7.32 provides a schematic of the allowed transitions between the different conformational states and states of ligand occupancy and their corresponding rate constants. There are many beautiful symmetries in the MWC model that allow us to relate the kinetic parameters in various ways. First, we note that regardless of how many ligands are already bound, the on and off rates for the ligands are the same. When the hemoglobin is in the R state, the rate for binding an oxygen molecule is k_R^+, and the rate of an oxygen molecule falling off is k_R^-. Similarly, when the hemoglobin is in the T state, the rate for binding an oxygen molecule is k_T^+, and the rate of an oxygen molecule falling off is k_T^-.

Our work on the equilibrium treatment of hemoglobin provides a series of constraints on our kinetic rate constants. We saw this philosophy unfold in the context of ion-channel dynamics in section 3.6 (p. 98). Building on these ideas and appealing to the equilibrium between binding and unbinding, we have the condition that

$$p_T^{(i)} k_T^+ [O_2] = p_T^{(i+1)} k_T^-, \tag{7.49}$$

where $p_T^{(i)}$ refers to the probability of being in the T state with i ligands bound, and $p_T^{(i+1)}$ refers to the probability of being in the T state with $i+1$ ligands bound. A similar constraint applies to binding and unbinding of ligands in the R state as

$$p_R^{(i)} k_R^+ [O_2] = p_R^{(i+1)} k_R^-. \tag{7.50}$$

From the perspective of the states and weights of Figure 7.11 we can relate these binding and unbinding rate constants through the dissociation constants as

$$K_R = \frac{k_R^-}{k_R^+} \tag{7.51}$$

and

$$K_T = \frac{k_T^-}{k_T^+}. \tag{7.52}$$

Although we have introduced the four rate constants k_R^-, k_R^+, k_T^-, and k_T^+, the equilibrium constants K_R and K_T reduce these to only two new parameters.

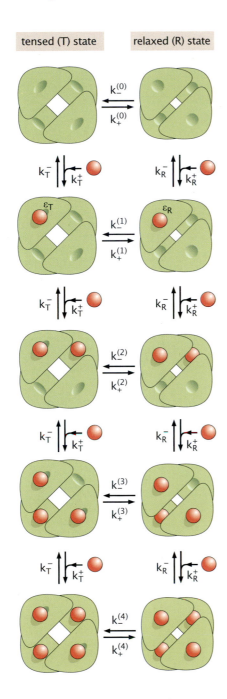

tensed (T) state   relaxed (R) state

Figure 7.32

Kinetics of hemoglobin in the MWC model. There are 10 distinct states: two conformational states T and R, each of which can have 0, 1, 2, 3 or 4 oxygens bound. The rate constants can be separated into two categories. There are those rate constants that characterize the gain and loss of an oxygen molecule within a given conformational state. The second set of rate constants characterize the switching between the T and R states at fixed ligand number.

Thus far, we have considered the rate constants shown in Figure 7.32 that concern binding and unbinding of ligands (vertical arrows). We must also consider the rate constants that describe the kinetics of conformational changes between the T and the R states for different degrees of ligand occupancy (horizontal arrows).

As seen in Figure 7.32, we have an additional 10 rate constants that tell us the rate of switching between the T and R states. To be specific, we introduce the notation $k_-^{(i)}$ and $k_+^{(i)}$ for the rate of switching from T to R and R to T for the states with i ligands, respectively. The equilibrium between the states can be formulated in both the language of statistical mechanics (i.e., states and weights) and the language of kinetics (i.e., detailed balance), and the relation between these two viewpoints allows us to reduce the number of free kinetic parameters. Assuming a separation of time scales such that we can treat the conformational change as being in quasi-equilibrium, we have the general condition

$$p_T^{(i)} k_+^{(i)} = p_R^{(i)} k_-^{(i)}, \tag{7.53}$$

where this condition expresses the balance between transitions from T_i to R_i and R_i to T_i. We can now implement that condition for each state of occupancy. For example, for the T_0 and R_0 states we have

$$\frac{[T_0]}{[R_0]} = \frac{e^{-\beta \varepsilon_T}}{e^{-\beta \varepsilon_R}} = L = \frac{k_-^{(0)}}{k_+^{(0)}}. \tag{7.54}$$

Similar conditions hold for the different levels of ligand occupancy as

$$\frac{[T_1]}{[R_1]} = \frac{e^{-\beta \varepsilon_T}}{e^{-\beta \varepsilon_R}} \frac{[O_2]/K_T}{[O_2]/K_R} = L \frac{K_R}{K_T} = \frac{k_-^{(1)}}{k_+^{(1)}}, \tag{7.55}$$

and

$$\frac{[T_2]}{[R_2]} = \frac{e^{-\beta \varepsilon_T}}{e^{-\beta \varepsilon_R}} \frac{([O_2]/K_T)^2}{([O_2]/K_R)^2} = L \left(\frac{K_R}{K_T} \right)^2 = \frac{k_-^{(2)}}{k_+^{(2)}}, \tag{7.56}$$

and

$$\frac{[T_3]}{[R_3]} = \frac{e^{-\beta \varepsilon_T}}{e^{-\beta \varepsilon_R}} \frac{([O_2]/K_T)^3}{([O_2]/K_R)^3} = L \left(\frac{K_R}{K_T} \right)^3 = \frac{k_-^{(3)}}{k_+^{(3)}}, \tag{7.57}$$

and finally,

$$\frac{[T_4]}{[R_4]} = \frac{e^{-\beta \varepsilon_T}}{e^{-\beta \varepsilon_R}} \frac{([O_2]/K_T)^4}{([O_2]/K_R)^4} = L \left(\frac{K_R}{K_T} \right)^4 = \frac{k_-^{(4)}}{k_+^{(4)}}. \tag{7.58}$$

This now gives us five additional constraints on the 10 new kinetic parameters.

In the early days of the MWC model (see the papers by Hopfield, Shulman, and Ogawa; and Szabo and Karplus in Further Reading), once it was clear that the hemoglobin binding curves could be self-consistently interpreted in terms of the MWC model and even reconciled with the structural insights of Perutz and coworkers, the next step in trying to break the model was to see how well it could explain kinetic data. One exciting category of experiments that had already reached the point of being a fine art were photolysis experiments, in which light is used to tune off rates, as we saw in the context of CO in Figure 7.28 (p. 260). In a very nuanced and thoughtful treatment of the difference between fitting data and interpreting data in a way that the precision of the experiments allows, Hopfield and coworkers used the constraints just described to make a best guess at parameter values and to show that the resulting theoretical predictions of key quantities were consistent with the MWC model. Figure 7.33 shows the result of using the model we have just sketched to interpret one class of kinetics experiments on oxygen binding in hemoglobin.

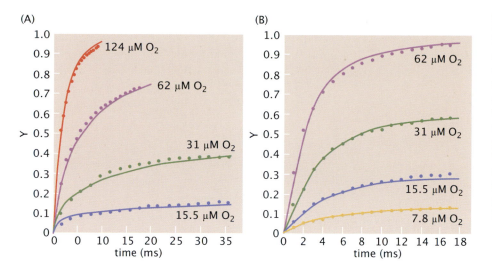

Figure 7.33

Kinetics of hemoglobin as described by the MWC model. The data points are the result of measuring the fractional occupancy of hemoglobin as a function of time at different concentrations of oxygen. The curves are parameter-free predictions using the model of MWC kinetics developed in this section. Adapted from Hopfield, Shulman, and Ogawa (1971); see Further Reading.

As we have found repeatedly throughout the book, often as our experimental acumen increases, so, too, must the sophistication of our models. The case of hemoglobin is no exception. Because of the celebrity status of the hemoglobin molecule as a poster child for many facets of structural biology, biochemistry and physiology, a huge array of data of different types have been accumulated over the years and present increasing demands on our theories. In the case of hemoglobin, there is an abundance of structural data, equilibrium binding data, and kinetic data, all of which have been greatly enhanced by being viewed through the prism of allosteric frameworks. In this final section of the chapter, we give a few hints of the beautiful experiments that have been performed to challenge allosteric models of proteins in general and hemoglobin in particular.

We begin with the binding data themselves, the primary consideration of this chapter. A very revealing set of experiments beyond those already considered that helped tease out the applicability of MWC thinking were based on measuring the binding properties of hemoglobin when constrained to be exclusively in the T state. In this case, the states and weights correspond to only those from the left column of Figure 7.11 (p. 240). The partition function simplifies to

$$Z = (1 + \frac{[O_2]}{K_T})^4. \quad (7.59)$$

If we recall our notation $c = K_R/K_T$, and $x = [O_2]/K_R$, we can write the occupancy in the form

$$Y = \frac{\langle N \rangle}{4} = \frac{cx(1+cx)^3}{(1+cx)^4} = \frac{cx}{(1+cx)}. \quad (7.60)$$

This implies in turn that the binding of oxygen to this stabilized crystalline version of the T state should exhibit no cooperativity. This idea was confirmed experimentally, as seen in Figure 7.34.

One of the biggest problems with the way this subject has unfolded philosophically is that often the focus has been on a retrospective basis, referring to how well the data are fit as opposed to making polarizing predictions. Even

Noncooperative binding of hemoglobin when constrained to be exclusively in the *T* state. Adapted from Eaton et al. (1999); see Further Reading.

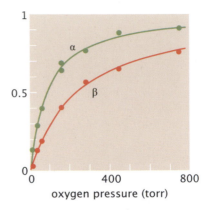

Kinetics of binding of CO to hemoglobin. Adapted from Eaton et al. (1999).

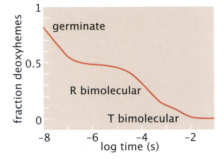

so, the story of hemoglobin gives an inspiring account of what it looks like to poke and prod a model in as many different ways as we can think of. The clever use of crystals and gels to freeze proteins in either the *T* or the *R* conformation and then to measure the occupancy curves provides a category of tests of the different conceptual frameworks.

Similarly, the kinetics of photodissociation experiments provided another class of stringent tests on our thinking. Early data on hemoglobin binding kinetics such as those shown in Figure 7.33 were ambiguous about whether different allosteric models could describe all the properties of hemoglobin. In another experimental and modeling tour de force, kinetics experiments with higher temporal resolution yielded data like those shown in Figure 7.35. Though it is beyond the scope of this book, the dialogue between theory and experiment was extremely enriching (see the papers by Eaton et al. in Further Reading) and led to a self-consistent picture of the physical behavior of hemoglobin.

All told, hemoglobin constitutes a compelling success story in the power of quantitative thinking in truly uncovering how a biological system works that is quite consistent with the broader definition of understanding in other domains of quantitative science.

7.8   Summary

In its original formulation, the allosteric model was introduced as a way to think about the nuanced behavior of proteins. This chapter showed how the allosteric framework can be used for a soluble protein whose main role is as a physiological carrier of oxygen. Indeed, the long and fascinating history of hemoglobin has been a signature event in modern biochemistry and biology by teaching us about the structure and function of this and many other molecules. This chapter explored a tiny part of that story by showing how the MWC framework and its extensions have been used to study the function of hemoglobin in the form of its oxygen-binding curves and the kinetics of ligand binding and unbinding, as well as how those molecular properties give rise to the physiological and evolutionary adaptation that is part of the fascinating world around us in contexts ranging from deep divers to high-mountain fliers.

7.9   Further Reading

Brunori, M. (2013) "Variations on the theme: Allosteric control in hemoglobin." *FEBS J.* 281:633. This excellent article explains many of the insights into hemoglobin over the last half century using the MWC model.

Eaton, W. A., E. R. Henry, J. Hofrichter, S. Bettati, C. Vaippiani, and A. Mozzarelli (2007) "Evolution of allosteric models for hemoglobin." *IUBMB Life* 59: 586–599. This paper does a thorough job of what I tried to do in section 7.7, namely, to explore the various bells and whistles that have been added to the simplest MWC model, with the ambition of exploring how well these different allosteric models respond to the entirety of the available data.

Eaton, W. A., E. R. Henry, J. Hofrichter, and A. Mozzarelli (1999) "Is cooperative oxygen binding by hemoglobin really understood?" *Nat. Struct. Biol.* 6:351. This wonderful article balances historical insight, philosophical insights into the nuanced meaning of understanding, and the technical details of how we know what we know about oxygen binding in hemoglobin. Should be read by everyone interested in hemoglobin.

Hopfield, J. J., R. G. Shulman, and S. Ogawa (1971) "An allosteric model of hemoglobin: I, Kinetics." *J. Mol. Biol.* 61: 425–443. The authors of this paper aim to assess the ability of the MWC model to account for both binding and kinetics experiments, finding at the time of their writing no disturbing anomalies.

Judson, H. F. (2014) *The Eighth Day of Creation.* Cold Spring Harbor, NY: Cold Spring Harbor Laboratory Press. A monumental classic in the history of science. For the purposes of the present chapter, there is wonderful material on the history of the study of hemoglobin, especially on the relentless pursuit of its structure and function by Max Perutz.

Mirceta, S., A. V. Signore, J. M. Burns, A. R. Cossins, K. L. Campbell, and M. Berenbrink (2013) "Evolution of mammalian diving capacity traced by myoglobin net surface charge." *Science* 340:1234192. This paper addresses the interesting challenge of elite divers and oxygen transport.

Schorr, G. S., E. A. Falcone, D. J. Moretti, and R. D. Andrews (2014) "First long-term behavioral records from cuviers beaked whales (*Ziphius cavirostris*) reveal record-breaking dives." *PLoS ONE* 9 (3): e926333. This article reveals the current record holder for whale diving times.

Scott, G. R., L. A. Hawkes, P. B. Frappell, P. J. Butler, C. M. Bishop, and W. K. Milsom (2015) "How bar-headed geese fly over the Himalayas." *Physiology* 30: 107–115. This excellent article gives an introduction to the challenges faced by high-flying birds and the many tricks they use to make such flying possible. Changes to the oxygen-binding affinity of hemoglobin are only a part of the rich story.

Szabo, A., and M. Karplus (1972) "A mathematical model for structure-function relations in hemoglobin." *J. Mol. Biol.* 72: 163–197. This classic paper reconciles the allosteric models for oxygen binding to hemoglobin with the structural knowledge that came from the work of Perutz.

Tanford, C., and J. Reynolds (2004) *Nature's Robots*. Oxford: Oxford University Press. This great book takes on a voyage of discovery about the history of proteins. One of the centerpieces of that history is the study of hemoglobin.

Wyman, J., and S. J. Gill (1990) *Binding and Linkage: Functional Chemistry of Biological Macromolecules*. Sausalito, CA: University Science Books. This book tackles the question of cooperative binding in the macromolecules of life, with hemoglobin serving as one of the main case studies.

7.10   REFERENCES

Bohr, C., K. Hasselbalch, and A. Krogh (1904) "Über einen in biologischer Beziehung wichtigen Einfluss, den die Kohlensäurespannung des Blutes auf dessen Sauerstoffbindung übt." (Concerning a biologically important relationship: The influence of the carbon dioxide content of blood on its oxygen binding.) *Skand. Arch. Physiol.* 16:401–412.

Brunori, M., J. Bonaventura, C. Bonaventura, E. Antonini, and J. Wyman (1972) "Carbon monoxide binding by hemoglobin and myoglobin under photodissociating conditions." *Proc. Natl. Acad. Sci.* 69:868–871.

Crick, F. H. C. (1958) "On protein synthesis." *Symp. Soc. Exp. Biol.* 12:138–163, p. 142.

Darwin, C. R. (1859) *On the Origin of Species by Means of Natural Selection, or the Preservation of Favoured Races in the Struggle for Life*, 1st ed. London: John Murray.

Funke, O. (1853) *Atlas der Physiologischen Chemie*. Leipzig: Engelmann, Table X.

Holmes, F. L. (1995) "Crystals and Carriers: The Chemical and Physiological Identification of Hemoglobin." Chapter 30 in edited by A. L. Kox and D. M. Siegel. *No Truth Except in the Details*. Dordrecht, The Netherlands: Kluwer Academic, 191–243.

Jendroszek, A., H. Malte, C. B. Overgaard, K. Beedholm, C. Natarajan, R. E. Weber, J. F. Storz, and A. Fago (2018) "Allosteric mechanisms underlying the adaptive increase in hemoglobin-oxygen affinity of the bar-headed goose." *J. Exp. Biol.* 221:1.

Maxwell, R. A. (2002) *Maxwell Quick Medical Reference*. Grass Valley, CA: Maxwell Publishing.

Melville, H. (1851) *Moby Dick; or the Whale*. New York: Harper & Brothers.

Milo, R., J. H. Hou, M. Springer, M. P. Brenner, and M. W. Kirschner (2007) "The relationship between evolutionary and physiological variation in hemoglobin." *Proc. Natl. Acad. Sci.* 104:16998–17003.

Mirceta, S., A. V. Signore, J. M. Burns, A. R. Cossins, K. L. Campbell, and M. Berenbrink (2013) "Evolution of mammalian diving capacity traced by myoglobin net surface charge." *Science* 340 (6138): 1234192.

Rapp, O., and O. Yifrach (2017) "Using the MWC model to describe heterotropic interactions in hemoglobin." *PLoS ONE* 12(8):e0182871.

Riggs, A. (1960) "The nature and significance of the Bohr effect in mammalian hemoglobins" *J. Gen. Physiol.* 43:737.

Scott, G. R., L. A. Hawkes, P. B. Frappell, P. J. Butler, C. M. Bishop, and W. K. Milson (2015) "How bar-headed geese fly over the Himalayas." *Physiology* 30:107–115.

Stokes, G. G. (1863–1864) "On the reduction and oxidation of the colouring matter of the blood." *Proc. R. Soc. London* 13:355–364.

Yonetani, T., S. Park, A. Tsuneshige, K. Imai, and K. Kanamori (2002) "Global allostery model of hemoglobin." *J. Biol. Chem.* 277:34508.

Zuckerkandl, E., R. T. Jones, and L. Pauling (1960) "A comparison of animal hemoglobins by tryptic peptide pattern analysis." *Proc. Natl. Acad. Sci.* 46:1349–1360.

8

HOW CELLS DECIDE WHAT TO BE: SIGNALING AND GENE REGULATION

> Probability theory is nothing more than common sense reduced to calculation.
>
> —Pierre Simon de Laplace

Cells make decisions all the time. Some of those decisions are fast, taking place on subsecond time scales, as we saw in chapter 4 (p. 124) in our discussion of chemotaxis. However, another important class of decisions unfolds on much slower time scales (tens of minutes). These cellular decisions dictate how genes are turned on and off, or tuned up and down, both in space and time. Of course, thinking about these questions takes us right to the heart of the processes of the central dogma. This chapter focuses on the all-important process of transcription. Specifically, we explore the control exercised in deciding when and where to read out the information carried in the genomic DNA. As we will see, our understanding of the control of gene expression is built on precisely the same allosteric concepts already highlighted throughout the book.

At the same time they were engaged in thinking about allostery, Jacob and Monod were also laying the foundations of the modern theory of transcriptional regulation (and indeed the two subjects were connected). The fascinating story has been told many times before, so we will resist the temptation to enter into the history in detail, but there is much about the tortured path to our understanding of transcription that is deeply instructive. Essentially, the same questions arose in the context of genes as had already manifested themselves in the context of enzymes. How do genes switch back and forth between inactive and active states? In the context of some proteins, the answer is they are made only when they are needed, and that decision is controlled in turn by molecules known as transcription factors that can in some cases be fruitfully described as MWC molecules, since their activity in controlling gene expression is altered by the binding of various signaling molecules.

In this chapter, we explore how the ideas already set forth earlier in the book can now be tailored to help us understand how repressors and activators work to alter the level of gene expression. Though our primary examples will center on bacterial transcription, since that is the forum where the quantitative underpinnings of allosteric control of transcription factors have been most carefully measured, the chapter will close by examining how these ideas might find their way into thinking about eukaryotic transcription, and parts of that story are

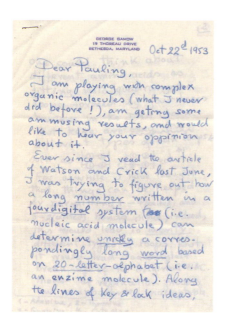

Figure 8.1

The mystery of the processes of the central dogma. Letter from George Gamow to Linus Pauling explaining the puzzle of how the language of DNA is related to the language of proteins. From "Linus Pauling and the race for DNA: A documentary history," OSU Libraries Special Collections & Archives Research Centers. Used with permission of the author's estate.

further elaborated when we discuss nucleosomes and chromatin in chapter 10 (p. 316).

8.1   Of Repressors, Activators, and Allostery

On October 22, 1953, in a letter directed to Linus Pauling, George Gamow acknowledged a deep mystery left in the wake of the Watson and Crick paper that had revealed the structure of DNA three short months earlier. That letter, reproduced here in Figure 8.1, articulated the puzzle of the "coding problem," namely, how does the information contained in the strings of four distinct nucleotides get converted into the strings of 20 distinct amino acids that make up proteins? Gamow's version of the challenge: "Ever since I read the article of Watson and Crick last June, I was trying to figure out how a long number written in a four digital system (i.e., nucleic acid molecule) can determine uniquely a correspondingly long word based on 20-letter alphabet (i.e., an enzyme molecule)."

After more than a decade of labor in the 1950s and 1960s, Gamow's challenge was answered in the form of the central dogma (indicated schematically in Figure 8.2) and culminated in the determination of the genetic code. The coding problem itself is resolved mechanistically in the process of translation, where transfer RNAs serve as decoders that relate nucleotides and amino acids, essentially providing a molecular way of speaking the nucleotide language and translating it into the amino acid language.

In this chapter, our central focus will be on the process of transcription, illustrated in Figure 8.2. Transcription is mediated by RNA polymerase (RNAP), which translocates along the DNA while synthesizing a messenger RNA molecule. A beautiful set of experiments made it possible to see the process of transcription structurally. Figure 8.3 shows Miller spreads, named

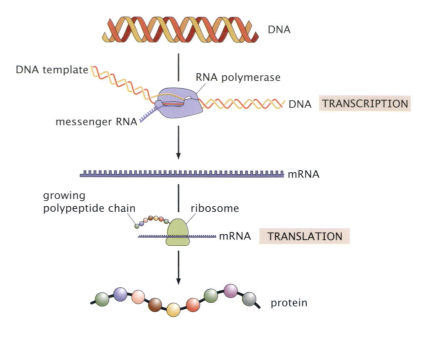

Figure 8.2

The processes of the central dogma. The genetic information in a gene is read out by RNA polymerase, which produces a messenger RNA copy of that gene. The ribosome catalyzes the formation of the correct polypeptide bonds in the nascent protein chain.

Figure 8.3

Electron microscopy image of the process of transcription. In classic experiments culminating in Miller spreads, genes that are in the process of being transcribed can be seen as Christmas-tree structures. The schematic illustrates how simultaneous transcription from single genes gives rise to such structures. Adapted from Gotta, L. S., et al. (1991), *J. Bacteriol.* 20: 6647.

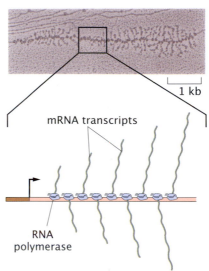

in honor of Oscar Miller, who pioneered techniques to perform electron microscopy on DNA of genes such as the highly transcribed ribosomal genes while they are being transcribed.

The process of transcription as schematized in Figure 8.2 tells us nothing about the decision about when and where those genes will be turned on or off. In that sense, Figure 8.2 needs to be amended to account for the regulatory processes discovered by Jacob, Monod, and Englesberg and shown in Figure 8.4. For the purposes of the present chapter, the key point we need to take away from this schematic is the idea that the action of RNA polymerase in the making of mRNAs is often under strict control by batteries of transcription factors that

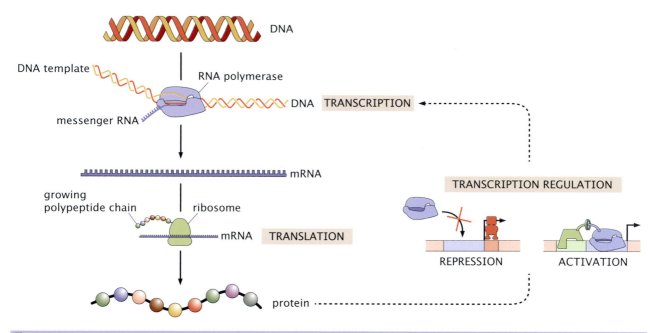

Figure 8.4

The processes of the central dogma including transcriptional regulation. The process of transcription is often controlled by transcription factors which can either tune the rate of transcription down (repression) or up (activation).

bind the DNA and up- or downregulate the transcription rate relative to its basal levels. In particular, we see there are classes of molecules known as repressors that inhibit the ability of RNA polymerase to make RNA. In addition, there are molecules that "recruit" the polymerase to some gene that are subject to upregulation.

The original hypotheses about the mechanism of transcriptional regulation focused on what is sometimes known as the repressor-operator model. This simple model envisioned *all* genes as being controlled by the binding of repressors. The gene would be off when the repressor was bound and on when the repressor was absent. Interestingly, many of the insights of the early days of molecular biology focused on those features of biological systems that were in some sense universal, such as the processes of the central dogma and the ubiquitous usage of similar metabolic pathways. In the repressor-operator model, Monod suspected that once again there would be a universal mechanism, this time for how genes are regulated, namely, by the binding of repressors to promoters at sites known as the operator and which shut those genes down for business. Over time, mounting evidence suggested a different kind of transcriptional control than the repressor-operator mechanism, namely, positive control by activators. The idea of activation is that a gene would be "on" only in the presence of positive control, in which the gene of interest was licensed to transcribe because of activator binding.

In a revealing article about the operon model, Beckwith (2011) notes: "It was not until the work of Ellis Englesberg and his coworkers on the genes determining arabinose metabolism that a true challenge to the universality of repressor-operator control appeared. While others presented genetic evidence

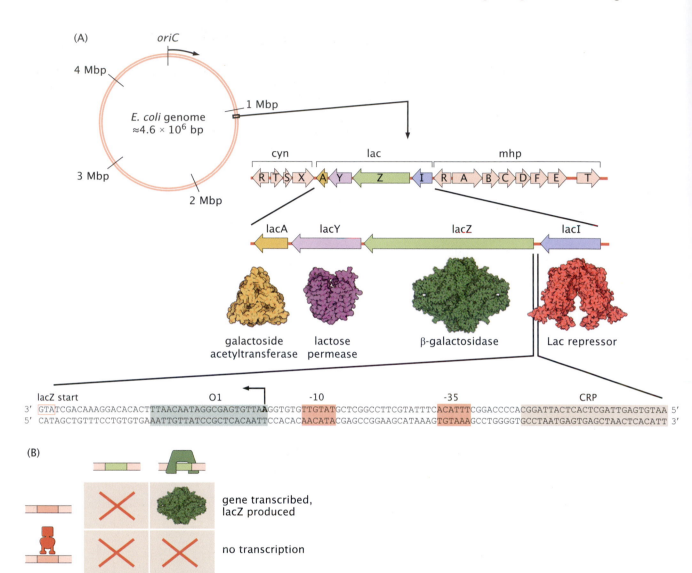

Figure 8.5

The *lac* operon regulatory architecture. (A) The *E. coli* genome revealing the position of the *lac* operon and the structural organization of the operon itself. The bottom image shows the nucleotide-resolution picture of the binding sites for repressor, RNAP, and activator (from left to right). (B) The logic table revealing how repressor and activator conspire to determine the level of gene activity. The enzyme β-galactosidase is produced only when activator CRP is present and repressor is not.

that was consistent with positive control operating in their systems, Englesberg was the first to hold tenaciously to such views until they were accepted." It is a great irony of the subject that the fundamental genetic circuit used by Monod to formulate the repressor-operator model is itself subject to the kind of positive regulation discovered by Englesberg. The present chapter follows that historical order as it focuses first on repression in the *lac* operon (see Figure 8.5) and then turns to the question of how those very same genes are activated by the broadly acting activator known as CRP.

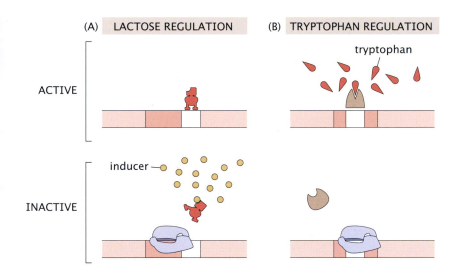

Figure 8.6

Inducers and allostery. (A) Induction in the *lac* operon. In the presence of inducer, the repressor has a low probability of being on the DNA, and the expression level increases. (B) In the regulatory circuit for tryptophan synthesis, corepression occurs in the *presence* of tryptophan.

But how does transcriptional regulation connect to our main subject of the molecular switch and allostery? The transcription of many genes is altered by environmental perturbations. That is, if we change the concentration of some chemical signal such as a metabolite, various genes in turn will change their level of expression. One of the classic mechanisms of such inducers is by their action on allosteric transcription factors that transition back and forth between states that are inactive and active in their role as regulators, as indicated schematically in Figure 8.6. In keeping with what we have already seen throughout the book, it will not be surprising that, once again, the mechanism of action for many of these transcription factors is through binding of some ligand that tilts the free-energy balance between two conformational states that have different activity in their role as transcriptional regulators. Working out the statistical mechanics of this process will form the central focus of this chapter.

As a result of various experimental advances, the study of transcription has reached a point at which the standards of the field are best directed toward constructing an understanding of the process that is predictive. In particular, the view offered here is that if people are going to go to all the trouble of making careful quantitative measurements of the level of transcription at the single-cell level as a function of various tuning parameters that vary transcription, then the field should be satisfied with nothing less than theoretical understanding that can predict the outcome of those quantitative experiments. In the remainder of this chapter, we will explore a small subset of theoretical ideas that have been set forth to explain transcription in just that way by invoking the concepts of the MWC model.

8.2 Thermodynamic Models of Gene Expression

Our starting point for thinking about the transcription process quantitatively is to consider the kinetics of mRNA production. In particular, we are interested in the time evolution of the mRNA census as a result of the key dynamical processes such as transcription itself and mRNA degradation. Given that the

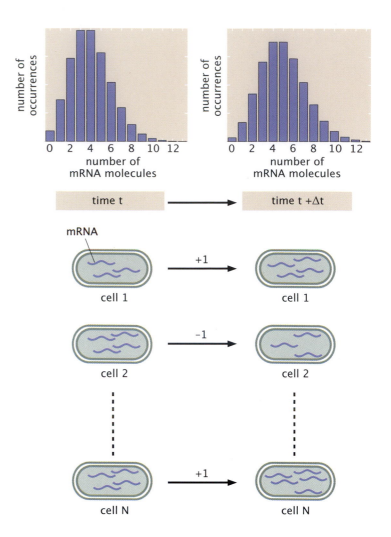

Figure 8.7

Transcription process resulting in change in mRNA census between times t and $t + \Delta t$. The histograms show the counts of different mRNA numbers in each cell at times t and $t + \Delta t$. The different cell schematics show each cell with its own number of mRNA, which changes over time.

amount of mRNA in the cell at time t is $m(t)$, the amount an instant Δt later is given by

$$m(t + \Delta t) = m(t) - \text{number of mRNA degraded}$$
$$+ \text{number of mRNA produced.} \tag{8.1}$$

Note that, as shown in Figure 8.7, when we write an equation for $m(t)$, it is the average value of m over all cells that we have in mind. To make further progress, we have to be able to write the degradation and production terms explicitly. To that end, we exploit the fact that in a simplest model of degradation, at every instant, each mRNA molecule has the same probability $\gamma \Delta t$ of decaying. Similarly, for each of the transcriptionally active states, there is a rate r_i of mRNA production. Given these ideas, the amount of mRNA at time $t + \Delta t$ can be written more formally as

$$m(t + \Delta t) = m(t) - \gamma \Delta t m + \sum_i (r_i \Delta t) p_i, \tag{8.2}$$

(A)   CONSTITUTIVE ARCHITECTURE

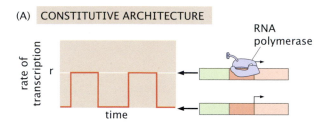

(B)   SIMPLE ACTIVATION ARCHITECTURE

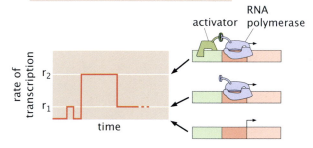

Figure 8.8

Transcriptional time series. (A) A constitutive promoter switches between the active and inactive configuration. When active, the promoter transcribes at a rate r. (B) The case of an activated promoter. Over time, the occupancy of the regulatory region, and hence the rate of transcription, changes. The state without an activator has a lower transcription rate than the state with an activator.

where p_i is the probability of being in the ith transcriptionally active state, and γ is the degradation rate and has units of time^{-1}. This idea is indicated schematically in Figure 8.8.

But what is the nature of the individual states and their probabilities? Though their explicit calculation is the business of the thermodynamic models which will take center stage in the next section, here we note that, the probability of the ith transcriptionally active state can be thought of as

$$p_i = p_i([TF_1], [TF_2], \cdots), \tag{8.3}$$

where the notation indicates that this probability is a function that reflects the occupancy of the regulatory DNA by the various transcription factors, such as the green activator molecule shown in Figure 8.8(B), that interact with the regulatory apparatus of the gene of interest. Hence, each transcriptionally active state denoted by the label "i," corresponds to a different state of the promoter characterized by a different constellation of bound transcription factors.

It is more traditional to rewrite our equation for mRNA time evolution in the limit of small Δt in the form of a differential equation as

$$\frac{dm}{dt} = -\gamma m + \sum_i r_i p_i \tag{8.4}$$

where, as above, p_i is the probability of the ith transcriptionally active state, and r_i is the rate of transcription when the system is in that state. The individual p_is are computed based upon the occupancy of various collections of transcription factors in the promoter region of the gene of interest.

The simplest example of this kind of model is the constitutive promoter. In this case, there are only two states: the state in which the promoter is empty and

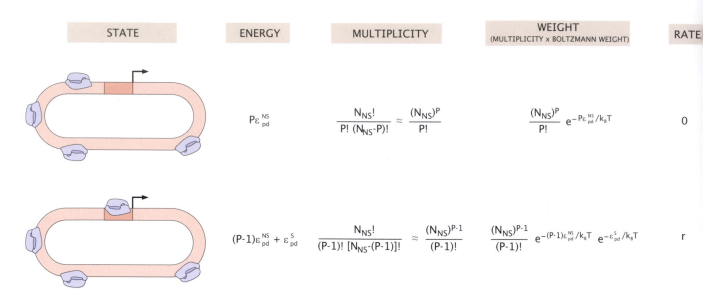

STATE	ENERGY	MULTIPLICITY	WEIGHT (MULTIPLICITY x BOLTZMANN WEIGHT)	RATE
	$P\varepsilon_{pd}^{NS}$	$\dfrac{N_{NS}!}{P!\,(N_{NS}-P)!} \approx \dfrac{(N_{NS})^{P}}{P!}$	$\dfrac{(N_{NS})^{P}}{P!}\, e^{-P\varepsilon_{pd}^{NS}/k_{B}T}$	0
	$(P-1)\varepsilon_{pd}^{NS} + \varepsilon_{pd}^{S}$	$\dfrac{N_{NS}!}{(P-1)!\,[N_{NS}-(P-1)]!} \approx \dfrac{(N_{NS})^{P-1}}{(P-1)!}$	$\dfrac{(N_{NS})^{P-1}}{(P-1)!}\, e^{-(P-1)\varepsilon_{pd}^{NS}/k_{B}T}\, e^{-\varepsilon_{pd}^{S}/k_{B}T}$	r

Figure 8.9

States, weights, and rates for a constitutive promoter. For the constitutive promoter, there are only two states: empty or occupied by RNAP.

thus transcriptionally silent and the state in which the promoter is occupied by RNA polymerase and hence transcriptionally active, with rate r. We can write the dynamical equation for mRNA production in this case as

$$\frac{dm}{dt} = -\gamma m + r p_{bound}. \tag{8.5}$$

The central preoccupation of the thermodynamic class of models is how to determine the quantity p_{bound}, the probability that RNA polymerase is bound to our promoter of interest. But how do we compute this probability using the tools of statistical mechanics?

The details of this kind of analysis have appeared elsewhere (see Further Reading at the end of the chapter), so here we resort to simply conveying an impression of the key points in the analysis. As was already discussed in chapter 2 in developing the statistical mechanical protocol (see Figure 2.1, p. 36), there are several key steps to carrying out a statistical mechanical analysis of some new problem. First, we have to enumerate the different microstates of the system. For the problem of the constitutive promoter, Figure 8.9 shows the two allowed categories of states: polymerase not bound to the promoter and polymerase bound to the promoter. When no polymerase molecule is bound at the promoter of our gene of interest, we assume that all P polymerase molecules are bound nonspecifically to the rest of the genome, as indicated schematically in the figure. By way of contrast, when the promoter is occupied by a polymerase molecule, there are then $P-1$ polymerases bound nonspecifically to the rest of the genome.

As seen in the figure, once we have identified the different states of the promoter of interest, our next task is to assign statistical weights to those separate states, which is a prerequisite to our being able to write the probabilities of

STATE	RENORMALIZED WEIGHT

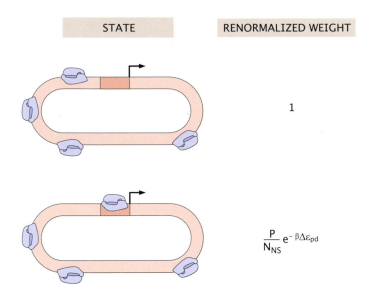

$$1$$

$$\frac{P}{N_{NS}}e^{-\beta\Delta\varepsilon_{pd}}$$

Figure 8.10
Renormalized states and weights for a constitutive promoter.

those states themselves. To compute the statistical weights, we need to evaluate the energies of the different states, as shown in Figure 8.9. In the simplest of models, we imagine that nonspecifically bound polymerases have an energy $\varepsilon_{PD}^{(NS)}$, and a polymerase bound to its promoter has the corresponding energy is $\varepsilon_{PD}^{(S)}$. Boltzmann tells us that the statistical weight of some set of configurations all having the same energy is

$$w_i = g_i(E_i)e^{-\beta E_i}, \tag{8.6}$$

where E_i is the energy of that state, and $g_i(E_i)$ tells us *how many* states there are with that energy. The quantity $g_i(E_i)$ is the multiplicity of the state with energy E_i.

The statistical weights shown in Figure 8.9 can be simplified by realizing that we can always multiply *all* statistical weights by any factor we please. The reason for this is that the probability of the state labeled with index i can always be written in the form

$$p_i = \frac{w_i}{\sum_j w_j}. \tag{8.7}$$

From this equation, we see that we can always multiply all weights by the common factor α and that the resulting probability will be of the form

$$p_i = \frac{\alpha w_i}{\sum_j \alpha w_j} = \frac{w_i}{\sum_j w_j}. \tag{8.8}$$

Hence, we see that by multiplying all statistical weights in Figure 8.9 by $(P!/N_{NS}^P)e^{P\varepsilon_{pd}^{NS}/k_BT}$, we have a much cleaner set of renormalized statistical weights, as shown in Figure 8.10.

Given the statistical weights of the two different states, the probability of the transcriptionally active state is given as the ratio

$$p_{bound} = \frac{\frac{P}{N_{NS}}e^{-\beta\Delta\varepsilon_{PD}}}{1 + \frac{P}{N_{NS}}e^{-\beta\Delta\varepsilon_{PD}}}. \tag{8.9}$$

Figure 8.11

States and weights for simple repression. For the simple repression motif considered here, there are three states of the promoter: empty, occupied by RNA polymerase, occupied by repressor. Only the state occupied by polymerase leads to transcription, as indicated by the rate r in the right column.

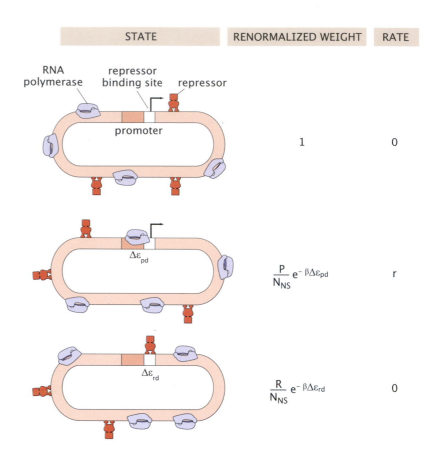

STATE	RENORMALIZED WEIGHT	RATE
	1	0
	$\frac{P}{N_{NS}}e^{-\beta\Delta\varepsilon_{pd}}$	r
	$\frac{R}{N_{NS}}e^{-\beta\Delta\varepsilon_{rd}}$	0

Essentially, this is simply a traditional noncooperative binding curve. As the number of RNAP molecules increases, the probability of binding saturates.

For the case of a promoter regulated by a single transcription factor that represses that gene, the repressor competes for the same region of the promoter as RNAP, and when it is bound, transcription is shut down. The various allowed microstates of the simple repression architecture are shown in Figure 8.11. Using precisely the same reasoning, we can compute the probability that the promoter is occupied as

$$p_{bound} = \frac{\frac{P}{N_{NS}}e^{-\beta\Delta\varepsilon_{PD}}}{1 + \frac{R}{N_{NS}}e^{-\beta\Delta\varepsilon_{RD}} + +\frac{P}{N_{NS}}e^{-\beta\Delta\varepsilon_{PD}}}. \tag{8.10}$$

If we introduce the simplifying notation $p = (P/N_{NS})e^{-\beta\Delta\varepsilon_{PD}}$, and $r = (R/N_{NS})e^{-\beta\Delta\varepsilon_{RD}}$, then we can rewrite our expression for the gene being on as

$$p_{bound} = \frac{p}{1+r+p}. \tag{8.11}$$

From the perspective of experimental measurements on transcription, often it is most convenient to characterize the level of expression by a comparative metric known as the *fold-change*. The idea of the fold-change is to compare the

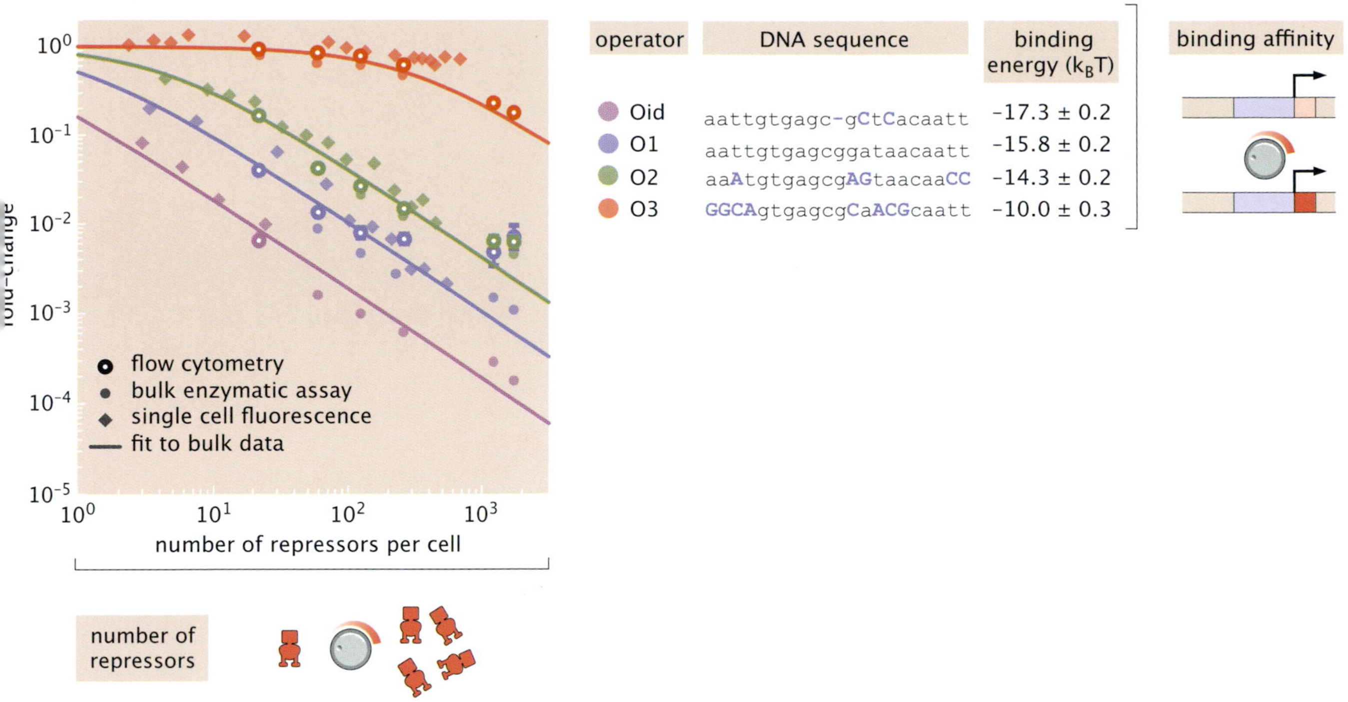

operator
DNA sequence
binding energy (k_BT)
binding affinity
Oid aattgtgagc-gCtCacaatt -17.3 ± 0.2
O1 aattgtgagcggataacaatt -15.8 ± 0.2
O2 aaAtgtgagcgAGtaacaaCC -14.3 ± 0.2
O3 GGCAgtgagcgCaACGcaatt -10.0 ± 0.3
flow cytometry
bulk enzymatic assay
single cell fluorescence
fit to bulk data
number of repressors per cell
number of repressors

Figure 8.12

Fold-change in the simple repression architecture. The four curves correspond to four different operators (binding sites for repressor). Highlighted bases indicate the sequence differences between O1 and all other operators. The fold-change depends upon the number of repressors. Measurements were made using a variety of different methods and readouts. Adapted from Brewster et al. (2014).

amount of expression in the presence of the regulatory machinery with that in its absence. This ratio can be rewritten as

$$\text{fold-change} = \frac{p_{bound}(R \neq 0)}{p_{bound}(R = 0)}. \tag{8.12}$$

For the simple repression architecture considered here, the fold-change simplifies to the form

$$\text{fold-change} = (1 + r)^{-1}, \tag{8.13}$$

where we have invoked the so-called weak-promoter approximation, in which $p \ll 1$. In this case, RNAP binds sufficiently weakly to the promoter that for the majority of the time, the promoter is not occupied by polymerase. What this tells us is that as the number of repressors gets larger (R increases), the fold-change becomes smaller and smaller, as expected since the repressor competes with the RNAP for binding to the promoter region. Detailed experimental exploration of the response of the simple repression architecture is shown in Figure 8.12, where we see that precisely the kind of scaling with repressor number (x-axis) expected from the thermodynamic framework is observed experimentally. A second way to effect differences in the fold-change is by changing the strength of the repressor binding sites, and those kinds of effects are also shown in the figure (different-colored curves).

The real subject of this chapter, however, is how the activity of a given promoter is controlled by some external agent such as an inducer molecule, as shown in Figure 8.6. To that end, we have to figure out what fraction of our transcription factors are active as a function of the inducer concentration, a subject we turn to now.

8.3 Induction of Genes

Though the model described in the previous section tells us that a very useful control parameter is the number of transcription factors, often both in the lab and in the "real world," cells respond not by changing the number of transcription factors but, rather, by changing the number of *active* transcription factors. How does this work?

How do we turn these ideas into a statistical mechanical framework for thinking about induction? Figure 8.13 shows one approach, which is to imagine that our collection of repressors can be separated into two populations, one of which is active and the other of which is "induced" and hence less active. This means we need to generalize our original picture of simple repression to include two distinct classes of repressed state, one corresponding to the active repressor (R_A) binding and the other to the reduced binding that accompanies binding of the inducer to the repressors (R_I). Clearly, we have a constraint on the total number of repressors of the form $R_A + R_I = R_{tot}$.

Using the states and weights shown in Figure 8.13 we can write the probability that the gene is on as usual as the ratio of the statistical weight of the state with polymerase bound to the sum over the statistical weights of all states, resulting in

$$p_{bound} = \frac{\frac{P}{N_{NS}} e^{-\beta \Delta \varepsilon_{PD}}}{1 + \frac{R_A}{N_{NS}} e^{-\beta \Delta \varepsilon_{RD,A}} + \frac{R_I}{N_{NS}} e^{-\beta \Delta \varepsilon_{RD,I}} + \frac{P}{N_{NS}} e^{-\beta \Delta \varepsilon_{PD}}}. \tag{8.14}$$

If we introduce the simplifying notation $p = (P/N_{NS})e^{-\beta \Delta \varepsilon_{PD}}$, $r_A = (R_A/N_{NS})e^{-\beta \Delta \varepsilon_{RD,A}}$, and $r_I = (R_I/N_{NS})e^{-\beta \Delta \varepsilon_{RD,I}}$, then we can rewrite our expression

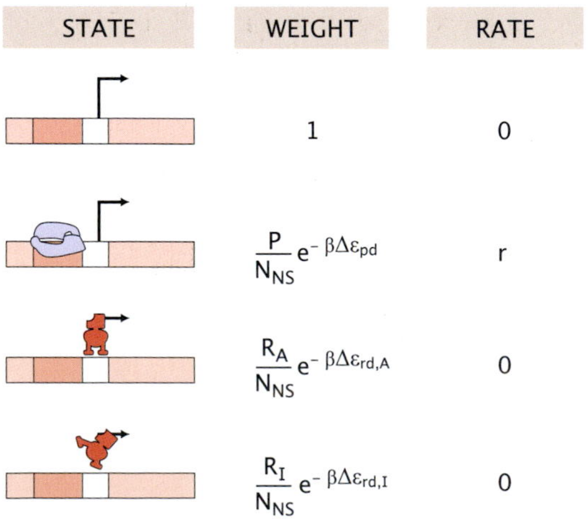

Figure 8.13

States and weights for a dual population of repressors. For the simple repression motif considered here, there are four states of the promoter: empty, occupied by RNA polymerase, occupied by uninduced repressor, and occupied by induced repressor. R_A denotes the number of active repressors, and R_I denotes the number of inactive repressors.

STATE	WEIGHT	RATE
	1	0
	$\frac{P}{N_{NS}} e^{-\beta \Delta \varepsilon_{pd}}$	r
	$\frac{R_A}{N_{NS}} e^{-\beta \Delta \varepsilon_{rd,A}}$	0
	$\frac{R_I}{N_{NS}} e^{-\beta \Delta \varepsilon_{rd,I}}$	0

for the gene being on as

$$p_{bound} = \frac{p}{1 + r_A + r_I + p}.$$ (8.15)

In many instances, as noted in the introduction to this chapter, it is most convenient for the purposes of comparison to experiments to reckon the fold-change, which for this case is given by

$$\text{fold-change} = \frac{1}{1 + r_A + r_I},$$ (8.16)

where we have once again invoked the weak-promoter approximation, $p \ll 1$.

We can simplify our analysis by making the assumption that the inactive repressors have their K_ds increased so much by inducer binding that they no longer bind the operator DNA at all (i.e., $r_I << 1$). In this case, the expression for the fold-change simplifies to

$$\text{fold-change} = \frac{1}{1 + r_A}.$$ (8.17)

This is more easily interpreted if written in terms of the actual concentrations of the constituents, such as the total number of repressors and the concentration of inducer. To that end, we note that we can rewrite the expression as

$$\text{fold-change} = \frac{1}{1 + \frac{p_A(c)[R]}{K}},$$ (8.18)

where we have used the fact that $R_A = p_A(c)R$, with $p_A(c)$ itself defined as the probability of the repressor being in the active state as a function of the concentration c of inducer. We have also introduced the notation $K = (N_{NS}/V)e^{\beta \Delta \varepsilon_{RD,A}}$ as the effective K_d that describes the binding of the active repressor to operator DNA. Now to make further progress, we have to determine how $p_A(c)$ depends upon the concentration of inducer, and this is where the allostery models that are the subject of this book come in.

Although the Lac repressor we will consider is a dimer, we begin by considering a repressor with only one inducer binding site, which means that four states are available to the repressor molecule, as shown in Figure 8.14. Specifically, there are two active states where the molecule can repress and two inactive states where it is assumed to no longer bind the operator DNA and hence cannot repress. The two states for each activity level correspond to the cases in which the inducer is either bound or not.

Given the states and weights shown in Figure 8.14, we follow suit with what we have already done repeatedly throughout the book to compute the activity of the repressor as a function of inducer concentration (c) in the form

$$p_A(c) = \frac{(1 + \frac{c}{K_A})}{(1 + \frac{c}{K_A}) + e^{-\beta \varepsilon}(1 + \frac{c}{K_I})},$$ (8.19)

where, as usual, we introduce distinct dissociation constants K_A and K_I for the active and inactive states, respectively, and we have introduced the notation

Figure 8.14

States and weights for binding of inducer to repressor. The two states on the left are active (i.e., can bind DNA), and the pair of states on the right correspond to inactive repressor.

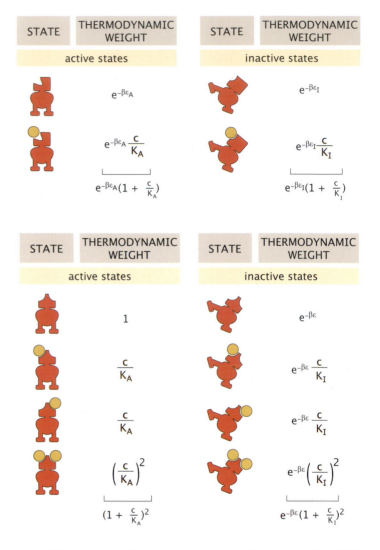

$\varepsilon = \varepsilon_I - \varepsilon_A$. We can now substitute this result back into equation 8.18, which then yields an equation for the fold-change as a function of the concentration of inducer. Specifically, we have

$$\text{fold-change} = \cfrac{1}{1 + \cfrac{(1+\frac{c}{K_A})}{(1+\frac{c}{K_A})+e^{-\beta\varepsilon}(1+\frac{c}{K_I})}\cfrac{[R]}{K}}, \tag{8.20}$$

where we recall that K_A and K_I are dissociation constants that characterize the binding of inducer to the repressor, and K is a dissociation constant that characterizes the binding of the active repressor to DNA.

Many transcription factors are more complicated than the monomeric repressors considered in Figure 8.14. For example, consider a repressor such as the Lac repressor, like that shown in Figure 8.15 that has two binding sites for inducers. In this case, our states and weights need to be expanded to account for the four distinct bound states corresponding to both the inactive and active states of the molecule.

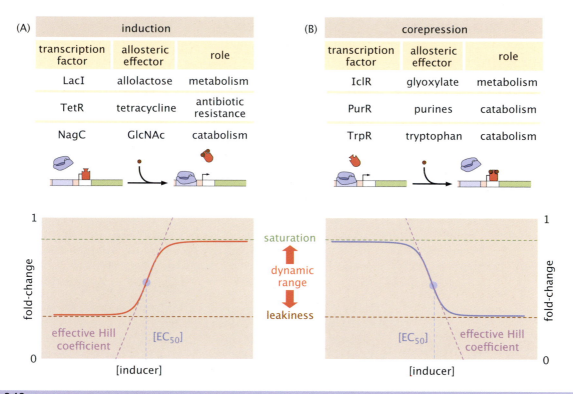

Figure 8.16

Allosteric induction in the simple repression architecture. (A) Examples where inducer causes the repressor to fall off the DNA. The response curve at the bottom showcases the key phenotypic parameters. (B) Examples where inducer stabilizes repressor binding (corepression). Similarly, the response curve at the bottom showcases the key phenotypic parameters for this case.

As has become habitual at this point, given the states and weights, we can write the probability of the active state as

$$p_A(c) = \frac{(1+\frac{c}{K_A})^2}{(1+\frac{c}{K_A})^2 + e^{-\beta\varepsilon}(1+\frac{c}{K_I})^2}. \tag{8.21}$$

As before, we can also use this in the context of equation 8.18 to write the fold-change as

$$\text{fold-change} = \frac{1}{1 + \frac{(1+\frac{c}{K_A})^2}{(1+\frac{c}{K_A})^2 + e^{-\beta\varepsilon}(1+\frac{c}{K_I})^2}\frac{[R]}{K}}, \tag{8.22}$$

where the formula is nearly identical to that provided in equation 8.20 with the difference that the terms involving the inducer concentration are squared, reflecting the presence of two binding sites for inducer.

Of course, although interesting as a statistical mechanical exercise, all these abstractions are really compelling only if they can tell us something about real experimental situations. The story of modern gene regulation owes much to the story of lactose metabolism in *E. coli*, and it serves as a convenient jumping-off point for our analysis of allostery in transcription factors. Figure 8.16 shows examples beyond just that of lactose metabolism in *E. coli*, where we see there

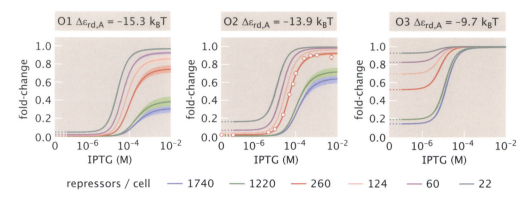

Figure 8.17

Predictions of allosteric response of simple repression motif as a function of inducer concentration. Each figure corresponds to a different repressor-DNA binding energy, shown above the curve. The series of curves correspond to different numbers of repressors per cell. The middle curve shows the one strain used to fit the MWC parameters. Adapted from Razo-Mejia et al. (2018).

are examples of conventional induction but also examples of corepression. As seen elsewhere in the book, the phenotypic parameters highlighted in Figure 8.16 can be computed explicitly in the MWC framework, making it possible to say a priori what values we expect for the leakiness, the dynamic range, the EC_{50}, and the effective Hill coefficient.

A recent study used the same strains used to generate Figure 8.12, which harbor different numbers of repressors and different operator strengths, to explore the validity of the model posited in equation 8.22. The different strains used for the experiments shown in Figure 8.12 have different values of K and $[R]$ in equation 8.22. As such, it is interesting to determine the other three parameters K_A, K_I, and ε using only one such strain, as shown in the middle panel of Figure 8.17. There we see that one induction curve serves to establish the remaining unknown parameters (K_A, K_I, and ε).

Once we have determined the parameters in the MWC model from our single induction curve, as shown in Figure 8.17, the model become strictly predictive, as seen in the various curves that attend the fit curve in the figure. That is, each of those curves is a prediction about what will happen when either the number of repressors per cell is changed or the strength of the repressor binding site on the DNA is changed. The beauty of having strains in which both the number of repressors and the strength of the binding sites have been varied is that it gives us the chance to see the parameter-free performance of the model. Each graph corresponds to a different operator strength, with the collection of curves corresponding in turn to the different repressor counts.

The predictions of Figure 8.17 are put to the test in Figure 8.18. The idea of these experiments is that induction was measured in each of the different strains, resulting in a fold-change as a function of the inducer concentration, as plotted in Figure 8.18. Globally, it is clear that for a series of parameter-free predictions, the agreement between theory and experiment is qualitatively satisfying. At the same time, it is perfectly reasonable to question the validity of the model in characterizing the data. Indeed, it is just such a change of conversation that is my central interest in developing models like this. Instead of having some reasonable qualitative story about the data, we are now faced with the sharper

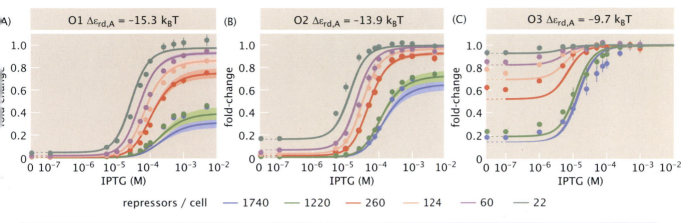

Figure 8.18
Experimental test of predictions of allosteric response of simple repression motif. (A) Fold-change for the O_1 operator for different repressor counts. (B) Fold-change for the O_2 operator for different repressor counts. (C) Fold-change for the O_3 operator for different repressor counts. Adapted from Razo-Mejia et al. (2018).

question of deciding whether nuances such as those seen in the context of the O_3 operator, where the blue and green sets of data points exhibit a systematic trend that is different from the theory, demand further reflection.

Though the predictions for the induction curves themselves provide a window onto the merits of the model, it is also interesting to consider the various phenotypic parameters that characterize the input-output curves, as shown in Figure 8.16. As demonstrated in earlier chapters, there are simple analytic formulas for some of these phenotypic parameters. Figure 8.19 shows predictions and corresponding measurements for the leakiness, the saturation, and the dynamic range as a function of the number of repressors per cell for three different operator strengths. Though ultimately these phenotypic parameters are implicitly present in the dose-response curves of Figure 8.18, recasting them in the language of the biophysical parameters such as leakiness, EC_{50}, and so on, holds up the allostery phenomenon to a different light and may provide a more transparent connection to the way physiological and evolutionary adaptation are achieved.

To make contact with the ideas on data collapse already considered on several occasions (see sec. 3.5, p. 94), we recall that we can rewrite the fold-change in the form

$$\text{fold-change} = \frac{1}{1 + e^{-\beta F(c)}}, \tag{8.23}$$

where by virtue of equation 8.22, $F(c)$ is defined as

$$F(c) = -k_B T (\log \frac{[R]}{K} + \log p_A). \tag{8.24}$$

We can rewrite this as

$$F(c) = -k_B T (\log \frac{[R]}{K} + \log \frac{(1 + \frac{c}{K_A})^2}{(1 + \frac{c}{K_A})^2 + e^{-\beta \varepsilon}(1 + \frac{c}{K_I})^2}), \tag{8.25}$$

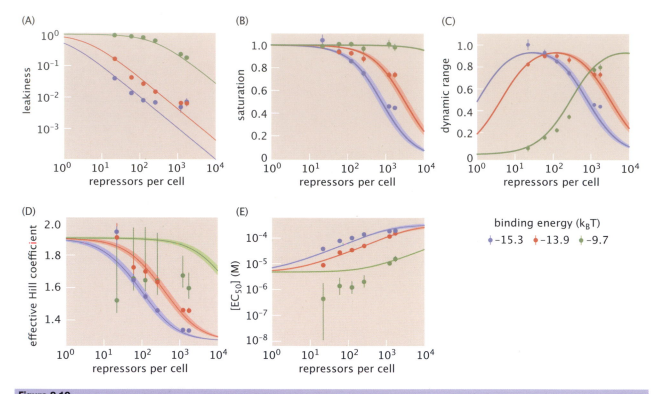

Figure 8.19

Experimental test of predictions of allosteric response for phenotypic parameters for different operator strengths and repressor copy numbers. (A) Predicted and measured leakiness. (B) Predicted and measured saturation. (C) Predicted and measured dynamic range. (D) Predicted and measured values of effective Hill coefficient. (E) Predicted and measured values of EC_{50}. Adapted from Razo-Mejia et al. (2018).

which describes the free-energy difference between the inactive and active states at concentration c. Using this free-energy difference as the natural variable that the full repression system "cares about," we see in Figure 8.20 that the entirety of the data plotted in Figure 8.18 falls onto one master curve. We argue that this data collapse is a powerful indication of a high-level understanding of induction in the regulatory setting of the original repressor-operator model of Jacob and Monod. In the next section, we expand our horizons to consider induction in the context of gene activation.

8.4 Activation

There is more to transcriptional regulation than repression. The original formulation of the repressor-operator model of Jacob and Monod asserted that regulation was inhibitory. Many of the experiments this idea of regulation by repression was based on took place in the context of lactose metabolism. As noted earlier in the chapter, in a delightful twist of history, the very same example of lactose metabolism considered by Jacob and Monod offered up overwhelming proof that genes are not only repressed but activated as well. As

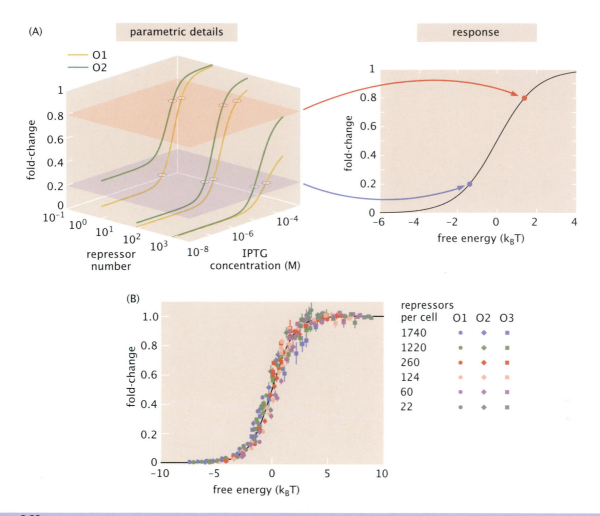

Figure 8.20

Data collapse of induction data. (A) Planes of constant fold-change show that there are many different points in parameter space with the same fold-change. (B) The data from all 18 strains can all be collapsed onto the same master curve when plotted using the free-energy difference between the active and inactive states as given by equation 8.25 as the "natural variable" of the system. Adapted from Razo-Mejia et al. (2018).

shown in Figure 8.4 (p. 275), the process of transcription can be tuned down by the binding of repressors (shown in red in that figure) and can be tuned up by the binding of activators (shown in green in that figure). In this section, we explore how the allostery idea can be used to understand transcriptional activation in the context of CRP, a broadly acting transcription factor in bacteria that is central to lactose metabolism. The role of CRP in the *lac* operon is shown in more detail in Figure 8.5 (p. 276).

8.4.1 Binding of Inducer to Activator

We begin by examining the probability that the activator will be active as a function of the concentration of its effector, cAMP. To that end, as usual, we set up

Figure 8.21

States and weights for transcriptional activation. The activator has two binding sites for its effector, resulting in four states of occupancy for both the active (left) and inactive (right) states of the molecule.

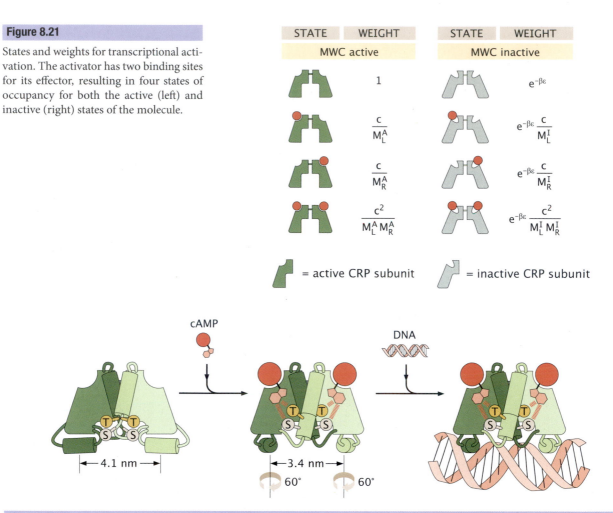

STATE	WEIGHT	STATE	WEIGHT
MWC active		MWC inactive	
	1		$e^{-\beta\varepsilon}$
	$\dfrac{c}{M_L^A}$		$e^{-\beta\varepsilon}\dfrac{c}{M_L^I}$
	$\dfrac{c}{M_R^A}$		$e^{-\beta\varepsilon}\dfrac{c}{M_R^I}$
	$\dfrac{c^2}{M_L^A M_R^A}$		$e^{-\beta\varepsilon}\dfrac{c^2}{M_L^I M_R^I}$

= active CRP subunit   = inactive CRP subunit

Figure 8.22

Illustration of how the allosteric transition works in the context of the transcriptional activator CRP. As a result of the conformational change, the α-helixes in the DNA-binding domain undergo large-scale rotations, as shown in the middle panel. Once these rotations have taken place, the activator is poised to bind the DNA, as shown in the right-hand panel. Adapted from Popovych et al. (2009).

the states and weights as shown in Figure 8.21. As in the Lac repressor example of Figure 8.15 (p. 286), our activator has two binding sites for its effector. As usual, we have a different dissociation constant for the active and inactive states, as shown in the figure, and we attribute an energy difference ε between the inactive and active states in the absence of effector. However, unlike in our earlier examples with repression, here we anticipate experiments on heterogeneous dimer mutants of CRP and allow for different dissociation constants for the "left" and "right" subunits, labeled M_L and M_R, respectively. With these states and weights in hand, we can compute a variety of statistical quantities of interest, including the probability that the transcription factor is active, and the fractional occupancy of CRP by cAMP, as will be highlighted further.

A higher-resolution view of the structural consequences of cAMP binding is shown in Figure 8.22. Upon binding of cAMP (shown here with more structural

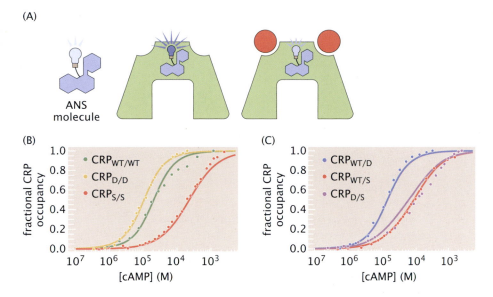

(A)

ANS molecule

(B)

fractional CRP occupancy

- CRP$_{WT/WT}$
- CRP$_{D/D}$
- CRP$_{S/S}$

[cAMP] (M)

(C)

fractional CRP occupancy

- CRP$_{WT/D}$
- CRP$_{WT/S}$
- CRP$_{D/S}$

[cAMP] (M)

Figure 8.23

CRP occupancy by cAMP. (A) Measurement of occupancy of CRP by cAMP by measuring the fluorescence of ANS. When ANS is bound to CRP, its fluorescence increases. When cAMP binds to CRP it ejects the ANS molecule, and fluorescence is reduced. Fraction bound is inferred from the fluorescence data. (B) Occupancy for wild-type and symmetric mutant versions of CRP. The curves are the results of fitting the MWC parameters. (C) Predicted and measured fractional occupancy curves for asymmetric mutants. Adapted from Einav, Duque, and Phillips (2018). See this work for the precise parameter values used for the wild-type and mutant activators. Experimental data from Lanfranco, et al. (2017).

detail than usual thus far in the book), two primary amino acid residues interact with the cAMP ligand, resulting in a large scale reorientation of the α-helices responsible for DNA binding, as shown in the middle panel of the figure. When DNA is present, those α-helices nestle into the major groove of the DNA double helix. This is an especially pleasing example in which the cartoons used in our statistical mechanical states-and-weights diagram are directly inspired by deep structural insights into the molecular actions that attend allosteric conformational change.

Recent in vitro biochemical experiments examined the occupancy of the CRP molecule as a function of the concentration of cAMP, measuring a key probabilistic quantity directly related to the states and weights presented in Figure 8.21. The concept of the experiment is introduced in Figure 8.23(A). As seen in the figure, a reporter molecule known as ANS (8-anilinonaphthalene-1-sulfonic acid) is used to read out the state of occupancy of the CRP. In particular, when the ANS molecule is bound to CRP, its fluorescence is increased. When cAMP binds to CRP the ANS molecule is less likely to bind CRP. The fluorescence as a function of cAMP concentration is used to infer the occupancy. As a result, as shown in Figure 8.23(B) and (C), the occupancy of CRP by cAMP can be measured as a function of cAMP concentration.

As a result of these experiments, we are in the unusual and pedagogically useful position of being able to disentangle the separate effects of binding of cAMP from DNA binding, which the CRP activator ultimately does (primarily when in the active state), to upregulate gene expression. To compare these experimental results, we have to use our states and weights to compute the fractional occupancy. Note that this probabilistic result is *different* from our usual probabilistic question, which is the probability of the molecule being active. The average number of cAMP binding sites that are occupied, related to the fractional occupancy of the CRP, is given as

$$\langle n_{bound} \rangle = \sum_{n=0}^{2} np(n), \tag{8.26}$$

where $p(n)$ is a shorthand notation for the probability of n ligands being bound. The value $p(0)$ is given by the sum of the weights in the first row of Figure 8.21, $p(1)$ is given by the sum of the weights in the second and third rows of Figure 8.21, and $p(2)$ is given by the sum of the weights in the fourth row of Figure 8.21. Implementing these injunctions using our states and weights we find

$$\langle n_{bound} \rangle = \frac{\left(\frac{c}{M_L^A} + \frac{c}{M_R^A}\right) + 2\frac{c}{M_L^A}\frac{c}{M_R^A} + e^{-\beta\varepsilon}\left(\frac{c}{M_L^I} + \frac{c}{M_R^I}\right) + 2e^{-\beta\varepsilon}\frac{c}{M_L^I}\frac{c}{M_R^I}}{\left(1 + \frac{c}{M_L^A}\right)\left(1 + \frac{c}{M_R^A}\right) + e^{-\beta\varepsilon}\left(1 + \frac{c}{M_L^I}\right)\left(1 + \frac{c}{M_R^I}\right)}.$$

$$(8.27)$$

The denominator of this expression is the partition function and amounts to simply summing over *all* the states considered in Figure 8.21.

Figure 8.23 shows a comparison between the results of measuring the fractional occupancy of the CRP molecule by cAMP and the theoretical results emerging from the MWC model and captured in equation 8.27. Note that the fractional occupancy is defined as $\langle n_{bound} \rangle / 2$, since the average number of bound cAMP molecules is between 0 and 2, and the fractional occupancy is normalized between 0 and 1. As seen in the figure, not content to examine only the properties of the wild-type protein, researchers performed experiments for a series of mutants, as shown in Figure 8.24. The mutant versions of the CRP protein shown in Figure 8.24 included a series of mutant homodimers (symmetric mutants) and then a corresponding set of mutant heterodimers (asymmetric mutants). One of the interesting questions raised by these mutants is, which subset of the MWC parameters will change owing to the mutation? Throughout the book, we have asserted the fascinating and complex interplay of physiological and evolutionary adaptation. One of our repeated arguments is that the phenotypic parameters such as leakiness, dynamic range, EC_{50}, and the effective Hill coefficient are the molecular features related to function, and hence targets for sculpting by evolution. Studies like those presented in Figure 8.23 give us the opportunity to see how mutations are tied to these phenotypic properties and to the underlying microscopic parameters in the statistical mechanical model.

The fractional occupancy of the mutant molecules by cAMP as a function of the cAMP concentration are shown in Figure 8.23. To compute these fractional occupancies, we resort to our equation 8.27 but with the recognition that we will have to alter some of the parameters in the equation in correspondence with the schematic of Figure 8.24, where we see that different mutants can alter different subsets of the MWC parameters and their corresponding biophysical phenotypes.

8.4.2   Binding of Activator to DNA

One of the beautiful aspects of the in vitro experiments considered here was that they were able to disentangle the effects of binding of cAMP to CRP and the binding of CRP to DNA. The ANS fluorescence reported in Figure 8.23

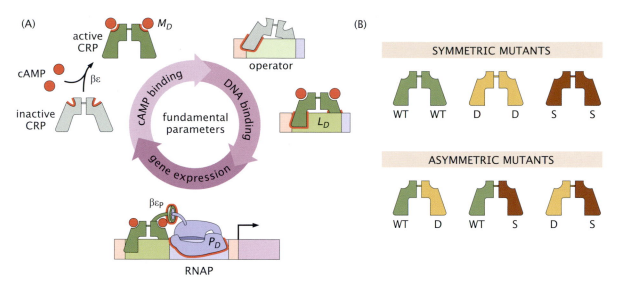

Figure 8.24

Parameters used in the treatment of activation by CRP. (A) The colored areas on molecules show the regions where mutations in the protein have been characterized quantitatively. (B) The collection of symmetric and asymmetric units examined in biophysical experiments of CRP.

determined the occupancy of CRP by cAMP. A second experiment, which works conceptually as seen in Figure 8.25(A), was used to see how DNA binding depends upon the concentration of CRP. The idea of the experiment is that labeled DNA fragments will tumble around in solution at a lower frequency when CRP is bound to DNA, and the tumbling rate reveals itself in the anisotropy of fluorescence. This anisotropy in fluorescence, was used to measure the probability of DNA binding as a function of CRP concentration, as shown in Figure 8.25(B). As seen in the figure, clearly the binding of the CRP to the DNA is strongly dependent upon whether there is cAMP present. In the absence of cAMP, the binding of CRP to the DNA is attenuated. By way of contrast, in the presence of 1 mM cAMP, the CRP molecules have a much higher probability of binding to the DNA.

For the purposes of comparing these experimental data with the results of statistical mechanical modeling of transcription factor activity, we have to compute the probability of DNA binding. To that end, we repeat our standard mantra and write the states and weights, as shown in Figure 8.26. The logic simply generalizes the states and weights already used in Figure 8.21 by noting that for every inactive and active state of the CRP molecule, that state can either be bound to DNA or not. We use the parameter A to characterize the copy number of the activators. There are now more parameters, since we have to account for the dissociation constant for CRP binding to DNA in the active (L_A) and inactive (L_I) states, and we adopt a strategy in which each of the two monomers of the activator has its own L_A or L_I. The reason for this notational complexity is that it now allows us to consider asymmetric mutants like those shown in Figure 8.24, with each of the asymmetric subunits having a distinct L_A.

Figure 8.25

Measurement of DNA binding as a function of concentration of CRP for different concentrations of cAMP. (A) Fluorescence anisotropy varies depending upon whether CRP is bound to DNA or not. The 32 bp fluorescein-labeled DNA molecule with a fragment of the *lac* promoter to bind CRP tumbles at a lower frequency when bound to CRP. (B) The measurements are compared with the results from the MWC model. Adapted from Einav, Duque, and Phillips (2018). Experimental data from Lanfranco et al. (2017).

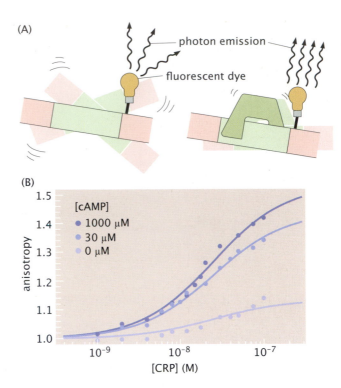

The price we have to pay for this complexity is the introduction of a parameter C_0 that has dimensions of concentration to balance the L_A^2 appearing in the statistical weights.

Given these states and weights, the probability that CRP will be bound to DNA is given by

$$p_{\text{bound}}^{\text{DNA}} = \frac{\frac{AC_0}{L_A^2}\left(1 + \frac{c}{M_R^A}\right)\left(1 + \frac{c}{M_L^A}\right) + e^{-\beta\varepsilon}\frac{AC_0}{L_I^2}\left(1 + \frac{c}{M_R^I}\right)\left(1 + \frac{c}{M_L^I}\right)}{\left(1 + \frac{AC_0}{L_A^2}\right)\left(1 + \frac{c}{M_R^A}\right)\left(1 + \frac{c}{M_L^A}\right) + e^{-\beta\varepsilon}\left(1 + \frac{AC_0}{L_I^2}\right)\left(1 + \frac{c}{M_R^I}\right)\left(1 + \frac{c}{M_L^I}\right)},$$

(8.28)

where c is the concentration of cAMP, and A is the concentration of CRP itself. Note that the numerator of this expression is the sum of the eight states at the bottom of Figure 8.26, while the denominator is the partition function and results from summing over all the states in Figure 8.26. Although this expression looks cumbersome, that is simply an inheritance of the large number of distinct states present in Figure 8.26. The relation between the probability of DNA binding and fluorescence anisotropy requires one more calculational step, because the measurement doesn't report the occupancy itself but, rather, the anisotropy. That next step in the calculation is beyond the scope of the present discussion, and the reader is invited to review the original work for that argument (see the paper by Duque et al. in Further Reading). The comparison between the measured and calculated values is consistent with the MWC model, as seen in Figure 8.25.

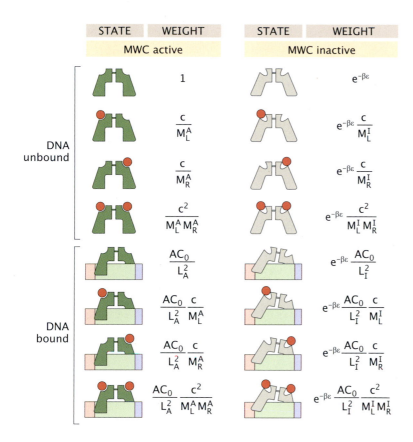

Figure 8.26
States and weights for activator binding to cAMP at concentration c and DNA. The activator can be in either active or inactive states, and the relative probabilities of those two classes of states are dictated by the concentration of effector (cAMP).

8.4.3 Activation and Gene Expression

So far, we have emphasized the use of the MWC framework for activation in the context of a series of in vitro biochemical experiments. But ultimately, transcriptional activators are part of the machinery of living cells and adjust the rate of gene expression in response to changes in the environment. We now turn to the question of how to use the model just developed to think about experiments in living cells.

As we have done throughout the chapter, we assume that gene expression is equal to the product of the RNAP transcription rate r and the probability that RNAP is bound to the promoter of interest. To compute the probability of RNAP binding, we resort to states and weights, this time as illustrated in Figure 8.27. The logic is entirely analogous to that carried out in the case of repressor earlier in the chapter. In particular, we divide the population of activators into two families, those that are active (A_A) and those that are inactive (A_I). Both families are illustrated in the figure. The probability that RNAP is bound is given by summing the statistical weights of the three states shown in the figure that have an RNAP present and dividing by the sum over all weights. The result for the level of transcription is then given by

$$\text{activity} = r\,\frac{\frac{P}{P_D}\left(1 + \frac{A_I}{L_I}e^{-\beta\varepsilon_{P,L_I}} + \frac{A_A}{L_A}e^{-\beta\varepsilon_{P,L_A}}\right)}{\frac{P}{P_D}\left(1 + \frac{A_I}{L_I}e^{-\beta\varepsilon_{P,L_I}} + \frac{A_A}{L_A}e^{-\beta\varepsilon_{P,L_A}}\right) + 1 + \frac{A_I}{L_I} + \frac{A_A}{L_A}}. \tag{8.29}$$

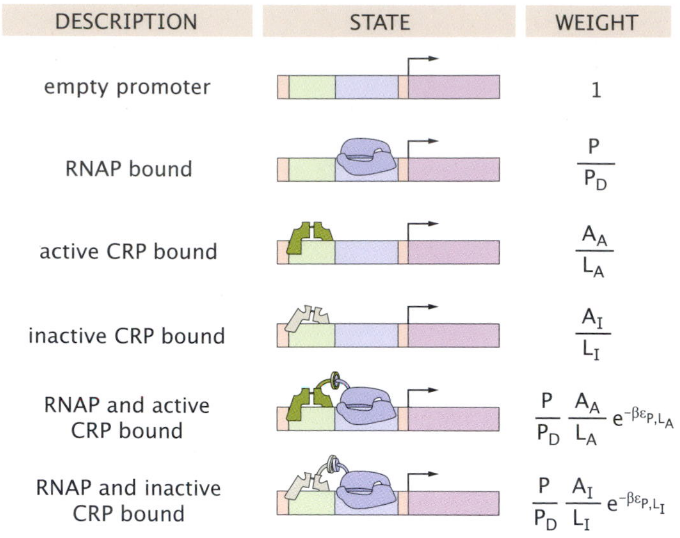

Figure 8.27

States and weights for a simple activation motif. Binding of RNAP (blue) to a promoter is facilitated by the binding of the activator CRP. Simultaneous binding of RNAP and CRP is facilitated by an interaction energy ε_{P,L_A} for active CRP (dark green) and ε_{P,L_I} for inactive CRP (light green). cAMP (not drawn) influences the concentration of active and inactive CRP.

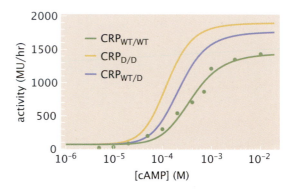

DESCRIPTION	STATE	WEIGHT
empty promoter		1
RNAP bound		$\dfrac{P}{P_D}$
active CRP bound		$\dfrac{A_A}{L_A}$
inactive CRP bound		$\dfrac{A_I}{L_I}$
RNAP and active CRP bound		$\dfrac{P}{P_D}\dfrac{A_A}{L_A}e^{-\beta\varepsilon_{P,L_A}}$
RNAP and inactive CRP bound		$\dfrac{P}{P_D}\dfrac{A_I}{L_I}e^{-\beta\varepsilon_{P,L_I}}$

Figure 8.28

Gene expression as a function of inducer concentration. Adapted from Einav, Duque, and Phillips (2018). Experimental measurements on gene expression as a function of cAMP concentration are adapted from Kuhlman et al. (2007).

Note that in this equation we have simplified the notation for CRP binding to DNA relative to Figure 8.26 by using a single dissociation constant (L_A or L_I) for a CRP molecule bound to DNA without distinguishing the two subunits.

As shown in Figure 8.28, the correspondence between the model predictions and the measured wild-type transcriptional activity as a function of cAMP concentration are quite satisfactory, though there are several subtleties that the reader is cautioned about, such as how one controls the *intracellular* concentration of cAMP. The reader is urged to review the original literature to appreciate these subtleties. Because we have gone to the trouble of determining the parameters for the various mutant CRP molecules, we can also make predictive statements about how we think the level of gene expression will depend upon cAMP concentration for these mutants, as also shown in the figure. Experiments to test these predictions, which have been only partially conducted thus far, would bring exciting closure to this story.

The key point of this exercise in activation was to show how the MWC framework can be used to try to understand a suite of different experiments, several in vitro biochemical assays that measure occupancy of activator by its effector and the probability of binding DNA as a function of the concentration of activator, and others that directly measure the level of gene expression due to activation in living cells. One of the philosophical pillars of that effort is that we demanded a single, minimal parameter set that could be used across the spectrum of all these different experiments and could be used to make predictions about future experiments.

This section on activation has been based thus far on a strict interpretation of the MWC model. However, as usual, we would like to adopt a more circumspect and critical attitude about the applicability of the model. One of the intellectual battles that enlivened the development of the allostery concept was the kinetic question of what sequence of structural events unfolds as a protein transitions from one state of activity to another. As we saw in Figure 1.28 (p. 29), one alternative framework to the all-or-nothing of the MWC model is the KNF model, which imagines sequential transitions as more ligands bind to a multivalent target. In fact, the KNF model does nearly as good a job of explaining the in vitro data as does the MWC model.

8.5   Janus Factors

Thus far, our discussion has focused primarily on bacterial transcription. But the idea of MWC molecules in the context of transcription has much broader reach. Just as in the bacterial case, there are many eukaryotic transcription factors that respond to environmental signals. Interestingly, some of these transcription factors have a Janus-like existence, in reference to the two-faced Roman deity, sometimes masquerading as repressors and at other times as activators, depending upon the presence or absence of some signaling molecule which causes changes in the activity of these factors. As already fleshed out in our discussion of the Lac repressor, the binding of a ligand to some transcription factor can tilt the free-energy balance in favor of either the active or inactive conformation, thus switching the protein between its inactive and active conformations. However, the conventional interpretation of allostery is too limiting to account for the diverse ways in which the activity of transcription factors is changed.

Figure 8.29 provides a gallery of several examples of signaling pathways that culminate in changes in transcriptional activity as a result of some change in the state of activity of the relevant transcription factors. The Notch-Delta signaling pathway features signaling between cells such that when the Notch receptor on one cell binds its Delta ligand on another cell, cleavage of the intracellular Notch domain allows it to translocate to the nucleus, where it then imposes transcriptional control as an activator, as indicated schematically in Figure 8.29(A).

The second example of a Janus transcription factor shown in Figure 8.29(B) is Gli3, a transcription factor in the Sonic Hedgehog pathway that functions both as an activator and as a repressor. Balance between the activator and repressor versions of this protein are thought to specify the identity and number of limb digits. The full-length transcription factor behaves as an activator once it has been phosphorylated and translocated to the nucleus, whereas the truncated form acts as a repressor. As we will explore in more detail in chapter 11, it is useful to entertain definitions of allostery that go beyond the simplest

Figure 8.29

Changes in activity of transcription factors due to signaling. (A) Notch-Delta signaling pathway and irreversible modification of activity of transcription factor. (B) Sonic Hedgehog signaling and repurposing of transcription factors as both activators and repressors.

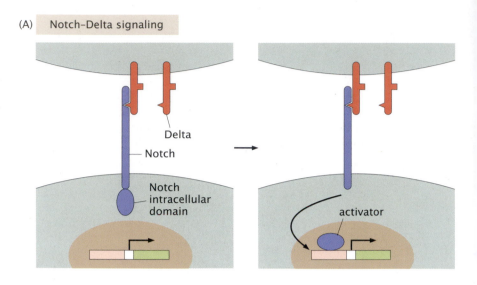

(A) Notch–Delta signaling

Delta
Notch
Notch intracellular domain
activator

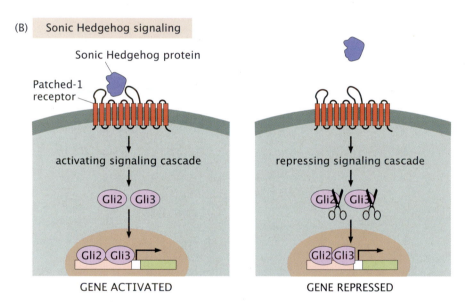

(B) Sonic Hedgehog signaling

Sonic Hedgehog protein
Patched-1 receptor
activating signaling cascade
Gli2 Gli3
Gli2 Gli3
GENE ACTIVATED

repressing signaling cascade
Gli2 Gli3
Gli2 Gli3
GENE REPRESSED

ligand-induced conformational changes that are the main focus of the book. Janus transcription factors are a fertile arena for continued efforts to build quantitative generalizations of the original MWC, KNF, and Eigen formulations of the statistical mechanics of proteins with more than one state of activity.

8.6 Summary

Transcription factors are undoubtedly one of the most important classes of allosteric molecules known in biology. The use of ligands of different types to alter the activity of transcription factors transcends any one class of organisms and is ubiquitous from Archaea to Eukaryotes. This chapter began on a historical note by considering the repressor-operator model of Jacob and Monod in the sense that we use the case study of simple repression as a tool to dig into all the details, both theoretical and experimental of what it looks like to design both approaches for a maximally constructive dialogue. That dialogue showed us how the statistical physics of allostery could be used to understand not only dose-response curves, but also key biophysical parameters such as leakiness, dynamic range and the EC_{50} in a predictive fashion. With the simple repression motif in hand, we then turned to an analysis of the case of activation with special emphasis on the paradigmatic example of the broadly acting transcriptional activator CRP. This chapter leaves the reader with a framework for exploring other problems in transcriptional regulation.

8.7   Further Reading

Alon, U. (2006) *An Introduction to Systems Biology*. Boca Raton, FL: Chapman & Hall/CRC Press. Alon's book gives a comprehensive and thoughtful discussion of regulation.

Beckwith, J. (1996) "The Operon: An Historical Account." In *Escherichia coli and Salmonella: Cellular and Molecular Biology*, 2nd ed., edited by F. Neidhardt et al. Washington, DC: ASM Press. (2011) "The operon as paradigm: Normal science and the beginning of biological complexity." *J. Mol. Biol.* 409 (1): 7–13. Beckwith's articles give a deep perspective on the origins of our thinking on gene regulation.

Bialek, W. S. (2013) *Biophysics: Searching for Principles*. Princeton: NJ: Princeton University Press. Bialek has many important things to say about how to think about gene expression as a predictive science.

Buchler, N. E., U. Gerland, and T. Hwa (2003) "On schemes of combinatorial transcription logic. *Proc. Natl. Acad. Sci.* 100: 5136. Excellent general discussion of thermodynamic models of gene regulation.

Echols, H. (2001) *Operators and Promoters*. Oakland, CA: University of California Press. Echols's book is a profoundly interesting and unique description of the history of much of molecular biology. In the context of the present chapter, it has a number of useful insights into the use of phage and *E. coli* as model systems, and describes topics such as gene regulation masterfully.

Judson, H. F. (1996) *The Eighth Day of Creation*. Cold Spring Harbor, NY: Cold Spring Harbor Laboratory Press. Judson's book, like that of Echols, recounts the history of the development of molecular biology and describes many of the experiments and advances discussed in the present chapter. This book is fascinating.

Müller-Hill, B. (1996) *The Lac Operon*. Berlin: Walter de Gruyter. Müller-Hill's book is a fascinating and idiosyncratic account of the development of thinking on gene regulation in general and the *lac* operon in particular. The book is full of interesting touches, such as Figure 3, which illustrates the ways in which synthetic analogues of lactose have played a role in the development of molecular biology.

Ptashne, M. (2004) *A Genetic Switch*. 3rd ed. Cold Spring Harbor, NY: Cold Spring Harbor Laboratory Press. A beautiful book that focuses on ideas as opposed to facts and paints a picture of how gene regulation works.

Ptashne, M. and A. Gann (2002) *Genes and Signals*. Cold Spring Harbor, NY: Cold Spring Harbor Laboratory Press. This book provides an excellent overview of transcriptional regulation.

Vilar, J and S. Leibler (2003) "DNA looping and physical constraints on transcriptional regulation." *J. Mol. Biol.* 331: 981–989. This paper gives a thorough thermodynamic treatment of the different facets of the *lac* operon.

8.8   REFERENCES

Beckwith, J. (2011) "The operon as paradigm: Normal science and the beginning of biological complexity." *J. Mol. Biol.* 409 (1): 7–13.

Brewster, R. C., F. M. Weinert, H. G. Garcia, D. Song, M. Rydenfelt, and R. Phillips (2014) "The transcription factor titration effect dictates level of gene expression." *Cell* 156: 1312–1323.

Einav, T., J. Duque, and R. Phillips, (2018) "Theoretical analysis of inducer and operator binding for cyclic-AMP receptor protein mutants." *PLoS ONE* 13 (9): e0204275.

Kuhlman, T., Z. Zhang, M. H. Saier, and T. Hwa (2007) "Combinatorial transcriptional control of the lactose operon of *Escherichia coli.*" *Proc. Natl. Acad. Sci.* 104 (14): 6043–6048.

Lanfranco, M. F., F. Gárate, A. J. Engdahl, and R. A. Maillard (2017) "Asymmetric configurations in a reengineered homodimer reveal multiple subunit communication pathways in protein allostery." *J. Biol. Chem.* 292:6086.

Popovych, N., S.-R. Shiou-Ru Tzeng, M. Tonelli, H. Richard, R. H. Ebright, and C. G. Kalodimos (2009) "Structural basis for cAMP-mediated allosteric control of the catabolite activator protein." *Proc. Natl. Acad. Sci.*, 106:6927.

Razo-Mejia M., S. L. Barnes, N. M. Belliveau, G. Chure, T. Einav, M. Lewis, and R. Phillips (2018) "Tuning transcriptional regulation through signaling: A predictive theory of allosteric induction." *Cell Syst.* 25:456–469.

BUILDING LOGIC FROM ALLOSTERY

9

It is particularly incumbent on those who never change their opinion, to be secure of judging properly at first.

—Jane Austen

Both cellular signaling and transcriptional regulation depend upon combinatorial control, in which multiple inputs determine the "choice" made by the signaling or regulatory circuit. In this chapter, we now ask ourselves whether the allosteric framework presented thus far provides a basis for constructing molecular input-output functions that mimic the classic two-input logic functions familiar from digital electronics. Specifically, in imitation of the interesting work on transcriptional regulatory logic in which combinatorial control has been thoroughly analyzed (see the paper by Buchler et al. in Further Reading), it is interesting to see how well individual molecules achieve different Boolean logic functions through binding to multiple distinct regulatory ligands. In Figure 1.9 (p. 11), we saw that in the context of actin polymerization in motile cells, multiple inputs are needed to activate the polymerization process. Similarly, section 5.4 (p. 192) showed how some ion channels are controlled by three distinct inputs. To understand the general principles of how individual molecules can lead to logic gates, in this chapter we undertake an analysis using the MWC model to explore the signaling response of allosteric molecules. We will examine the conditions under which such molecules can give rise to AND, OR, NAND, and NOR responses (see Figure 9.1 for the definitions of these logic gates). Interestingly, in the absence of explicit molecular cooperativity, the MWC model appears to be incapable of generating the XOR or XNOR logic functions.

9.1 Combinatorial Control and Logic Gates

A hallmark of cellular signaling and regulation is combinatorial control. Whether we are thinking about the enzymes of the glycolysis pathway or the transcription factors that control transcriptional networks, multiple inputs into the same pathway often give rise to a much richer response than can be achieved through single-input functions. That is, the response of a given signaling or regulatory architecture depends on the presence of more than one molecular

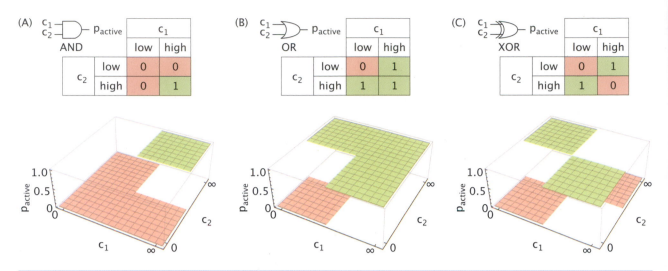

Figure 9.1

Logic gates and their target molecular responses. Different logic gates represented using Boolean logic and as the activity of an allosteric molecule (e.g., the probability an enzyme is active). (A) The AND gate. (B) The OR gate. (C) The XOR gate. In each case, the schematic beneath the logic table shows the target molecular response, indicated by the activity of the logic molecule as a function of the concentrations of the two input ligands.

species. For example, in the glycolysis pathway as shown in Figure 1.6 (p. 8), the enzyme phosphofructokinase has multiple-input ligands (see Figure 1.7, p. 10) for the array of different molecules that inhibit and activate this enzyme alone, some of which enhance enzyme activity and some of which inhibit it. In the context of transcriptional regulation, elegant earlier work (see the paper by Buchler et al. in Further Reading) explored the conditions under which transcriptional regulatory networks can give rise to the familiar Boolean logic operations like those shown in Figure 9.1. In the transcriptional setting, it was found that if we consider the example of two distinct transcription factors, their combined effect on the transcriptional activity of a given promoter as discussed in chapter 8 depends on their respective binding strengths, as well as cooperative interactions between each other and RNA polymerase. Indeed, it was found that by tuning the binding strengths and cooperativity parameters, one can generate a panoply of logic gates, such as the familiar AND, OR, NAND, and NOR gates known from the world of digital electronics.

We now ask whether a single molecule can yield combinatorial control in the same way in which transcriptional networks have already been shown to. Specifically, the question we address is the extent to which an allosteric molecule described by the MWC model can deliver input-output functions similar to the ideal logic gates described in Figure 9.1. To address that question, we must first determine how the input-output response function for such a molecule depends on the concentration of each regulatory ligand. For example in Figure 9.1(A), we see that there are four input conditions for the concentrations c_1 and c_2 of the two input molecules (i.e., low/low, low/high, high/low, high/high). Further, there are two output possibilities for each input set. Hence, the total number of output conditions is $2 \times 2 \times 2 \times 2 = 16$, as shown in Figure 9.2. Of course, this suite of possible logics includes familiar ones such as the AND, NAND, OR,

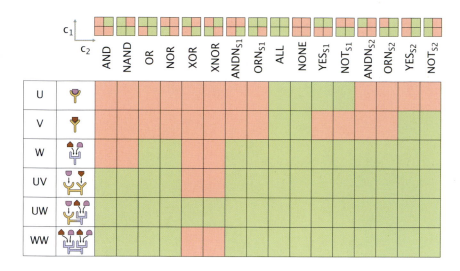

Figure 9.2

The 16 logic functions of a two-input system. Each column corresponds to one particular logical function. The rows correspond to different receptor types. For the squares above each logic's label, the red color signifies that the switch is in the "off" state, and the green color signifies that the switch is in the "on" state for that particular combination of c_1 and c_2. Adapted from de Ronde, ten Wolde, and Mugler (2012); see Further Reading.

and NOR functions. But it also indicates others that I had not thought about before such as the "trivial" NONE logic, in which all inputs lead to the same nonresponse. Nevertheless, it is a logic. The figure also shows cases in which one of the ligands overrides the other, such as the YES_{S1}, which is 0 when ligand 1 is absent, and 1 when ligand 1 is present.

As just noted, Figure 9.2 shows the entirety of the logical responses that can be achieved by a two-input circuit. However, the question of how a given molecule gives rise to different logic gates is more subtle than might first appear, because what can be achieved depends upon the molecular character of the receptor itself. The figure considers a number of different receptor types. Receptor type U responds to one ligand type we will call species S1, which has a concentration c_1. Receptor type V corresponds to a second ligand species of type S2 and characterized by concentration c_2. Receptor type W responds to both S1 and S2 but does so only through one binding site, such that both ligands cannot be present simultaneously. Receptor type UV is a heterodimer made up of a U receptor and a V receptor. This heterodimer can respond to both S1 and S2. UW is a heterodimer of the U receptor and the W receptor and is the only one that can implement all 16 logical functions. WW is a homodimer of the W receptor. Each of these different receptor types will have its own unique states and weights, and it is only by evaluating these states and weights that we can compute the activity ($p_{active}(c_1, c_2)$) of the receptor as a function of the two inputs.

To that end, Figure 9.3 shows how statistical mechanics can be used to formulate the statistical weights of the different molecular states of the UV receptor of Figure 9.2. As already shown exhaustively throughout the book, the states and weights characterize the probability that the MWC molecule is active or inactive, as well as whether it is occupied by ligands or not. The individual statistical weights depend upon the the energy difference between the active and inactive states in the absence of ligands (denoted here by $\Delta\varepsilon$), the concentrations of the ligands of interest, and the binding strength of each ligand to the allosteric molecule, which in general depends upon the activity state of the molecule. Given these states and weights, we are now poised to explore the ability of MWC molecules to achieve combinatorial control.

Figure 9.3

States and weights for the UV allosteric enzyme. The two different ligands (blue and red) are present with concentrations c_1 and c_2, respectively. Ligand dissociation constants in the active A and active I states of the enzyme are $K_{A,i}$ and $K_{I,i}$ ($i = 1, 2$), respectively. The energetic difference between the inactive and active states of the enzyme is denoted by $\Delta\epsilon = \epsilon_I - \epsilon_A$. Total weights of the active and inactive states are shown at the bottom of each column.

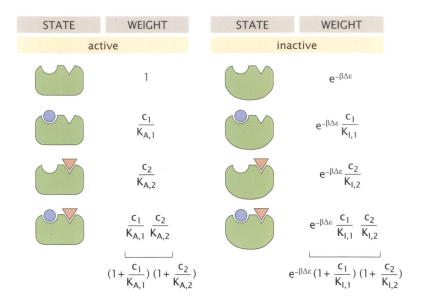

9.2 Using MWC to Build Gates

Consider an MWC molecule represented by the states and weights of Figure 9.3. We are interested in the activity of this molecule as a function of the concentrations of the two regulatory ligands, c_1 and c_2. The probability of the active state, p_{active}, is given by the ratio of the statistical weights of all the states that are active to the sum of the weights of all states. This yields the activity function

$$p_{active}(c_1, c_2) = \frac{\left(1 + \frac{c_1}{K_{A,1}}\right)\left(1 + \frac{c_2}{K_{A,2}}\right)}{\left(1 + \frac{c_1}{K_{A,1}}\right)\left(1 + \frac{c_2}{K_{A,2}}\right) + e^{-\beta\Delta\epsilon}\left(1 + \frac{c_1}{K_{I,1}}\right)\left(1 + \frac{c_2}{K_{I,2}}\right)}. \quad (9.1)$$

Here we have introduced the notation $K_{A,i}$ as the dissociation constant for ligand type S_i binding the active state of the allosteric molecule, and $K_{I,i}$ as the dissociation constant for ligand type S_i binding the inactive state. We also introduced the parameter $\Delta\epsilon$ to capture the energy difference between the inactive and active states in the *absence* of all ligands.

Our goal is to examine the extent to which allosteric molecules can serve as molecular logic gates. To that end, we need analytical tools for evaluating how well the behavior of an allosteric molecule reproduces a desired logic function. A useful way to explore this question is to examine the activity of our MWC molecule in the limits of very low and very high concentrations ($c_i \rightarrow 0$ and $c_i \rightarrow \infty$, respectively). These limits in the context of a single-ligand MWC molecule help us discern key phenotypic features such as the leakiness and dynamic range of our allosteric molecule. Here we generalize those ideas to two regulatory ligands in Figure 9.4(A). We introduce a simplifying parameter, γ_i, that is defined as the ratio of the K_ds for ligand c_i binding to the allosteric molecule in the active and inactive states, $\gamma_i \equiv K_{A,i}/K_{I,i}$. The key point is that the activity of the MWC molecule as given by equation 9.1 takes very simple forms in the limits of very low and very high concentrations. For example, the

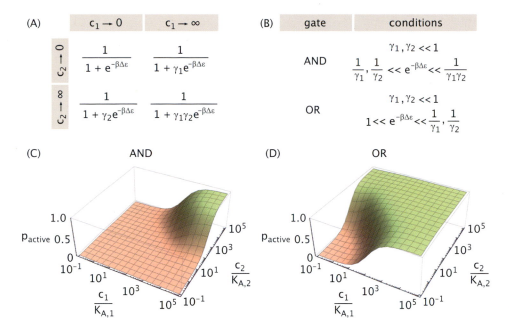

Figure 9.4

Constructing AND and OR logic from allosteric molecules. (A) Probability of activity, p_{active}, in different limits of ligand concentrations as determined by equation 9.1. We define $\gamma_i \equiv K_{A,i}/K_{I,i}$. (B) Conditions on the model parameters required for the operation of AND and OR gates. (C and D) Realizations by an MWC molecule of AND and OR logic gates, respectively. Parameters used were $K_{A,1} = K_{A,2} = 2.5 \times 10^{-8}$ M, $K_{I,1} = K_{I,2} = 1.5 \times 10^{-4}$ M ($\gamma_1 = \gamma_2 = \frac{5}{3} \times 10^{-4}$), and either $\Delta\varepsilon = -14.2\, k_B T$ in (C) or $\Delta\varepsilon = -5.0\, k_B T$ in (D).

upper left corner of Figure 9.4(A) tells us that the probability of the active state in the absence of all ligands, also termed the leakiness of the MWC molecule, depends only on $\Delta\varepsilon$.

The entire suite of conditions in Figure 9.4(A) can be analyzed by comparing the relative values of p_{active} in each concentration limit to arrive at parameter conditions that constrain behavior to the desired logic gate, as shown in Figure 9.4(B). As seen in that figure, there are parameter regimes on the γ_i and $\Delta\varepsilon$ in which we can produce both AND and OR logic (and hence, by definition, NAND and NOR by altering which states are favored at zero concentration). Interestingly, no viable parameter conditions exist for the XOR and XNOR logic gates.

9.2.1 Making Logic

We now delve more deeply into the precise nature of the conditions that give rise to each kind of logical architecture. We begin by introducing simplifying notation for the limits of very small and large concentrations of c_1 and c_2. The point of this notation is that we are going to repeatedly consider the output of the MWC molecule when the input concentrations take the extremes of $c \to 0$ and $c \to \infty$. Thus, we define

$$p_{active}(c_1 \to 0, c_2 \to 0) = p_{0,0} \tag{9.2}$$

for the case in which both inputs are at zero concentration, giving us the leakiness. We then define the two cases in which only one of the concentrations is high as

$$p_{active}(c_1 \to \infty, c_2 \to 0) = p_{\infty,0}, \tag{9.3}$$

and by symmetry,

$$p_{active}(c_1 \to 0, c_2 \to \infty) = p_{0,\infty,}. \tag{9.4}$$

Finally, we consider the limit of which both concentrations are high as

$$p_{active}(c_1 \to \infty, c_2 \to \infty) = p_{\infty,\infty,}. \tag{9.5}$$

These four extreme conditions will give us the opportunity to probe the logical behavior of the MWC molecule.

Given the notational convenience just introduced, we begin by analyzing the conditions under which we can realize the AND gate logic. From the ideal logic gate behavior as visualized in Figure 9.1, we see that we require the conditions $p_{0,0} \approx 0$, $p_{0,\infty} \approx 0$, $p_{\infty,0} \approx 0$, and $p_{\infty,\infty} \approx 1$, which correspond in turn to the following inequalities,

$$e^{-\beta \Delta \varepsilon} \gg 1, \tag{9.6}$$

$$\gamma_1 e^{-\beta \Delta \varepsilon} \gg 1, \tag{9.7}$$

$$\gamma_2 e^{-\beta \Delta \varepsilon} \gg 1, \tag{9.8}$$

$$\gamma_1 \gamma_2 e^{-\beta \Delta \varepsilon} \ll 1. \tag{9.9}$$

Combining equations 9.7–9.9, we obtain the condition for an AND gate, namely,

$$\frac{1}{\gamma_1}, \frac{1}{\gamma_2} \ll e^{-\beta \Delta \varepsilon} \ll \frac{1}{\gamma_1 \gamma_2}. \tag{9.10}$$

Note, that the outer inequalities imply

$$\gamma_1, \gamma_2 \ll 1, \tag{9.11}$$

meaning that both ligands bind more tightly to the protein in the active than the inactive state. With these conditions in hand, we are now poised to examine specific parameter regimes that satisfy them, as we will do in section 9.2.2 (p. 309).

We can make a similar analysis to explore the conditions under which we can realize the OR gate. This time, the logical structure of the OR gate requires that we have the relationships $p_{0,0} \approx 0$, $p_{0,\infty} \approx 1$, $p_{\infty,0} \approx 1$, and $p_{\infty,\infty} \approx 1$. This requires that the parameters obey

$$e^{-\beta \Delta \varepsilon} \gg 1, \tag{9.12}$$

$$\gamma_1 e^{-\beta \Delta \varepsilon} \ll 1, \tag{9.13}$$

$$\gamma_2 e^{-\beta \Delta \varepsilon} \ll 1, \tag{9.14}$$

$$\gamma_1 \gamma_2 e^{-\beta \Delta \varepsilon} \ll 1. \tag{9.15}$$

Combining equations 9.12–9.14, we obtain a constraint on the free-energy difference,

$$1 \ll e^{-\beta \Delta \varepsilon} \ll \frac{1}{\gamma_1}, \frac{1}{\gamma_2}. \tag{9.16}$$

As with the AND gate, the outer inequalities imply that the ligands prefer binding to the protein in the active state,

$$\gamma_1, \gamma_2 \ll 1. \tag{9.17}$$

Because the NAND and NOR gates are the logical complements of AND and OR gates, respectively, the parameter constraints under which they are realized are the opposites of those for AND and OR gates. In particular, the NAND gate is intimately related to the AND gate, since now we demand that quantities that were much greater than 1 be much smaller than 1. By framing these conditions in terms of the MWC probabilities following from the states and weights of Figure 9.3 (p. 306), we can rewrite them as

$$\frac{1}{\gamma_1 \gamma_2} \ll e^{-\beta \Delta \varepsilon} \ll \frac{1}{\gamma_1}, \frac{1}{\gamma_2}, \tag{9.18}$$

while the conditions for NOR gates are

$$\frac{1}{\gamma_1}, \frac{1}{\gamma_2} \ll e^{-\beta \Delta \varepsilon} \ll 1. \tag{9.19}$$

We note that in both cases, the outer inequalities imply that both ligands bind more tightly to the protein in the inactive state than in the active state, $\gamma_1, \gamma_2 \gg 1$.

Because of its unusual outcome, we finally turn to the example of the XOR logic and leave the rest of the examples in Figure 9.2 as exercises for the reader. Here, we show that the XOR gate (and by symmetry the XNOR gate) are not achievable with the form of p_{active} given in equation 9.1. An XOR gate satisfies $p_{0,0} \approx 0$, $p_{0,\infty} \approx 1$, $p_{\infty,0} \approx 1$, and $p_{\infty,\infty} \approx 0$, which necessitates the parameter conditions

$$e^{-\beta \Delta \varepsilon} \gg 1, \tag{9.20}$$

$$\gamma_1 e^{-\beta \Delta \varepsilon} \ll 1, \tag{9.21}$$

$$\gamma_2 e^{-\beta \Delta \varepsilon} \ll 1, \tag{9.22}$$

$$\gamma_1 \gamma_2 e^{-\beta \Delta \varepsilon} \gg 1. \tag{9.23}$$

However, these conditions cannot all be satisfied, as the left-hand side of equation 9.23 can be written in terms of the left-hand sides of equations 9.20–9.22, resulting in

$$\gamma_1 \gamma_2 e^{-\beta \Delta \varepsilon} = \frac{\left(\gamma_1 e^{-\beta \Delta \varepsilon}\right)\left(\gamma_2 e^{-\beta \Delta \varepsilon}\right)}{e^{-\beta \Delta \varepsilon}} \ll 1, \tag{9.24}$$

contradicting equation 9.23. To achieve this particular logic requires introducing cooperativity explicitly.

9.2.2   A Tour of Parameter Space

In light of the conditions summarized in Figure 9.4(A) and (B), we can now choose parameters satisfying the conditions shown in the figure and examine the activity functions themselves. Figures 9.4(C) and (D) give examples

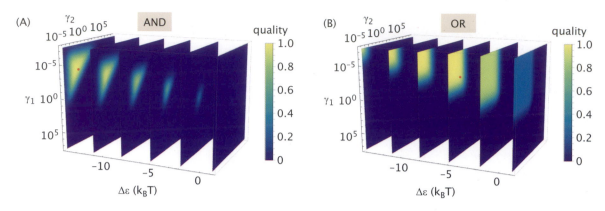

Figure 9.5

Quality of logic gates and MWC parameters. (A) Quality of the AND gate as a function of the three parameters γ_1, γ_2, and $\Delta\varepsilon$. (B) Quality of the OR gate as a function of the three parameters γ_1, γ_2, and $\Delta\varepsilon$. The red dot in the two figures signifies the parameter values used in Figure 9.4.

of molecular activity of the AND and OR character as a function of the concentrations of the regulatory ligands c_1 and c_2, exhibiting logic-gate behavior with reasonable values of the three key MWC parameters. We assess the reasonableness of the parameters by noting that these parameters have values comparable to those seen earlier in the book in the context of both ion channels and transcription factors.

To explore how the activity changes across parameter space and hence the extent to which the response function satisfies one of the logic architectures, we define a quality metric for how closely p_{active} matches its target value at different concentration limits for a given idealized logic gate. Specifically, we advance the definition

$$Q(\gamma_1, \gamma_2, \Delta\varepsilon) = \prod_{\lambda_1 = 0, \infty} \prod_{\lambda_2 = 0, \infty} (1 - \left| p_{\lambda_1, \lambda_2}^{ideal} - p_{\lambda_1, \lambda_2} \right|), \tag{9.25}$$

where $p_{\lambda_1, \lambda_2} = p_{active}([L_1] \to \lambda_1, [L_2] \to \lambda_2)$. A value of 1 for this entire product (high-quality gate) implies a perfect match between the target function and the behavior of the allosteric molecule, while a value near 0 (low-quality gate) suggests that the response behavior deviates from the target function in at least one limit.

From equation 9.25, the quality metric for the AND gate is defined as

$$Q_{AND} = (1 - p_{0,0})(1 - p_{\infty,0})(1 - p_{0,\infty})p_{\infty,\infty}, \tag{9.26}$$

while for the OR gate it takes on the form

$$Q_{OR} = (1 - p_{0,0})\, p_{\infty,0}\, p_{0,\infty}\, p_{\infty,\infty}. \tag{9.27}$$

Figure 9.5 shows the regions in parameter space where the protein exhibits these gating behaviors.

More specifically, for a fixed $\Delta\varepsilon$, the AND behavior is achieved in a finite triangular region in the γ_1-γ_2 plane which grows larger as $\Delta\varepsilon$ decreases. The

OR gate, on the other hand, is achieved in the region defined by $\gamma_1, \gamma_2 \lesssim e^{\beta \Delta \varepsilon}$ In either case, a high-quality gate can be obtained only when the base activity is very low ($\Delta \varepsilon \lesssim 0$) and when both ligands are strong activators ($\gamma_1, \gamma_2 \ll 1$), in agreement with the derived conditions shown in Figure 9.4(B). Lastly, we note that the quality metrics for AND/OR and their complementary NAND/NOR gates obey a simple relation, namely, $Q_{AND/OR}(\gamma_1, \gamma_2, \Delta \varepsilon) = Q_{NAND/NOR}\left(\frac{1}{\gamma_1}, \frac{1}{\gamma_2}, -\Delta \varepsilon\right)$, which follows from the functional form of equation 9.25 and the symmetry between the two gates.

9.3   Beyond Two-Input Logic

Our treatment of Phosphofructokinase in section 6.4.1 (p. 223) showed us that enzymes in the glycolysis pathway can have more than two inputs. As a reminder, the reader is urged to reexamine Figure 1.7 (p. 10), which shows the molecular inhibitors and activators of the enzymes of the glycolysis pathway. Similarly, in chapter 5, we saw how the GIRK ion channels are gated by a combination of G proteins, PIP_2, and Na^+ ions. Motivated by examples like these, we now seek to generalize our thinking about molecular logic to cases with more than two inputs.

The result of equation 9.1 may be generalized to the case in which there are N distinct input ligands which can give rise to an even larger space of signaling responses. In particular, if we imagine there are n_i binding sites for the ith input ligand, then the activity of the multi-input MWC molecule is the generalized case of equation 9.1 and becomes

$$p_{active} = \frac{\prod_{i=1}^{N}\left(1 + \frac{c_i}{K_{A,i}}\right)^{n_i}}{\prod_{i=1}^{N}\left(1 + \frac{c_i}{K_{A,i}}\right)^{n_i} + e^{-\beta \Delta \varepsilon}\prod_{i=1}^{N}\left(1 + \frac{c_i}{K_{I,i}}\right)^{n_i}}, \tag{9.28}$$

where c_i is the concentration of the ith ligand. Note that each ligand i has its own $K_{I,i}$ and $K_{A,i}$. For example, in the case of a three-input molecule with concentrations c_1, c_2, and c_3, respectively, with only one binding site for each of these inputs, we have

$$p_{active}(c_1, c_2, c_3) =$$

$$\frac{\left(1 + \frac{c_1}{K_{A,1}}\right)\left(1 + \frac{c_2}{K_{A,2}}\right)\left(1 + \frac{c_3}{K_{A,3}}\right)}{\left(1 + \frac{c_1}{K_{A,1}}\right)\left(1 + \frac{c_2}{K_{A,2}}\right)\left(1 + \frac{c_3}{K_{A,3}}\right) + e^{-\beta \Delta \varepsilon}\left(1 + \frac{c_1}{K_{I,1}}\right)\left(1 + \frac{c_2}{K_{I,2}}\right)\left(1 + \frac{c_3}{K_{I,3}}\right)}. \tag{9.29}$$

The generalization to multiple-input ligands inherent in equation 9.28 makes it possible to see how increasing the complexity of MWC models gives rise to yet further modes of combinatorial control of allosteric molecular activity. In the case of three ligand types, each with a single binding site ($N = 3$, $n_1 = n_2 = n_3 = 1$), simple extensions of the two-ligand case are easily achieved. To be concrete, Figure 9.6 gives an assessment of the diversity of three-input molecular logics. As shown in the figure, there are eight possible input configurations (HHH, HHL, HLH, LHH, HLL, LHL, LLH, LLL), where H represents

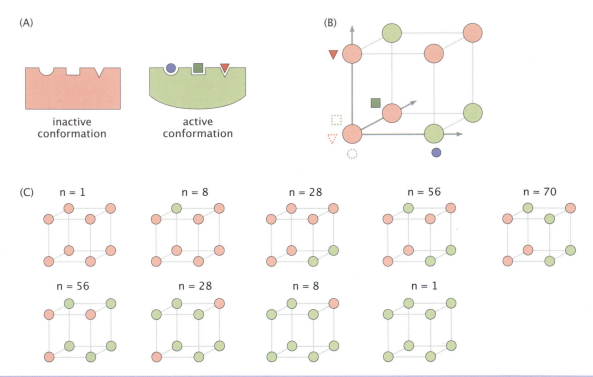

Figure 9.6

Three-input logic. (A) Schematic of the three-input allosteric molecule in both the inactive and active conformations. (B) The eight input conditions into a three-input molecule. Ligands 1, 2, and 3 can all be at either low or high concentration. (C) The different classes of molecular outputs in which none of the input conditions lead to high output (i.e., NEVER), where one of the inputs gives rise to high output (i.e., EXCLUSIVE$_i$), two of the inputs give rise to high output, and so on.

"high," and L represents "low," and our notation means $c_1 = H, c_2 = H, c_3 = H$ (HHH). Since for each of the eight input possibilities, there are two output possibilities, either low or high, this means that by analogy with Figure 9.2, there are $2^8 = 256$ distinct logics for the three-input case. However, by symmetry, not all these different logics are actually different.

We consider two responses to be functionally identical if one can be obtained from another by relabeling the ligands, for example, $(1, 2, 3) \rightarrow (3, 1, 2)$. Eliminating all redundant responses leaves 80 unique cases out of the 256 possibilities. In addition, since the molecule's activity in the eight ligand concentration limits is determined by only four MWC parameters, namely, $\{\Delta\varepsilon, \gamma_1, \gamma_2, \gamma_3\}$, we expect the space of possible three-input gates to be constrained (analogous to XOR/XNOR gates being inaccessible to two-input MWC proteins). Imposing the constraints leaves 34 functionally unique logic responses that are compatible with the MWC framework, as shown in Figure 9.7.

In addition to expanding the scope of combinatorial control relative to the two-input case, we can think of the role of the third ligand as a regulator whose presence switches the logic performed by the other two ligands. We illustrate this behavior in Figure 9.7(C) by first focusing on the leftmost cubic diagram. The gating behavior on the left face of the cube (in the absence of c_1) exhibits NONE logic, while the behavior on the right face of the cube (in the presence

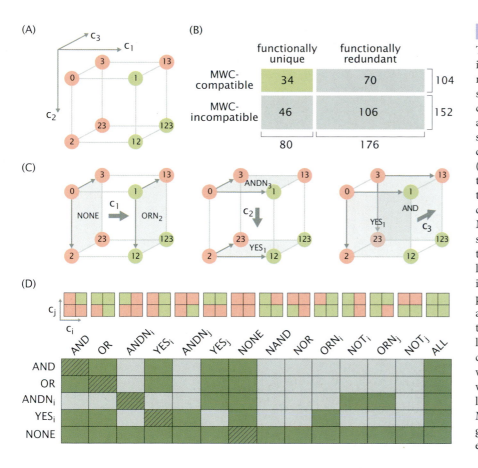

Figure 9.7

Three-input logic. (A) The eight input conditions into a three-input molecule. A red vertex means that the switch is off in that combination of concentrations of the three ligands, and a green vertex means that the switch is on at that combination of concentrations of the three ligands. (B) Breakdown of the 256 possible three-input logic gates into those that are functionally unique and can or cannot be realized by the MWC model. (C) Single inputs switch the logical character of a given two-input logic. In the first cube, ligand 1 switches a NONE logic into a ORN_2 logic. (D) Table of all possible logic transitions inducible by a third ligand. The transition switches the logic shown in the row to the logic shown in the column for those cases with a green cell. Those cells with hashed lines show the cases in which the third ligand switches a logic into itself. Schematics of the 14 MWC-compatible two-ligand logic gates corresponding to each column entry are shown on top.

of saturating c_1) is the ORN_2 logic. In this way, adding c_1 switches the logic of the remaining two ligands from NONE → ORN_2. In a similar vein, adding c_2 changes the logic from $ANDN_3$ → YES_1, while adding c_3 causes a YES_1 → AND switch. Figure 9.7(D) shows the entirety of allowed logics.

We repeat the same procedure for all functionally unique three-ligand MWC gates and obtain a table of all possible logic switches that can be induced by a third ligand, as shown in Figure 9.7(D), which tabulates row → column logic switches. As we can see, a large set of logic switches are feasible, the majority of which (the left half of the table) do not involve a change in the base activity (i.e., activity in the absence of the two ligands). Comparatively fewer transitions that involve flipping of the base activity from OFF to ON are possible (the right half of the table).

Figure 9.8 shows how these generalizations of equation 9.28 play out concretely at the level of the activity curves (or surfaces in this higher-dimensional case). Figures 9.8(A) and (C) show the conditions on the MWC parameters that convert an AND into an OR logic either by increasing the number of binding sites per ligand or by increasing the number of species of ligand. Evidence from the case of phosphofructokinase hints to me that this kind of multi-input logic is in fact ubiquitous, and as shown in this section, a rich variety of behaviors can come from this kind of logic.

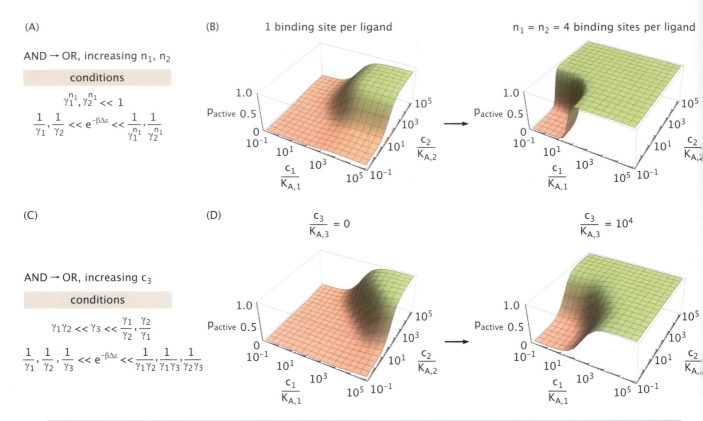

(A)

AND → OR, increasing n_1, n_2

conditions

$$\gamma_1^{n_1}, \gamma_2^{n_1} \ll 1$$

$$\frac{1}{\gamma_1}, \frac{1}{\gamma_2} \ll e^{-\beta\Delta\varepsilon} \ll \frac{1}{\gamma_1^{n_1}}, \frac{1}{\gamma_2^{n_1}}$$

(B)   1 binding site per ligand $n_1 = n_2 = 4$ binding sites per ligand

(C)

AND → OR, increasing c_3

conditions

$$\gamma_1\gamma_2 \ll \gamma_3 \ll \frac{\gamma_1}{\gamma_2}, \frac{\gamma_2}{\gamma_1}$$

$$\frac{1}{\gamma_1}, \frac{1}{\gamma_2}, \frac{1}{\gamma_3} \ll e^{-\beta\Delta\varepsilon} \ll \frac{1}{\gamma_1\gamma_2}, \frac{1}{\gamma_1\gamma_3}, \frac{1}{\gamma_2\gamma_3}$$

(D) $\frac{c_3}{K_{A,3}} = 0$ $\frac{c_3}{K_{A,3}} = 10^4$

Figure 9.8

Beyond the simplest two-input model. (A) Conditions for switching from an AND to an OR by tuning the number of binding sites. (B) Comparison of logical output for the case where there is one binding site per ligand and the case where there are four binding sites per ligand, revealing that the AND logic is converted into OR logic. (C) Conditions for switching from an AND to an OR by increasing the concentration of a third species of ligand. (D) Comparison of logical output for the cases where $c_3 = 0$ and the case where $c_3 \gg K_{A,3}$, revealing that the AND logic is converted into OR logic.

9.4 Summary

Combinatorial control is one of the primary facts of life. The networks of biological systems of all kinds have multiple inputs, and the resulting output depends upon all these inputs. For the simplest case in which there are two inputs, we can enumerate all 16 different possible logics, as shown in Figure 9.2. For a three-input case, there are 256 distinct logics. The statistical mechanics of the MWC model allows us to systematically explore the regions in MWC parameter space where these different logics can be realized, if at all. We focused on several key examples such as the AND and OR gates and their negations, such as the NAND and NOR, which can be implemented by flipping the relative stability of the active and inactive states. Certain logics such as XOR require contradictory conditions on the MWC parameters and hence cannot be realized. In fact, many biochemical pathways are subject to more than two inputs at a time, with the case of phosphofructokinase serving as a canonical example. In the latter part of the chapter, we gave a flavor of how the simplest MWC framework can be extended to account for multi-input combinatorial control.

9.5 Further Reading

Agliari, E., M. Altavilla, A. Barra, L. Dello Schiavo, and E. Katz (2015) "Notes on stochastic (bio)-logic gates: Computing with allosteric cooperativity." *Sci. Rep.* 5:9415. This paper reviews the kinds of results found in de Ronde et al. and shows how such logics can be made experimentally with enzyme interactions.

Buchler, N. E., U. Gerland, and T. Hwa (2003) "On schemes of combinatorial transcription logic." *Proc. Natl Acad. Sci.* 100:5136. This paper gives a thorough treatment of combinatorial control in transcriptional regulation.

de Ronde, W., P. R. ten Wolde, and A. Mugler (2012) "Protein logic: A statistical mechanical study of signal integration at the single-molecule level." *Biophys. J.* 103:1097–1107. This excellent paper gives a thorough and insightful treatment of the subject of this chapter. Some of my figures are drawn directly from this paper.

Galstyan, V., L. Funk, T. Einav, and R. Phillips (2019) "Combinatorial control through allostery." *J. Phys. Chem.* B123:2792–2800.

Graham, I., and T. Duke (2005) "The logical repertoire of ligand-binding proteins." *Phys. Biol.* 2:159–165. This paper shows how dual-input allosteric molecules can serve as logic gates.

10

DNA PACKING AND ACCESS: THE PHYSICS OF COMBINATORIAL CONTROL

If you hold a cat by the tail you learn things you cannot learn any other way.

—Mark Twain

The study of allostery has traditionally been focused on conformational changes in proteins that transition back and forth between inactive and active states, with the binding of some ligand tipping the balance between those two states of activity. In this chapter, we generalize our thinking on allostery by considering the protein-DNA complexes that mediate the packing of DNA into the eukaryotic nucleus. Interestingly, when DNA is wrapped up in these structures, known as *nucleosomes*, it is no longer as accessible to binding by DNA-binding proteins and it is the transition between these inaccessible and accessible states that is the focus of this chapter.

10.1 Genome Packing and Accessibility

An estimate in eukaryotic biology that should be known to all working scientists focuses on the incredible mismatch between the length of genomic DNA if fully stretched out and the size of the nucleus where that DNA is stored. Let's explore those numbers.

Though it is beyond the scope of the present discussion, polymer physics teaches us ways to think about the cost in free energy to take a semiflexible polymer such as DNA and to fold it up in such a way that it can fit in the nucleus. One of the sources of the free energy to effect such packing is DNA-protein interactions. At the several-nanometer scale, 147 bp segments of the DNA are wrapped $1\frac{3}{4}$ times around an octamer of histone proteins to make a nucleosome, as shown in Figure 10.1. The free-energy budget for this wrapping has both a cost, owing to the elastic bending of the DNA, and an advantage, owing to the favorable interactions between the negative charges on the DNA and the positive charges on the histones. Together, these costs result in a net advantage, leading to spontaneous formation of nucleosomes, as has been demonstrated in in vitro experiments with mixtures of DNA and histone proteins.

ESTIMATE

Estimate: DNA Compaction in the Eukaryotic Nucleus

With its 3×10^9 bp–long genome sequence, when fully stretched out, the DNA of the human genome is a meter in length, as seen by using the length per base pair of 1/3 nm,

$$L_{genome} = \text{number of bp} \times \frac{\text{length}}{\text{bp}} = 3 \times 10^9 \text{ bp} \times \frac{1}{3} \text{ nm/bp} = 1 \text{ m}. \quad (10.1)$$

This length is to be contrasted with the linear dimensions of the typical nucleus, which has a size of a few microns. This estimate is a pointed reminder of the enormous compaction to which genomes are subjected when squeezed into the confines of the eukaryotic nucleus. Further, there are two copies of the genome in every such nucleus, making this feat all the more impressive.

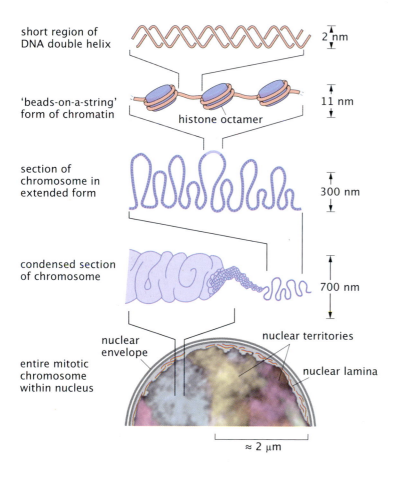

short region of DNA double helix — 2 nm

'beads-on-a-string' form of chromatin — 11 nm
histone octamer

section of chromosome in extended form — 300 nm

condensed section of chromosome — 700 nm

nuclear envelope
nuclear territories
entire mitotic chromosome within nucleus
nuclear lamina

≈ 2 μm

Figure 10.1

Hierarchical packing of eukaryotic DNA. Much of the DNA in eukaryotes is wrapped up in protein-DNA complexes involving 147 bp of DNA and an octamer of histone proteins to form a complex known as the nucleosome. These nucleosomes are spaced by a few tens of base pairs. The DNA is condensed into higher-order structures culminating in the nuclear territories seen in the bottom panel. Bottom adapted from Steven et al. (2016).

Figure 10.1 goes beyond the single-nucleosome picture to provide a schematic of current thinking on how eukaryotic DNA is folded up hierarchically. In ways that are still not fully understood, this hierarchy continues all the way up in scale from several nanometers until the fully packed nucleus is reached, as shown in the bottom schematic of Figure 10.1.

The geometric and energetic story of chromosome organization across scales is one of the greatest stories of modern biology. For our purposes we permit ourselves the more modest but still fascinating goal of trying to understand the implications of such DNA packing for how genes are turned on and off. Chapter 8 (p. 272) gave us an introduction to the way that DNA-protein interactions conspire to give rise to transcriptional regulation. In this chapter, we now extend that discussion to include the question of genome architecture in eukaryotes (sometimes characterized as chromatin), and the synergy between genome packing, binding of transcription factors and polymerases to DNA, and transcriptional regulation itself.

10.2   The Paradox of Combinatorial Control and Genomic Action at a Distance

One of the obvious ways in which our study of transcriptional regulation needs to be extended beyond the story offered in chapter 8 (p. 272) is to account for the fact that genes are subject to combinatorial control; that is, the state of activity of a given gene is determined by the concentrations of entire batteries of different transcription factors. In a now-classic paper, Britten and Davidson (1969) offered a panorama of regulatory biology that went far beyond the original, beautiful work of Jacob and Monod, who had introduced the repressor-operator model, in which transcription of a gene is controlled by the binding of a repressor. In Figure 10.2 we show the original conception of Britten and Davidson, in which each gene has many inputs and many outputs. That is to say, a given gene can be controlled by the binding of a constellation of distinct transcription factors. Concretely, Britten and Davidson envisioned that there would be regulatory genes, dubbed "sensor genes" by them and labeled as S_1, S_2, S_3, and S_4 in the figure, whose gene products would then control producer genes. In the figure, the producer genes are labeled by the letters A through K, with the notation $P_E(1, 2, 3, 4)$ signifying that gene E is controlled by regulatory genes 1, 2, 3, and 4. Since the original formulation of this regulatory hypothesis about networks of connected genes the basic picture has been confirmed over and over again.

If we adopt an overly simplistic view in which a promoter is represented as harboring some region with a size of at most several hundred base pairs where the basal transcription apparatus binds as shown in Figure 10.3(A), then we see that the number of sites on the immediately adjacent DNA where a transcription factor can bind and "touch" the basal transcription apparatus is small. To be concrete, in the absence of some sort of DNA looping, if we imagine the footprint of each transcription factor as being of order 10 bp in length (or 1/3 nm/bp $\times$ 10 bp $\approx$ 3 nm in size), then several binding sites already take us 10 nm or more away from the promoter region itself. Geometric constraints limit the extent of combinatorial control, as shown in Figure 10.3(A).

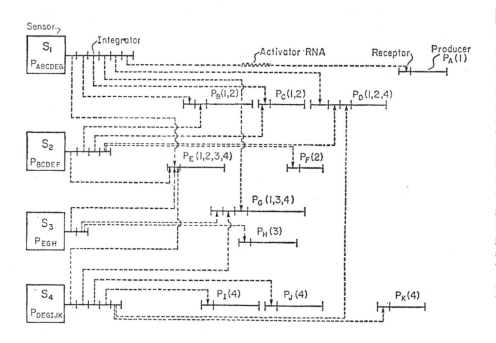

Figure 10.2

Beyond the operon model. The original conception of combinatorial control by Britten and Davidson showing how a collection of sensor (regulatory) and producer genes are linked together in networks. The numbers in parentheses after each producer gene's label reveal combinatorial control by showing the multiple inputs that control that gene. For example, producer gene E (middle of the figure) is controlled by all four sensor genes. Adapted from Britten, R. J. and E. H. Davidson (1969), "Gene regulation for higher cells: A theory," *Science* 349. Reprinted with permission from AAAS.

(A)
COMPACT BACTERIAL PROMOTER

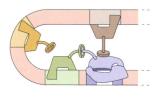

(B)
BACTERIAL ACTION AT A DISTANCE

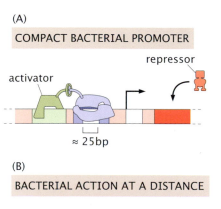

(C)
FULL COMBINATORIAL CONTROL

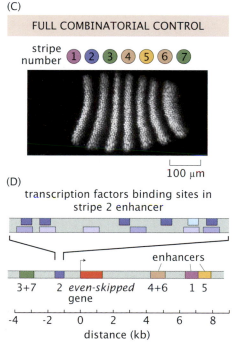

(D)
transcription factors binding sites in stripe 2 enhancer

Figure 10.3

The paradox of combinatorial control. (A) There is insufficient genomic real estate to harbor sufficient binding sites for combinatorial control. (B) The concept of action at a distance. Bending of the DNA from some distant site to come in contact with the promoter opens the door to combinatorial control. (C) and (D) An example of the regulatory landscape with enhancers. Note that for the *even-skipped* stripe 2 enhancer, there is a constellation of binding sites for transcription factors to exercise combinatorial control. (C) adapted from Surkova S., et al. (2008), "Characterization of the Drosophila segment determination morphome," *Dev. Biol.* 313(2):844, with permission from Elsevier. (D) adapted from Surkova et al. (2008) and Davidson (2006).

The paradox of combinatorial control is answered in part by action at a distance, as shown in Figure 10.3(B). The idea here is that parts of the genome not adjacent to the promoter of interest can still interact with the promoter by looping of the DNA such that these distal parts of the genome are in close *physical* proximity to the promoter. As shown in the figure, such combinatorial control is a key part of the regulatory machinery in prokaryotes. The situation becomes even more interesting in the context of eukaryotes.

As seen in Figure 10.3(C) and (D) in the context of the body plan of the fruit fly, combinatorial control and action at a distance go hand in hand. The various stripes (stripes 1–7) that appear in the anterior-posterior patterning of the fly embryo are controlled by a variety of enhancers. As seen in the detailed schematic of Figure 10.3(D), each of these enhancers itself comprises a number of binding sites for transcription factors such as Bicoid, Hunchback, Kruppel, and Giant. These enhancers can be more than hundreds of base pairs in length and can be tens of thousands of base pairs away from the promoter (and even more in mammals). Though enhancers are a powerful example of action at a distance, the rules of how sequences in these enhancer regions actually work remain unknown.

In the remainder of this chapter, our objective is see how the ideas developed throughout the book can help us understand the transactions between proteins and eukaryotic genomes and how those transactions can be understood quantitatively within the two-state framework. We begin in the next section with the question of binding-site accessibility for sites that are buried at different depths within a single nucleosome. With those results in hand, we then consider how the addition of chemical groups, either to the DNA or the histone proteins, can modify the accessibility of nucleosomal DNA in a way that is analogous with the Bohr effect (see Figure 7.2, p. 233) we have already seen in the context of hemoglobin. We then turn to the more general case of transitions in chromatin between accessible and inaccessible states mediated by the binding of multiple distinct transcription factors. In light of this analysis, we show how combinatorial control can be used to implement key logic functions such as and and or gates.

10.3   Nucleosomes and DNA Accessibility

Figure 10.1 gave us our first view of the geometry of chromatin. The "beads-on-a-string" pattern of repeating nucleosomes shown in that figure raises a very interesting question about binding-site accessibility. A more detailed structural view of the nucleosome assembly is offered in Figure 10.4, where we see with amazing detail the interaction between the DNA, with its dense array of negative charges on the phosphate groups, nestled up against the histone proteins. But this figure also raises the question of how binding sites within the 147 base pairs of DNA wrapped within the nucleosome are accessed.

As a first foray into DNA accessibility problems from the MWC perspective, we consider the accessibility of a DNA segment wrapped within a single nucleosome. To give a first impression of the kinds of molecular states of interest and how they can be described using statistical mechanics, Figure 10.5 shows a hypothetical eukaryotic promoter bound with a specific arrangement relative

(A)

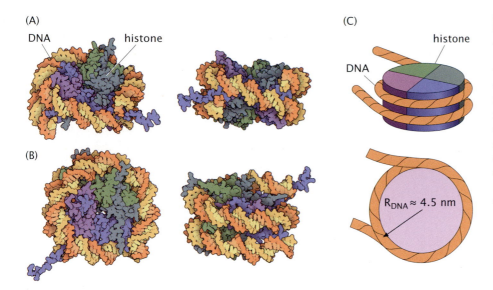

(B)

(C)

Figure 10.4

Structure of the nucleosome. (A) Atomic-level depiction of the nucleosome core particle as obtained using crystallography, revealing both the histone octamer and the encircling DNA and histone tail. (B) Alternative atomic-level view of the nucleosome. (C) Cartoon representation of the nucleosome core particle. (A) and (B) courtesy of David Goodsell.

to a nucleosome. We note from the outset that because of the rules of nucleosome positioning the real situation is more subtle than this example, which is intended only to illustrate the "indirect regulation" that could be exercised by the presence of nucleosome-bound DNA. As seen in the figure, the DNA segment of interest harbors both a promoter and a binding site for a transcription factor. When the promoter is wrapped within the nucleosome, the gene of interest is inactive. The four states of this promoter in this simple model then correspond to inactive and active configurations of the promoter and the transcription factor binding site either unoccupied or occupied, with the transcription factor serving as the ligand in much the same way as we saw other ligands do in previous examples throughout the book.

As usual, once we have assigned the states and weights, our work is effectively done. In this case, if we are interested in the probability that the promoter is in the active state, we add up the statistical weights of the two open states and divide by the sum over all statistical weights, resulting in

$$p_{active} = \frac{e^{-\beta \varepsilon_o}(1 + \frac{[P]}{K_O})}{e^{-\beta \varepsilon_o}(1 + \frac{[P]}{K_O}) + e^{-\beta \varepsilon_c}(1 + \frac{[P]}{K_C})}. \tag{10.2}$$

Note that this expression is the now-familiar MWC form and tells us the probability of the active state as a function of the concentration of activator $[P]$, as shown in Figure 10.5(B).

The model presented in Figure 10.5 is overly naive in that its statistical mechanical interpretation acknowledges only an "open" configuration and a "closed" configuration, as though there were only two distinct DNA conformations. The energy of the open and closed conformations are captured in the parameters ε_o and ε_c, respectively. However, in principle, there is a continuum of degrees of nucleosomal opening for which we have not accounted. Each such state will have a different energy because we change both the elastic energy of bending the DNA and the interaction energy between DNA and histones by different amounts depending upon how much we open the nucleosome.

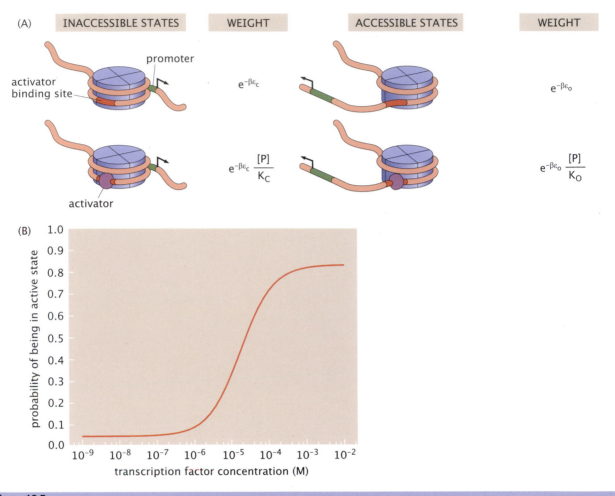

Figure 10.5

MWC model of nucleosome accessibility. (A) States-and-weights diagram for a toy model of nucleosome accessibility which illustrates how transcription factors could alter the equilibrium of nucleosome-bound DNA. ε_c and ε_o refer to the conformational energies of the closed (inaccessible) and open (accessible) states, respectively, and K_C and K_O are the dissociation constants for transcription factor binding in those two states. (B) Probability of a nucleosome being in the open state as a function of the transcription factor concentration as described by the states and weights shown in part (A) and resulting in equation 10.2. Parameters used are $K_O = 10^{-6}$ M, $K_C = 10^{-4}$ M, and $\Delta\varepsilon = \varepsilon_c - \varepsilon_o = -3\,k_BT$.

To address the question of continuous rather than discrete states of nucleosomal opening, a beautiful physical experiment was conceived. The idea of the experiment is shown in Figure 10.6, where we see three DNA constructs, each harboring a restriction enzyme binding site at different burial depths within the nucleosome. For the restriction enzyme to act, the DNA must be in the open conformation, as shown in the right panel of the figure. That is, when the DNA is in the closed conformation (left panel), the restriction site is not accessible. By way of contrast, when the nucleosomal DNA has lifted off to at least a distance d, the binding site becomes accessible and hence can be cut. To compute the probability of such nucleosomal opening, we need to know the energies of the different states and how those energies depend upon the distance d.

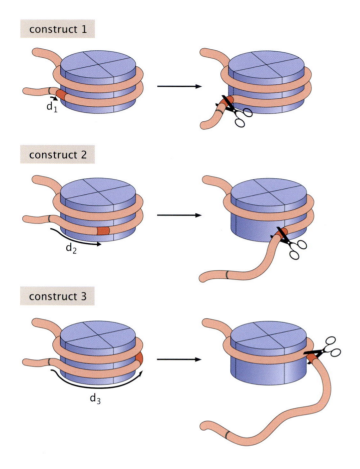

construct 1

d_1

construct 2

d_2

construct 3

d_3

Figure 10.6

Experiment to measure accessibility of binding sites in nucleosomal DNA. The restriction enzyme cut site is labeled on the DNA with a darker color than the surrounding DNA. The depth of burial of the restriction site for the ith construct is given by d_i.

To assess the energy cost to bend DNA into the circular conformation of the nucleosome, we recall that the persistence length for DNA (ξ_p) is of the order of 50 nm. The *persistence length* gives a rough measure of the scale over which an elastic rod is stiff, implying that the energy cost to bend that elastic rod is energetically very costly. In contrast, note that the radius of curvature associated with the DNA wrapped around the histone octamer is roughly 4.5 nm, as shown in Figure 10.4. There are clearly elastic consequences to such highly deformed DNA fragments. One of the goals of this section is to examine the free-energy balance associated with the assembly of the nucleosome core particle. In addition to the cost of elastic deformation, we need to consider the favorable electrostatic interactions between the DNA and the histone octamer, since the DNA itself is negatively charged, and the histone octamer surface is covered with lysine and arginine residues that present a compensating positive charge. Within this simple picture, the free energy of formation for a nucleosome can be written as

$$G_{nucleosome} = G_{bend} + G_{DNA\text{-}histone}. \qquad (10.3)$$

G_{bend} is the elastic energy cost of taking some segment of the DNA and making it adopt the circular conformation of the nucleosome assembly. $G_{DNA-histone}$ is essentially a one-dimensional adhesion energy between DNA and histones.

To construct a preliminary estimate of the deformation energy associated with the formation of the nucleosome core particle, we treat the DNA as a featureless rod subject to a uniform state of deformation. The energy stored in each turn of the DNA by virtue of its deformation into a circle of radius R_{DNA} is given by

$$G_{bend} = \frac{\pi \xi_p k_B T}{R_{DNA}}, \tag{10.4}$$

where R_{DNA} is the radius of curvature of the wrapped DNA. This is a classic result of the elastic theory of beams that is worked out elsewhere (see Further Reading at the end of the chapter), with this being one of those rare cases where we ask the reader's indulgence for a result that is quoted without explicit derivation. The key aspect is that the bending energy of a circular hoop scales as $1/R$, where R is the radius of that loop, estimated to be roughly 4.5 nm for nucleosomal DNA, as shown in Figure 10.4.

To proceed with our estimate, we need to compute the curvature associated with the 147 bp segment when it is wrapped around the histone octamer. As shown in Figure 10.4, the DNA wraps around the histone octamer in a helical pattern (we are not speaking here of the DNA double helix itself) with a small pitch. For simplicity of calculation, we use an approximate expression for the curvature, in which the helical pitch is neglected. In addition, we assume that the DNA is wrapped fully around the histone octamer two times, resulting in a total bending free-energy cost of $G_{bend} = 2\pi \xi_p k_B T / R_{DNA} \approx 70\,k_B T$.

The second key contribution to the free energy of formation of the nucleosome is dictated by the interactions between the DNA and the positively charged residues on the histones. As a simplest model, we characterize this interaction energy via an adhesive energy γ_{ad}, which has units of energy/length. In light of this model, the adhesive contribution to the total free energy is

$$G_{DNA-histone} = 4\pi R_{DNA} \gamma_{ad}, \tag{10.5}$$

where we have assumed that the DNA is twice wrapped fully around the histone octamer. In the next section, we show how this adhesive energy can be deduced from experimental data on the equilibrium accessibility of nucleosomes.

10.3.1   Equilibrium Accessibility of Nucleosomal DNA

As noted previously, DNA molecules in eukaryotic cells wind around histone octamers to form the nucleosome. In this state the DNA is not directly accessible to regulatory proteins, since their binding sites are occluded by the nucleosome. Key experiments on nucleosome accessibility were performed in vitro by assessing the susceptibility of particular sites on the DNA to cleavage by restriction enzymes as a function of the distance of these sites from the unwrapped ends of the nucleosomal DNA. Figure 10.7 gives a view of the transitions that need to take place for a cutting event by the restriction enzyme to occur. These restriction enzymes are proteins that cleave DNA at specific recognition sites and serve as a convenient readout for assessing nucleosome accessibility. Effectively, these experiments provide a position-dependent equilibrium constant that depends upon the distance of the site of interest from the unwrapped ends of the nucleosomal DNA.

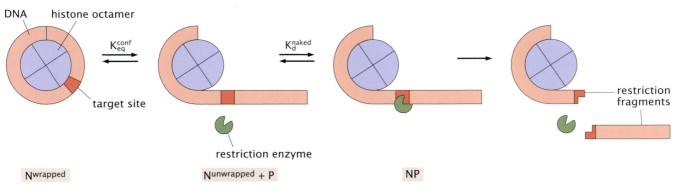

Figure 10.7

Experiment to measure equilibrium accessibility of nucleosomes. Nucleosomal DNA is prepared with a binding site for a restriction enzyme. A wrapped nucleosome, $N^{wrapped}$ can transiently unwrap ($N^{unwrapped}$) and interact with a restriction enzyme, P, upon exposure of its target binding site, forming the NP complex. A measurement is made of the rate of restriction digestion as a function of the distance of the target site from the unwrapped ends of the nucleosomal DNA. This rate is compared with the rate of restriction digestion for nonnucleosomal DNA, and the relative rates in these two cases yield the configurational equilibrium constant.

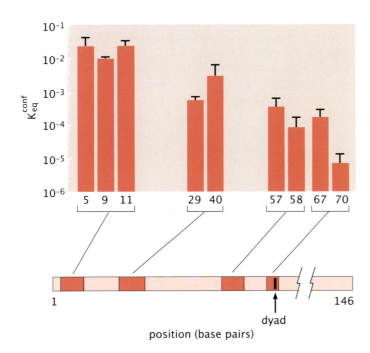

Figure 10.8

Equilibrium constant for site exposure as a function of the distance of the site from the unwrapped end of the nucleosomal DNA. The approximate positions of the binding sites are shown. The demarcated dyad is the center of symmetry of the DNA molecule when it is wrapped around the histone octamer. Adapted from Polach and Widom (1995).

The model put forward to interpret these results envisions the binding of a DNA-binding protein (for the experiment in question, restriction enzymes were used as the protein of interest) to its target site as a two-step process: first the DNA unwraps from the histones simply as a result of thermal fluctuations, and then the restriction enzyme binds to its specific site, which is no longer occluded by the nucleosome. This process is shown schematically in Figure 10.7. In Figure 10.8 we present the results from the experimental concept of Figure 10.7 for the equilibrium constant of DNA unwrapping from the nucleosome as a

Geometry of site accessibility of the nucleosome. The coordinate x refers to how much the DNA is unwrapped, and x_{re} refers to the minimum distance the DNA needs to be unwrapped to access the target site.

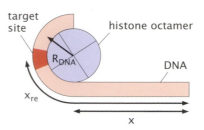

function of burial depth of the target site within the nucleosome. The simple observation is that this equilibrium constant decreases the farther the binding site is from the unwrapped ends of the nucleosome. To address these data in a quantitative way we turn once again to statistical mechanics and compute the probability that an enzyme is bound to the target site as a function of the target site location. Note that we hypothesize that all the data shown in Figure 10.7 can be addressed using a simple generalization of the model of Figure 10.5, but with each different construct having its own effective opening energy $\varepsilon_o(d)$. Because there is actually a continuum of states, since the degree of opening of the nucleosome is a continuous variable, this model can be viewed as a kind of ensemble allostery.

The simple model we consider has two degrees of freedom, as shown in Figure 10.9. One of the relevant degrees of freedom is the length x of DNA unwound from the nucleosome, which is continuous and takes values between 0 and $L \approx 25$ nm; this length is half the total DNA length wound in a typical nucleosome. The other degree of freedom is discrete, and it simply keeps track of whether a restriction enzyme is bound or not to its specific site. The binding site for the enzyme is located at position x_{re}, which is somewhere along the DNA length between 0 and L. The quantity we wish to compute is the probability p_{bound} that the restriction enzyme is bound to its site, as a function of the restriction site location.

The probability that an enzyme is bound is readily computed by summing over all the possible conformations of the nucleosomal DNA. Each of these different conformations has a different energy depending upon how much of the DNA has been unwrapped from the nucleosome, as shown in Figure 10.9. We can use equation 10.4 to define an elastic energy per unit length γ_{bend} as

$$\gamma_{bend} = \frac{G_{bend}}{2\pi R_{DNA}} = \frac{\xi_p k_B T}{2R_{DNA}^2}. \tag{10.6}$$

This result follows because we defined G_{bend} as the energy of a single circular hoop of DNA with radius of curvature R_{DNA}. Hence, to find the energy per unit length, we simply divide G_{bend} by the length of a single circular hoop, namely, $2\pi R_{DNA}$.

Because there is a continuum of states corresponding to all the possible amounts that the nucleosome is unwound, our sum over states becomes an integral such that the partition function is

$$\mathcal{Z} = \underbrace{\int_0^L e^{-\beta(\gamma_{bend}+\gamma_{ad})(L-x)}\frac{dx}{a}}_{\text{no enzyme bound}} + \underbrace{e^{-\beta(\varepsilon_{bind}-\mu)}\int_{x_{re}}^L e^{-\beta(\gamma_{bend}+\gamma_{ad})(L-x)}\frac{dx}{a}}_{\text{enzyme bound}}.$$

$$\tag{10.7}$$

The first term in $\mathcal{Z}$ is a sum over all possible values for the length x of DNA unwound from the nucleosome, with no restriction enzyme bound. The sum over discrete base pairs is for convenience replaced by an integral over the DNA length, in which case a is the distance between consecutive base pairs; this guarantees that the number of states, $\int_0^L dx/a$, is equal to the number of base pairs of DNA in the nucleosome. The statistical weight for a given x is determined by the energy per unit length associated with bending the DNA (γ_{bend}) and the energy per unit length for DNA binding to histones (γ_{ad}). Recall that the bending energy is positive, since it costs energy to bend the DNA into the wrapped conformation. By way of contrast, the adhesion energy is negative, since there is a favorable interaction between the positive charges on the histones and the negative charges on the DNA.

The second term accounts for unwound states of the DNA when there is a restriction enzyme bound to its specific binding site. For this binding to occur the DNA must unwind at least by an amount given by the position of the binding site x_{re}, which sets the lower bound of the integral. If the DNA is unwound by less than this amount, the binding site remains inaccessible. The weight associated with a given x now has two additional factors, one which takes into account the chemical potential of the restriction enzymes in solution, μ, and the other arising from the binding energy of the enzyme to its specific site on the DNA, ε_{bind}.

Note that the two integrals in equation 10.7 can both be evaluated, since they can be beaten into the form $\int e^x dx$. For the contribution to the partition function with no enzyme bound, we obtain

$$Z_{no\ enzyme} = \frac{1}{a\beta(\gamma_{bend} + \gamma_{ad})}\left(1 - e^{-\beta(\gamma_{bend}+\gamma_{ad})L}\right), \qquad (10.8)$$

and for the contribution to the partition function coming from the states with enzyme bound we have

$$Z_{enzyme} = \frac{e^{-\beta(\varepsilon_{bind}-\mu)}}{a\beta(\gamma_{bend} + \gamma_{ad})}\left(1 - e^{-\beta(\gamma_{bend}+\gamma_{ad})(L-x_{re})}\right). \qquad (10.9)$$

Now, the probability that the enzyme is bound to its restriction site is the ratio of the second term in the partition function in equation 10.7 to the total partition function, resulting in

$$p_{bound} = \frac{\frac{[P]}{[P]_0} e^{-\beta(\varepsilon_{bind}-\varepsilon_{sol})}\left(1 - e^{-\beta(\gamma_{bend}+\gamma_{ad})(L-x_{re})}\right)}{\frac{[P]}{[P]_0} e^{-\beta(\varepsilon_{bind}-\varepsilon_{sol})}\left(1 - e^{-\beta(\gamma_{bend}+\gamma_{ad})(L-x_{re})}\right) + \left(1 - e^{-\beta(\gamma_{bend}+\gamma_{ad})L}\right)}, \qquad (10.10)$$

where we have made use of the ideal-solution result for the chemical potential, $\mu = \varepsilon_{sol} + k_B T \ln([P]/[P]_0)$. Here $[P]$ is the concentration of proteins (restriction enzymes) in solution, while $[P]_0$ is the standard-state concentration.

To make contact with the experimental results we should translate the preceding expression into the language of dissociation constants. We rewrite equation 10.10 as

$$p_{bound} = \frac{[P]/K_d^{eff}}{1 + [P]/K_d^{eff}} , \tag{10.11}$$

where we have defined the effective dissociation constant K_d^{eff}. Comparing this equation with equation 10.10, we deduce

$$K_d^{eff} = [P]_0 e^{\beta(\varepsilon_{bind} - \varepsilon_{sol})} \frac{1 - e^{-\beta(\gamma_{bend} + \gamma_{ad})L}}{1 - e^{-\beta(\gamma_{bend} + \gamma_{ad})(L - x_{re})}} , \tag{10.12}$$

for the equilibrium dissociation constant.

We now wish to connect our model with the quantity measured experimentally, namely, the equilibrium constant for nucleosomal accessibility, K_{eq}^{conf}. It is defined as the ratio of the concentration of nucleosomes whose DNA is sufficiently unwrapped so as to allow the restriction enzymes to bind, to the concentration of nucleosomes whose DNA is wrapped such that no DNA binding is allowed. We call these concentrations $[N^{unwrapped}]$ and $[N^{wrapped}]$, respectively, as shown in Figure 10.7. Our claim is that K_{eq}^{conf} is embedded as a part of the total K_d^{eff}. To see this we multiply and divide K_d^{eff} by $[N^{unwrapped}]$, resulting in

$$K_d^{eff} = \frac{[P][N]}{[NP]} \times \frac{[N^{unwrapped}]}{[N^{unwrapped}]}. \tag{10.13}$$

Note that $[NP]$ is the same as $[N^{unwrapped}P]$, since to form the nucleosome-restriction-enzyme complex we need the nucleosomes to be sufficiently unwrapped. With this in mind we can express the dissociation constant as

$$K_d^{eff} = \underbrace{\frac{[N^{unwrapped}][P]}{[N^{unwrapped}P]}}_{K_d^{naked}} \frac{[N]}{[N^{unwrapped}]}, \tag{10.14}$$

where $K_d^{naked} = [P_0]e^{\beta(\varepsilon_{bind} - \varepsilon_{sol})}$ is the dissociation constant associated with naked DNA, where the equilibrium between restriction enzymes bound to the DNA and those in solution is established in the absence of histones. We can write the last term in equation 10.14 in terms of the wrapped and unwrapped nucleosome species. To do that we remind ourselves that $[N] = [N^{unwrapped}] + [N^{wrapped}]$, leading to

$$\frac{[N]}{[N^{unwrapped}]} = \frac{[N^{unwrapped}] + [N^{wrapped}]}{[N^{unwrapped}]} = 1 + \underbrace{\frac{[N^{wrapped}]}{[N^{unwrapped}]}}_{\left(K_{eq}^{conf}\right)^{-1}}, \tag{10.15}$$

which results in a dissociation constant as a function of K_d^{naked} and K_{eq}^{conf}

$$K_d^{eff} = K_d^{naked}\left(1 + \frac{1}{K_{eq}^{conf}}\right). \tag{10.16}$$

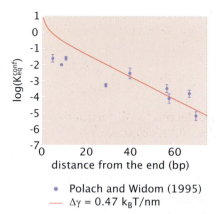

Figure 10.10

Equilibrium constant for DNA accessibility. The data points result from measurements described in this section. The curve is the result of the model worked out using statistical mechanics. We define the parameter $\Delta\gamma$ as the net energy cost per unit length for unwrapping of DNA from the histone octamer. Using a persistence length of 50 nm implies $\gamma_{bend} = 1.2\ k_BT/\text{nm}$. Data adapted from Polach and Widom (1995).

Comparing this with equation 10.12, we arrive at the expression

$$K_{eq}^{conf} = \frac{1 - e^{-\beta(\gamma_{bend}+\gamma_{ad})(L-x_{re})}}{e^{-\beta(\gamma_{bend}+\gamma_{ad})L} - e^{-\beta(\gamma_{bend}+\gamma_{ad})(L-x_{re})}}, \tag{10.17}$$

which depends only on the position of the restriction binding site along the DNA sequence and not on the strength of the binding site ε_{bind}.

We are now in the position to compare the theoretical results with the measurements discussed earlier. Figure 10.10 shows both the experimental data points as well as a best fit to these points using the adhesive energy as a free parameter. We conclude that K_{eq}^{conf} decreases by four orders of magnitude as the position of the binding site shifts from the end of the wound DNA to the dyad at $x = 70$ bp ≈ 25 nm, in good agreement with the measurements of Polach and Widom (1995). Note that the sharp increase at $x_{re} = 0$ seen in Figure 10.10 is an artifact of the continuum approximation assumed by the model; it leads to an infinite K_{eq}^{conf} when $x_{re} \to 0$. Deriving the discrete model that remedies this pathology of the continuous model is left as an exercise for the reader. The reader can also find a full treatment of the problem in the discrete approximation rather than the continuum description in the Further Reading.

Another way of thinking through the implications of the model we have set forth in this section is by considering the probability of being in the active state as a function of concentration of the DNA-binding protein and depth of burial of the binding site. Figure 10.11(A) shows a family of such activity curves for three different burial depths of the protein binding site within the nucleosome. By rewriting equation 10.2 in the form

$$p_{active} = \frac{1}{1 + e^{-\beta\Delta F}}, \tag{10.18}$$

and plotting the probability of the open state as a function of the parameter ΔF, as we have done throughout the book (see sec. 3.5, p. 94), we see that these three distinct curves collapse onto a single master curve, as shown in Figure 10.11(B).

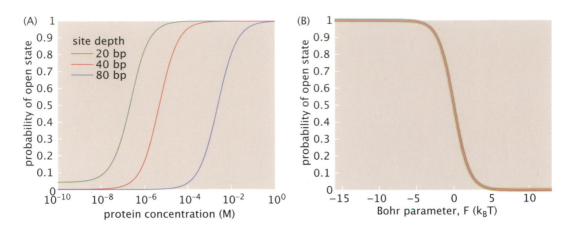

Probability of open state of the nucleosome. (A) Open-state probability as a function of transcription factor concentration for three different binding site depths within the nucleosome. (B) Data collapse for different binding site distances within the nucleosome.

10.4 MWC Model of Nucleosomes: Arbitrary Number of Binding Sites

Our treatment thus far has focused on the case in which there is one binding site within the nucleosome, and we broached the very interesting question of how the binding of a DNA-binding protein depends upon the distance within the nucleosome of that binding site. Now we consider a related problem, which is how to think about situations in which there are multiple binding sites within the nucleosome. To get a feel for the numbers, we note that typical transcription factor binding sites in eukaryotes are of order 10 bp in length, which means that given the 147 bp of DNA "buried" within a nucleosome, many distinct transcription factor binding sites are possible.

As in the previous section, our abstraction of the problem of nucleosomal binding imagines a nucleosome as having two states, a closed state and an open state. These microscopic states for the case of multiple binding sites and their corresponding states of occupancy and energies are shown in Figure 10.12. Using the states and weights presented in Figure 10.12, we can compute the partial partition function for the nucleosome state as

$$Z_{nucleosome} = \sum_{n=0}^{N} e^{-\beta \varepsilon_c} \frac{N!}{n!(N-n)!} e^{-\beta n(\epsilon_b^{(c)} - \mu)}. \tag{10.19}$$

We can effect this sum by using the fact that the binomial theorem tells us that

$$(x+y)^N = \sum_{n=0}^{N} \frac{N!}{n!(N-n)!} x^n y^{N-n}. \tag{10.20}$$

For our case, if we take

$$x = e^{-\beta(\epsilon_b^{(c)} - \mu)} \tag{10.21}$$

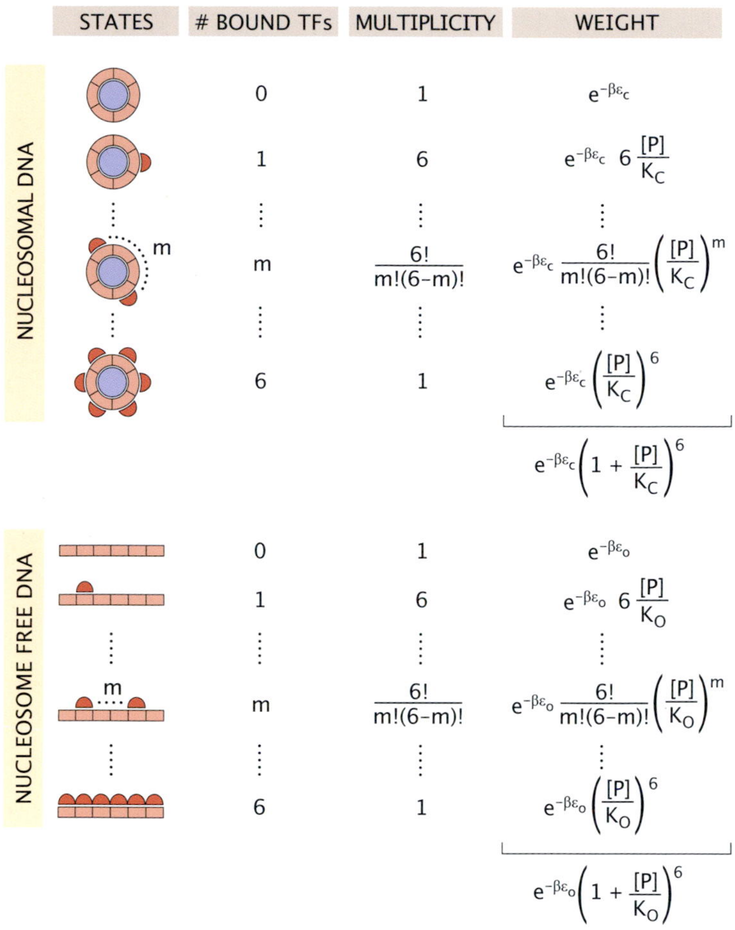

STATES	# BOUND TFs	MULTIPLICITY	WEIGHT
	0	1	$e^{-\beta\varepsilon_c}$
	1	6	$e^{-\beta\varepsilon_c}\, 6\dfrac{[P]}{K_C}$
m	m	$\dfrac{6!}{m!(6-m)!}$	$e^{-\beta\varepsilon_c}\dfrac{6!}{m!(6-m)!}\left(\dfrac{[P]}{K_C}\right)^m$
	6	1	$e^{-\beta\varepsilon_c}\left(\dfrac{[P]}{K_C}\right)^6$

$e^{-\beta\varepsilon_c}\left(1+\dfrac{[P]}{K_C}\right)^6$

NUCLEOSOMAL DNA

STATES	# BOUND TFs	MULTIPLICITY	WEIGHT
	0	1	$e^{-\beta\varepsilon_o}$
	1	6	$e^{-\beta\varepsilon_o}\, 6\dfrac{[P]}{K_O}$
m	m	$\dfrac{6!}{m!(6-m)!}$	$e^{-\beta\varepsilon_o}\dfrac{6!}{m!(6-m)!}\left(\dfrac{[P]}{K_O}\right)^m$
	6	1	$e^{-\beta\varepsilon_o}\left(\dfrac{[P]}{K_O}\right)^6$

$e^{-\beta\varepsilon_o}\left(1+\dfrac{[P]}{K_O}\right)^6$

NUCLEOSOME FREE DNA

Figure 10.12

States and weights for nucleosomal DNA with multiple binding sites. For concreteness, we consider the case of $N=6$ transcription factor binding sites.

and

$$y=1, \tag{10.22}$$

we are left with

$$Z_{nucleosome}=e^{-\beta\varepsilon_c}\left(1+e^{-\beta(\epsilon_b^{(c)}-\mu)}\right)^N. \tag{10.23}$$

Once again using the result for the chemical potential that $\mu=\varepsilon_{sol}+k_BT\ln\frac{[P]}{P_0}$, where $[P]$ is the concentration of the DNA-binding protein of interest, P_0 is the standard-state concentration, and that $K_C=P_0e^{\beta(\varepsilon_b^{(c)}-\varepsilon_{sol})}$, we can rewrite our result as

$$Z_{nucleosome}=e^{-\beta\varepsilon_c}\left(1+\frac{[P]}{K_C}\right)^N. \tag{10.24}$$

In light of the preceding analysis, we can similarly determine the partial partition function for the accessible open states as

$$Z_{open}=e^{-\beta\varepsilon_o}\left(1+\frac{[P]}{K_O}\right)^N. \tag{10.25}$$

by following precisely the same kind of argument we used to obtain $Z_{nucleosome}$. Given the partial partition functions, we can then write the probability that the chromatin is in the accessible open state as

$$p_{open} = \frac{Z_{open}}{Z_{nucleosome} + Z_{open}}. \tag{10.26}$$

Using our results for the partial partition functions, we can rewrite this as

$$p_{open}([P]) = \frac{e^{-\beta \varepsilon_o} \left(1 + \frac{[P]}{K_O}\right)^N}{e^{-\beta \varepsilon_o} \left(1 + \frac{[P]}{K_O}\right)^N + e^{-\beta \varepsilon_c} \left(1 + \frac{[P]}{K_C}\right)^N}. \tag{10.27}$$

We can rewrite equation 10.27 in a more familiar form by recourse to a few notational simplifications. We begin by defining an equilibrium constant that tells us about the relative frequency of the nucleosomal and open states as

$$L = e^{-\beta \Delta \varepsilon} = \frac{[N]}{[O]}, \tag{10.28}$$

where we have introduced the energy difference $\Delta \varepsilon = \varepsilon_c - \varepsilon_o$. We also parameterize the relative strength of transcription factor binding to those two states as

$$c = \frac{K_O}{K_C}, \tag{10.29}$$

and also define

$$x = \frac{[P]}{K_O}. \tag{10.30}$$

Given these definitions, we can now write the probability that the DNA will be wrapped in the nucleosomal state (inactive for transcription) as

$$Y_N = \frac{L(1 + xc)^N}{(1 + x)^N + L(1 + xc)^N}, \tag{10.31}$$

as plotted in Figure 10.13.

A second very interesting quantity is the fractional occupancy of the binding sites. We can write the number of bound DNA-binding proteins abstractly as

$$\langle n_{bound} \rangle = \sum_{n=0}^{N} n p(n), \tag{10.32}$$

where $p(n)$ is the probability of having n ligands bound. To compute this occupancy we invoke the states and weights shown in Figure 10.12 to write

$$\langle n_{bound} \rangle = \frac{1}{Z} \Big(\sum_{n=0}^{N} e^{-\beta \varepsilon_c} \frac{N!}{n!(N-n)!} e^{-\beta n (\epsilon_b^{(c)} - \mu)} n$$

$$+ \sum_{n=0}^{N} e^{-\beta \varepsilon_o} \frac{N!}{n!(N-n)!} e^{-\beta n (\epsilon_b^{(o)} - \mu)} n \Big), \tag{10.33}$$

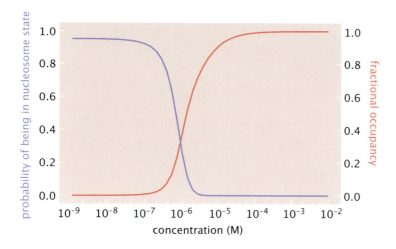

Figure 10.13

Results for MWC model of nucleosome. The probability of being in the nucleosomal state (Y_N) and the fractional occupancy ($Y = (\langle n_{bound} \rangle / N$) are both shown for comparison. The parameters used in this figure are $K_C = 10^{-4}$ M, $K_O = 10^{-6}$ M, $\varepsilon_c - \varepsilon_o \approx$ -3 $k_B T$ and $N = 6$ binding sites.

where the first term sums over all states of occupancy (i.e., 0 bound, 1 bound, 2 bound, $\cdots$, N bound) of the nucleosomal state, and the second term sums over all states of occupancy of the open state. We can rewrite this more conveniently as

$$\langle n_{bound} \rangle = \frac{1}{Z} \left(x_1 \frac{\partial}{\partial x_1} \sum_{n=0}^{N} e^{-\beta \varepsilon_c} \frac{N!}{n!(N-n)!} x_1^n + x_2 \frac{\partial}{\partial x_2} \sum_{n=0}^{N} e^{-\beta \varepsilon_o} \frac{N!}{n!(N-n)!} x_2^n \right),$$

(10.34)

where we have defined $x_1 = e^{-\beta(\epsilon_b^{(c)} - \mu)}$, and $x_2 = e^{-\beta(\epsilon_b^{(o)} - \mu)}$. These definitions are useful because we can then evaluate the sums indicated in equation 10.33 by taking derivatives with respect to x_1 and x_2 and obtaining

$$\langle n_{bound} \rangle = \frac{1}{Z} (x_1 N (1 + x_1)^{N-1} + x_2 N (1 + x_2)^{N-1}).$$

(10.35)

Using the notation already introduced, it is more convenient to write this as

$$Y = \frac{\langle n_{bound} \rangle}{N} = x \frac{(1+x)^{N-1} + Lc(1+cx)^{N-1}}{L(1+xc)^N + (1+x)^N},$$

(10.36)

as plotted in Figure 10.13. The key point is that as the fractional occupancy goes up, the probability of being in the nucleosomal state goes down.

One of the most mysterious parameters in the regulation of transcription in prokaryotes and eukaryotes is the number of binding sites controlling the expression of a given gene. Binding-site distributions associated with the regulatory architectures controlling different genes are very diverse, with some architectures having only a single binding site, and other architectures having many such binding sites. That said, the current state of the art in our understanding of binding-site copy number as a tunable knob controlling transcription is very primitive. The model just described gives us an opportunity to explore the implications of binding-site copy number as shown in Figure 10.14.

Figure 10.14

Figure 10.14

Probability of being in the nucleosomal (inactive) state as a function of the number of transcription factor binding sites. The plot shows the probability of being in the nucleosomal state as a function of the number of binding sites for two different concentrations of transcription factor.

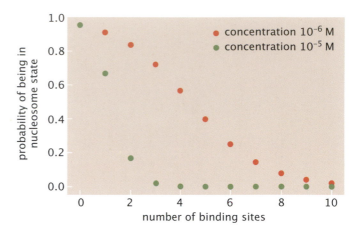

10.5 Nucleosome Modifications and the Analogy with the Bohr Effect

In a beautiful and important turn of phrase, Prabakaran (2012) and collaborators refer to posttranslational modifications as "nature's escape from genetic imprisonment." One of the archetypal examples of such posttranslational modifications is that of genome packing. As shown in Figure 10.15(A), the genomic DNA itself is subject to modifications through the addition of methyl groups which can alter the interaction energy between DNA and its protein partners. But these proteins, too, are liberated from their genetic imprisonment. As a result of the labors of a generation of scientists, there are now detailed maps of the way in which the histone proteins at the core of nucleosomes are modified by methylation, acetylation, phosphorylation, and ubiquitination, as shown in Figure 10.15(B). These so-called epigenetic modifications have a wide swath of functional consequences that make them an interesting twist on the story we have presented so far of nucleosomal stability and accessibility.

In the context of the two-state model of nucleosome accessibility, we can think of the action of posttranslational modifications as changing the parameter γ_{ad} in our statistical mechanical model. Recall that this parameter measures the stickiness between the DNA and the histones. As shown in Figure 10.16, the presence of such modifications can lead to a reduced interaction energy between the DNA and the histones. The model laid out for the probability of transcription factor occupancy of nucleosomal DNA can tell us how this probability will be altered by virtue of such posttranslational modifications.

Figure 10.17 shows an example of the shift in the transcription factor occupancy curve for different choices of the adhesion energy. Note that changing the adhesion energy effectively amounts to changing the parameter $\Delta\varepsilon = \varepsilon_c - \varepsilon_o$ in equation 10.27. Said differently, this parameter changes the relative probability of the nucleosomal and open states in the *absence* of any DNA-binding proteins. This encourages a fascinating analogy between the Bohr effect in hemoglobin (see Fig. 7.2, p. 233) and the role of histone modifications in the context of chromatin. A clear prediction of this model resulting from this analogy is that if one could set up an experiment in which the occupancy curves for a particular binding site buried within the nucleosome were measured for different sets of

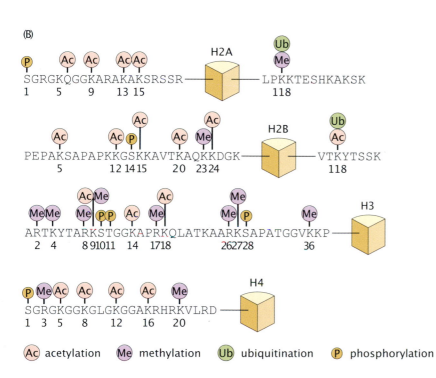

nuclear territories

chromosome

chromonema fiber

nucleosomes

H1

30 nm fiber

histones

nuclear envelope

DNA

nuclear pore

nuclear lamina

Figure 10.15

Sites of histone modifications. (A) Large-scale view of the nucleus and the compacted genome. The zoomed-out view provides different views of the underlying structure including the histone proteins. (B) High-resolution view of the four histone proteins and the sites of posttranslational modification each of them has available. Adapted from Steven et al. (2016).

(B)

H2A

SGRGKQGGKARAKAKSRSSR — LPKKTESHKAKSK
1   5   9   13 15   118

H2B

PEPAKSAPAPKKGSKKAVTKAQKKDGK — VTKYTSSK
      5   12 14 15   20 23 24   118

H3

ARTKYTARKSTGGKAPRKQLATKAARKSAPATGGVKKP
 2 4   8 9 10 11 14   17 18   26 27 28   36

H4

SGRGKGGKGLGKGGAKRHRKVLRD
1 3 5   8   12   16   20

Ac acetylation   Me methylation   Ub ubiquitination   P phosphorylation

Figure 10.16

Histone modifications and nucleosomal stability. The modified nucleosome is depicted as having a less stable closed state.

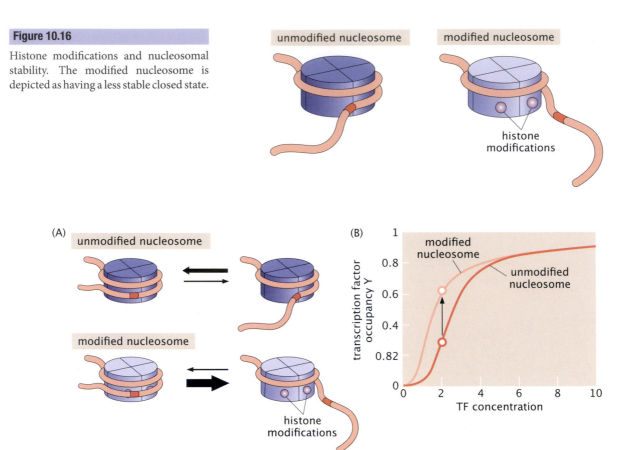

Figure 10.17

Analogy of Bohr effect in hemoglobin and the histone modification effect. (A) Kinetics of DNA unwrapping from nucleosomes in an unmodified nucleosome and a modified nucleosome. In the modified case, the inaccessible state is less stable. (B) Transcription factor occupancy on the nucleosomal DNA as a function of the concentration of transcription factor. The shift in the occupancy curves is analogous to the Bohr effect in hemoglobin. Adapted from Mirny (2010), see Further Reading.

histone modifications, these curves should form a one-parameter family that could be subjected to data collapse.

10.6 Stepping Up in Scales: A Toy Model of Combinatorial Control at Enhancers

One of the most compelling discoveries to emerge from the study of eukaryotic gene regulation, especially in multicellular organisms, is the existence of DNA-binding regions known as enhancers, which result in regulatory "action at a distance." Several examples of enhancers from the fruit fly *Drosophila melanogaster* that participate in the regulatory decisions behind embryonic development are shown in Figure 10.18. As seen in the figure, for such enhancers there are binding sites that are not in genomic proximity to the promoter they control. Further, depending upon the binding of transcription factors to these enhancers,

(A)

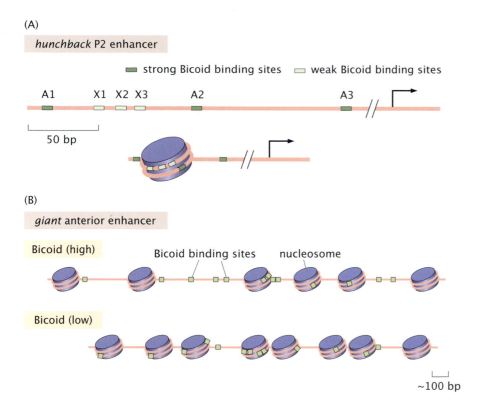

Figure 10.18

The nucleosome landscape of *Drosophila* genes. (A) The *hunchback* P2 enhancer has multiple Bicoid binding sites. A number of those binding sites (*X1*, *X2*, *X3*, and *A2*) are found within a nucleosome. (B) Bicoid dosage-dependent nucleosomal occupancy of the *giant* anterior enhancer. In high Bicoid concentrations, a subset of the nucleosomes are released. Data courtesy of Shelby Blythe.

the genes are expressed to differing extents. A particularly intriguing aspect of these enhancers from the point of view of more traditional mechanisms of gene regulation is their extreme flexibility—in some cases there seems to be a generic indifference to the number of binding sites, their specific position, and even their sequence. Interestingly, a generalization of the MWC concept introduced earlier in the chapter to characterize single nucleosomes is also useful for characterizing these ubiquitous eukaryotic regulatory architectures.

For example, the embryonic development of the fruit fly *Drosophila melanogaster*'s body plan is determined by the expression levels of a hierarchy of genes with single-cell resolution along the anterior-posterior axis of the embryo. Figure 10.19 shows a schematic representation of this hierarchy of genes and the successive steps in the development of the anterior-posterior part of the fly body plan. One such gene is *even-skipped*, which is expressed in seven stripes along the anterior-posterior axis of the embryo. Each one of these stripes is controlled by an individual enhancer (though some stripes, such as 3 and 7, use the same enhancer) located up to 8 kbp upstream or downstream of the actual *even-skipped* gene. The enhancer that controls stripe 2, for example, is located 1.5 kbp upstream from the gene and in its minimal form spans 480 bp, as indicated schematically in Figure 10.3(C). It contains several binding sites for two activators and two repressors. Specifically, it has three binding sites for the activator Bicoid, despite the fact that the deletion of any one of these sites does not usually cause any qualitative changes to the output pattern. Perhaps more revealing in terms of the flexibility of these regions is that this enhancer sequence has undergone significant changes throughout

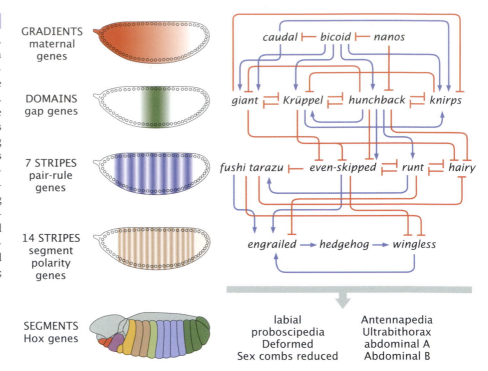

Figure 10.19

Establishing the fly body plan. The schematics of the embryo on the left show a succession of patterns of gene expression over time during embryonic development. The regulatory network on the right shows how the various genes involved in embryonic patterning activate (blue lines) and repress (red lines) other genes in the network. Our focus in this discussion is on the arrow connecting *bicoid* in the first row to *hunchback* in the second row. Adapted from Carroll, Grenier, and Weatherbee (2001); Edgar, Odell, and Schubiger (1989); Jaeger (2011); and von Dassow et al. (2000).

evolution while retaining its function. For example, in another species of fly called *D. pseudoobscura* the same enhancer has lost and gained binding sites while the remaining binding sites have changed in their affinities with respect to the *D. melanogaster* enhancer. The spacing between some of these sites has also changed in some cases by up to 80 bp. Nevertheless, when the *D. pseudoobscura* enhancer is introduced into *D. melanogaster* not only does it result in a very similar pattern of expression, it can even rescue mutations in the *even-skipped* gene.

An exciting hypothesis about the action enhancer of is that they control the level of gene expression by controlling chromatin accessibility. This is in stark contrast to a picture in which transcriptional cooperativity is attributed to direct interactions between transcription factors and the basal transcription apparatus. Figure 10.20(A) shows a schematic example of how the MWC concept can be applied to a model chromatin state. In the "closed" or inaccessible state, the DNA is wrapped up in a tight nucleosomal configuration, here indicated by one of many hypothetical higher-order chromatin structures (i.e., the putative 30 nm fiber). While in this state, the promoter of interest is hypothesized to be unavailable for transcription. The concept of the model is that RNA polymerase and transcription factors can bind more easily to DNA when it is in its open or accessible state, indicated schematically in Figure 10.20 by DNA that is freely available in the "open chromatin" configuration.

As usual, we can think about the system quantitatively by recourse to the statistical mechanical states and weights shown in Figure 10.20(B). In this case, we consider a situation in which there are two transcription factors R and B that bind to sites within this chromatin region and that their binding affinity is

(A)

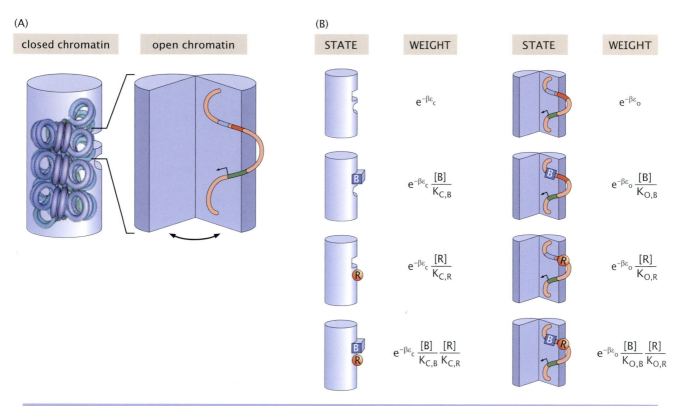

Figure 10.20

Schematic descriptions of MWC chromatin. (A) The genomic DNA exists in two classes of state: in the "closed" state of chromatin and the "open" state of chromatin, which permits transcription. Transcription factor binding controls the relative probability of these different eventualities. (B) States and weights for the binding of two transcription factors, here denoted by *R* and *B*, which occupy the open and closed conformations with different affinity.

higher for the open state. As a consequence, the transcription factors will shift the free-energy balance to the open chromatin conformation. Our goal is to compute the probability of the open state, since in this simplest of models, the gene is considered to be on when the chromatin is in the accessible state. Using the states and weights, we see that the probability of the open state is given as

$$p_{open} = \frac{e^{-\beta\varepsilon_o}(1 + \frac{[R]}{K_{O,R}})(1 + \frac{[B]}{K_{O,B}})}{e^{-\beta\varepsilon_o}(1 + \frac{[R]}{K_{O,R}})(1 + \frac{[B]}{K_{O,B}}) + e^{-\beta\varepsilon_c}(1 + \frac{[R]}{K_{C,R}})(1 + \frac{[B]}{K_{C,B}}).} \quad (10.37)$$

Even within the relatively simple scenario depicted in Figure 10.20(A), there is already a great deal of conceptual and quantitative flexibility to account for a host of different regulatory architectures. For example, one can imagine situations such as that shown in Figure 10.20(B) in which the transcription factors bind more favorably in either the closed or open conformations, thus stabilizing one state or the other. Similarly, we can imagine both positive and negative cooperativity between the transcription factors themselves through direct physical contacts, permitting the construction of various logic functions such as AND and OR functions (and many others). From the perspective of the MWC

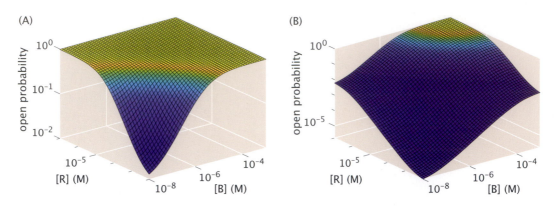

Figure 10.21

Plot of the probability of the on state of chromatin as a function of the concentrations of the two transcription factors. In both figures, the K_ds are given by $K_{O,R} = 2.5 \times 10^{-8}$ M, $K_{C,R} = 1.5 \times 10^{-4}$ M, $K_{O,B} = 2.5 \times 10^{-8}$ M, and $K_{C,B} = 1.5 \times 10^{-4}$ M. For the OR logical function (A), $\Delta\varepsilon = -5\,k_B T$, and for the AND logical function (B), $\Delta\varepsilon = -14\,k_B T$.

model itself, the key parameters that come into play are the difference in energy between the closed and open conformations, $\Delta\varepsilon = \varepsilon_c - \varepsilon_o$, the binding energies (or K_ds) for the relevant transcription factors in each of the states, and the "Hill coefficient," which can be tuned by changing the number of binding sites for the DNA-binding proteins in question. In Figure 10.21, we show an example of how the probability of being in the active state depends upon the concentrations of the two species of transcription factor.

10.7   An Application of the MWC Model of Nucleosomes to Embryonic Development

We can take the model of an enhancer with multiple binding sites like that described to the case of fly development. After the fly embryo begins its rapid journey of development, the body plan is specified through patterns of gene expression. For example, as shown in Figure 10.19, the long axis is specified by a hierarchy of genes resulting in the segmented body plan along the anterior-posterior axis.

As seen in Figure 10.19, one level in the hierarchy of developmental patterning plays out as a result of the input-output function connecting the morphogen Bicoid and the transcription factor Hunchback. The measured input-output function between Bicoid and Hunchback is shown in Figure 10.22. The figure shows that when Bicoid concentration is low, so, too, is the Hunchback concentration. By way of contrast, when the Bicoid concentration is high, the Hunchback concentration is high, as well. One key point evident in the figure is that at each level in the hierarchy, the patterns of gene expression become sharper. For example, Bicoid is characterized by an exponentially decaying gradient from the anterior to the posterior end of the embryo, as shown in the first level of the hierarchy. Patterns of Hunchback expression show a sharper interface between regions of high and low expression.

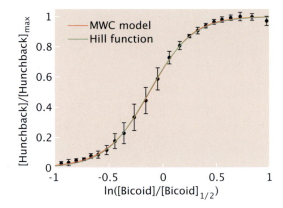

Figure 10.22

Hunchback-Bicoid input-output function. The data points show Hunchback protein expression as a function of Bicoid concentration normalized by the Bicoid concentration at which Hunchback expression reaches half its maximum ($[Bicoid]_{1/2}$). The fit to the MWC model from equation 10.39 yields $\Delta \varepsilon = 10.85 \ k_B T$ and $K_d/[Bicoid]_{1/2} = 0.17$, while the fit to the Hill function from equation 10.38 results in $K_d/[Bicoid]_{1/2} = 0.88$ and $N = 5.01$. Data adapted from Gregor et al. (2007).

This sharp input-output function has been speculated to be the result of protein-protein interactions between Bicoid molecules, leading to cooperativity in the binding of this transcription factor to the DNA. As seen in Figure 10.22, this makes it possible to fit the data using a Hill function

$$p_{active}([Bcd]) = \frac{(\frac{[Bcd]}{K_d})^N}{1 + (\frac{[Bcd]}{K_d})^N}. \tag{10.38}$$

However, the allosteric framework advocated throughout the book and specifically the ideas on nucleosomal accessibility offered in this chapter offer an alternative possible interpretive framework for thinking about these problems. Figure 10.23 shows how we can set up the states and weights for the binding-site configuration in this *hunchback* enhancer. We once again imagine that there are two classes of genome configurations, those in which the genome is folded in a way that renders our gene of interest inactive and those in which the genome architecture is open, allowing for the gene of interest to be on. In addition, we make the simplifying assumption that when folded, the genome is inaccessible, such that Bicoid binding cannot take place.

Given these states and weights, we have a prescription for computing the activity of the *hunchback* gene as a function of Bicoid concentration as

$$p_{active}([Bcd]) = \frac{e^{-\beta \Delta \varepsilon}(1 + \frac{[Bcd]}{K_O})^6}{e^{-\beta \Delta \varepsilon}(1 + \frac{[Bcd]}{K_O})^6 + 1}, \tag{10.39}$$

where $[Bcd]$ is the Bicoid concentration and where we have defined $\Delta \varepsilon = \varepsilon_{on} - \varepsilon_{off}$ as the difference between the energy of the chromatin open and closed states, respectively. Figure 10.22 shows an example of fits using both the MWC framework and Hill functions for the Bicoid-Hunchback input-output function. One key distinction between the MWC and Hill frameworks is that in the MWC approach, there is an effective cooperativity without invoking any explicit cooperative reactions, as are central to the Hill function. Further, the number of binding sites in the MWC approach is a real parameter, while in the Hill framework it is a fit parameter that cannot be interpreted as the number of binding sites. Note that the model presented here is provisional. Ultimately, the power of an analysis like this is that it suggests interesting parameters that

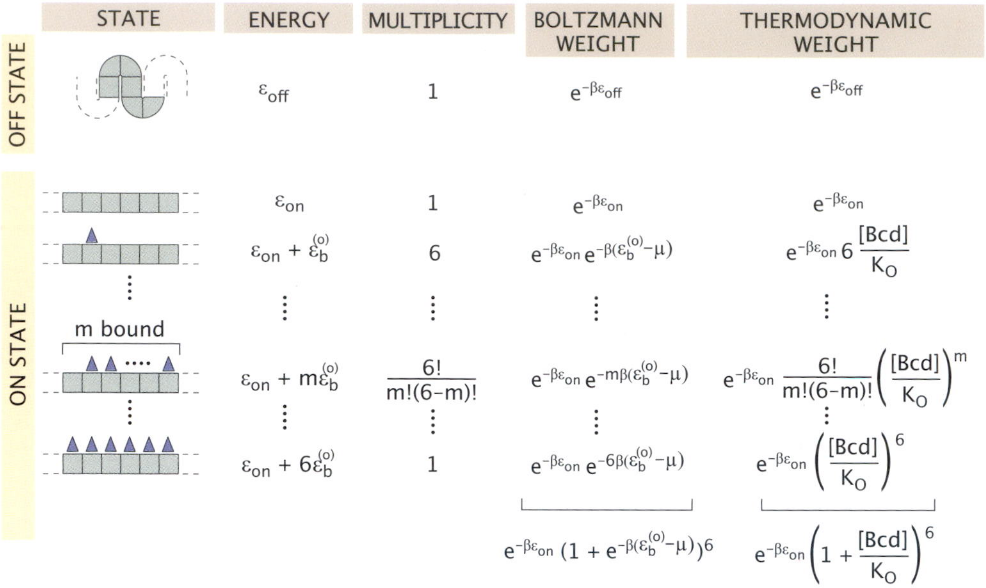

Figure 10.23

States and weights for MWC model of DNA binding by transcription factors. In the chromatin state, the DNA is imagined to be so highly folded that Bicoid proteins cannot bind. In the on state, the chromatin is unfolded, and six sites are accessible to the Bicoid protein.

can be tinkered with experimentally to find out whether our thinking is on track or not. As noted earlier one of the most interesting, sharp predictions that comes out of the MWC framework is that the parameter N characterizing the number of Bicoid binding sites is directly related to the coefficient seen in equation 10.39 (in that case, $N = 6$). This raises the experimental possibility of tuning the number of Bicoid binding sites and measuring how the input-output curve changes.

10.8 Summary

In its original formulation, the allosteric model was introduced as a way to think about the nuanced behavior of proteins. This chapter showed how the allosteric framework can be extended to other kinds of molecular switches. In particular, it has been hypothesized that genomic packing results in the emergence of a "chromatin switch" in which packaged DNA transitions between a compact chromatin state which is unavailable for transcription and an open state which is transcriptionally active. Note that the wide-ranging analysis carried out in this chapter leaves in its wake many currently untested predictions about the accessibility of nucleosomal DNA and the implications of this accessibility for gene expression. We argue that the great power of models like those presented in this chapter in contrast with phenomenological fits to activity curves such as are offered by Hill functions is that the allosteric models acknowledge key tuning variables that can serve as experimental knobs.

10.9 Further Reading

Culkin, J., L. de Bruin, M. Tompitak, R. Phillips, and H. Schiessel (2017) "The role of DNA sequence in nucleosome breathing." *Eur. Phys. J.* E40:106. This paper shows how to use a discrete model for nucleosome accessibility rather than the simplified continuum model provided in the chapter.

De Gennes, P. G. (1979) *Scaling Concepts in Polymer Physics*. Ithaca, NY: Cornell University Press. De Gennes gives a clear derivation of the role of excluded-volume effects in polymers.

Doyle, B., G. Fudenberg, M. Imakaev, and L. Mirny (2014) "Chromatin loops as allosteric modulators of enhancer-promoter interactions." *PLoS Comp. Biol.* 10:e1003867. This excellent article shows the ways in which the allostery concept can be generalized to the example of chromatin.

Mirny, L. A. (2010) "Nucleosome-mediated cooperativity between transcription factors." *Proc. Natl. Acad. Sci.* 107(52):22534–22539. This is one of the articles that really persuaded me to spend some years thinking about the broad reach of the allostery concept. Seeing it applied to DNA structure rather than to a strictly protein problem was very enlightening.

Narula, J., and Igoshin O. A. (2010) "Thermodynamic models of combinatorial gene regulation by distant enhancers." *IET Syst. Biol.* 4:393–408. This excellent article discusses combinatorial control in nucleosomes.

Phillips, R., J. Kondev, J. Theriot, and H. G. Garcia (2012) *Physical Biology of the Cell*. New York: Garland Press. This book gives a detailed explanation of the energy cost to elastically deform a polymer such as DNA into a circular hoop.

Polach, K. J., and Widom J. (1995) "Mechanism of protein access to specific DNA sequences in chromatin: A dynamic equilibrium model for gene regulation." *J. Mol. Biol.* 254:130. This important article shows how proteins access DNA wrapped in nucleosomes and inspires the allosteric model of that phenomenon shown in the chapter.

Schiessel, H. (2003) "The physics of chromatin." *J. Phys.: Condens. Matter* 15:R699. This excellent article describes the structure and energetics of nucleosomes.

10.10 REFERENCES

Britten, R. J., and E. H. Davidson (1969) "Gene regulation for higher cells: A theory." *Science* 165:349.

Carroll, S. B., J. K., Grenier, and S. D. Weatherbee (2001) *From DNA to Diversity: Molecular Genetics and the Evolution of Animal Design*. Malden, MA: Blackwell Science.

Davidson, E. H. (2006) *The Regulatory Genome: Gene Regulatory Networks in Development and Evolution*. Cambridge, MA: Academic Press.

Edgar, B. A., G. M. Odell, and G. Schubiger (1989) "A genetic switch, based on negative regulation, sharpens stripes in *Drosophila* embryos." *Dev. Genet.* 10:124–142.

Gregor, T., D. W. Tank, E. F. Wieschaus, and W. Bialek (2007) "Probing the limits to positional information." *Cell* 130:153–164.

Jaeger, J. (2011) "The gap gene network." *Cell Mol. Life Sci.* 68:243–274.

Polach, K. J., and J. Widom (1995) "Mechanism of protein access to specific DNA sequences in chromatin: A dynamic equilibrium model for gene regulation." *J. Mol. Biol.* 254:130.

Prabakaran, S., G. Lippens, H. Steen, and J. Gunawardena (2012) "Post-translational modification: Nature's escape from genetic imprisonment and the basis for dynamic information encoding." *Wiley Interdiscip. Rev. Syst. Biol. Med.* 4(6):565–583, doi: 10.1002/wsbm.1185.

Steven, A. S., W. Baumeister, L. N. Johnson, and R. N. Perlham (2016) *Molecular Biology of Assemblies and Machines.* New York: Garland Science.

Surkova, S., D. Kosman, K. Kozlov, Manu, E. Myasnikova, A. A. Samsonova, A. Spirov, C. E. Vanario-Alonso, M. Samsonova, and J. Reinitz (2008) "Characterization of the Drosophila segment determination morphome." *Dev. Biol.* 313(2):844–862.

von Dassow, G., E. E. Meir, M. Munro, and G. M. Odell (2000) "The segment polarity network is a robust developmental module." *Nature* 406:188–192.

PART III

BEYOND ALLOSTERY

ALLOSTERY EXTENDED

Things are not always what they seem; the first appearance deceives
many; the intelligence of a few perceives what has been carefully hidden.
—Phaedrus

Theoretical arguments and various structural probes have both indicated that the simple two-state framework that has served as the backdrop for the entire book overlooks a host of hidden dynamical processes within allosteric molecules. To that end, in this chapter we provide a deeper explanation of the ideas introduced in section 2.6 (p. 63) to see how unaccounted-for degrees of freedom behave as hidden variables. The key outcome is that by integrating over these hidden degrees of freedom, in some cases we can still describe the activity of some allosteric molecule of interest in the form

$$p_{active} = \frac{Z_{active}}{Z_{active} + Z_{inactive}} = \frac{e^{-\beta F_{active}}}{e^{-\beta F_{active}} + e^{-\beta F_{inactive}}}, \qquad (11.1)$$

where F_{active} and $F_{inactive}$ are both obtained by summing over the ensemble of active and inactive states, respectively.

11.1 Ensemble Allostery

The primary concern of this chapter is that the original MWC, KNF, and Eigen models all oversimplify the space of different conformations available to macromolecules by neglecting the full spectrum of motions available in each case. Figure 11.1 gives an indication of the classes of hidden degrees of freedom that can influence the free-energy balance between the inactive and active states of a molecule. As we proceed from left to right in the figure, we see increasingly rich internal dynamics. In the leftmost example, hemoglobin illustrates rigid-body rotations of the subunits of a protein. The point is that the static picture of the structure of hemoglobin that comes from X-ray crystallographic studies, for example, needs to be amended to account for the fact that the different subunits are incessantly jostling among different rotational states, and these rotations make an entropic contribution to the free energy that depends upon whether oxygen is bound. In the second example we use another player in one of our key case studies, namely, the bacterial transcriptional activator CRP, to argue that

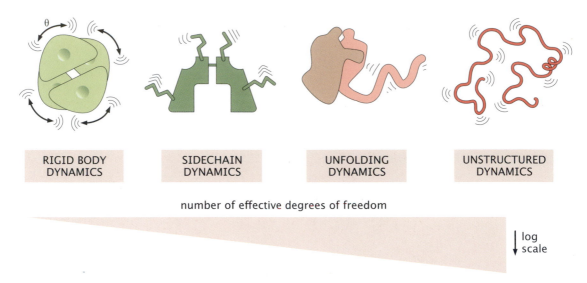

Figure 11.1

Levels of coarse graining of hidden degrees of freedom in allosteric transitions. From left to right, different classes of internal degrees of freedom and their corresponding dynamics are shown. Adapted from Motlagh et al. (2014); see Further Reading.

the side-chain dynamics of the amino acids can be different between the inactive and active states. The third example shows a dimeric protein with one of its subunits completely unfolded in one of the states of activity. The final figure shows an intrinsically disordered protein whose dynamics can be altered depending upon whether a ligand is bound or not. As we see from the schematic bar at the bottom of the figure, the number of effective degrees of freedom increases from left to right, with the number of hidden microscopic states increasing exponentially as the spatial scale of the internal degrees of freedom gets smaller. In this chapter, we consider how to go from the words and pictures used to characterize these internal degrees of freedom to explicitly calculating their role in the free-energy balance that determines when a macromolecule will play some key functional role and when it will not.

A beautiful example of the ensemble allostery concept that has come into focus over the last several decades is one of our old favorites, the transcriptional activator CRP, shown in Figure 11.2 and already described in detail in section 8.4 (p. 290). As with many transcription factors, the activity of CRP depends upon the presence or absence of some effector, in this case cAMP. Because of methods such as nuclear magnetic resonance (NMR), we now have a more refined structural view of this protein. Specifically, as shown in Figure 11.2(B) and (C), NMR has revealed that in the presence of cAMP, the internal dynamics of the protein (more specifically, its side chains) are inhibited. In the next section, we will make a toy model showing how changes in the vibrational states of a molecule change its vibrational entropy. This entropy change is inherited by the free-energy difference between the active and inactive states. We also estimate the magnitude of this effect in $k_B T$ units to show how it compares with the $\Delta \varepsilon$ parameter that determine the free-energy difference between the inactive and active states in the absence of ligand.

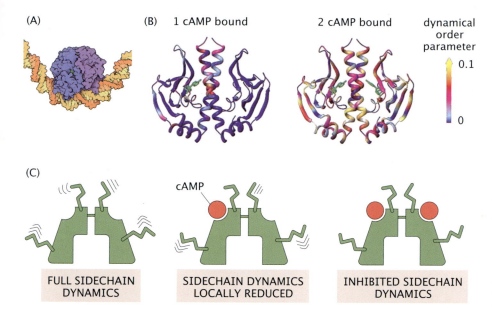

(A)

(B) 1 cAMP bound   2 cAMP bound   dynamical order parameter

0.1

0

(C)

cAMP

FULL SIDECHAIN DYNAMICS

SIDECHAIN DYNAMICS LOCALLY REDUCED

INHIBITED SIDECHAIN DYNAMICS

Figure 11.2

Internal dynamics of the transcriptional activator CRP. (A) Structure of CRP bound to DNA. (B) Dynamical order parameter for CRP when different numbers of cAMP molecules are bound. The dynamical order parameter provides a measure of the amplitude of the fast (ps-ns) internal motions of the protein. When the order parameter is small, it means there are larger motions. (C) Inhibition of sidechain dynamics of CRP resulting from the binding of cAMP. (A) courtesy of David Goodsell. (B) adapted by permission from Springer Nature via Copyright Clearance Center: Popovych, N., S. Sun, R. H. Ebright, and C. G. Kalodimos (2006), "Dynamically driven protein allostery," *Nat. Struc. Mol. Biol.*

11.1.1   Normal Modes and Mechanisms of Action at a Distance

What exactly is meant by the cartoons showing jiggling motions in Figures 11.1 and 11.2? An intuitive way to understand such jiggling motions is in terms of the so-called normal modes of vibration. In this section, we will explore the simplest models of such normal modes, because they will provide a natural language for our discussion of ensemble allostery. The point of the analysis is the recognition that an allosteric molecule is not a static entity. Rather, it is constantly jiggling around, and the nature of those jiggling motions can be different in the different conformational states of the protein. One objective of this section will be to show how these normal modes permit us to adjust the "bare" free energy of the allosteric system to include all these hidden vibrational (ensemble) degrees of freedom.

To be concrete, we consider the example of the mass-spring system shown in Figure 11.3. To describe the motions of this system, the natural first instinct is to assign a displacement x_1 to the first atom and a displacement x_2 to the second atom. We can write the equations of motion for the two masses by inspection using Newton's second law of motion ($F = ma$) as

$$m\frac{d^2x_1}{dt^2} = -kx_1 + k(x_2 - x_1). \tag{11.2}$$

The first term on the right expresses the force on mass 1 due to the spring on the left. The second term on the right gives the force on mass 1 due to the middle spring, whose amount of stretch depends upon the difference in the displacements of the two masses. The equation of motion for the second mass can be written similarly as

$$m\frac{d^2x_2}{dt^2} = -kx_2 + k(x_1 - x_2). \tag{11.3}$$

Figure 11.3

Normal modes of a two-atom molecule. (A) x_1 and x_2 are the coordinates of the two atoms. (B) Normal modes are the natural coordinates of the problem. The symmetric normal mode x_s involves motions of both atoms in a concerted fashion in the same direction. The asymmetric normal mode x_a involves concerted motions of the two-atoms with equal amplitudes in opposite directions.

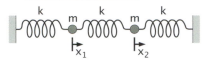

(A) BAD COORDINATES

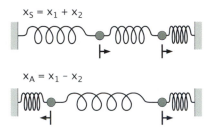

(B) NATURAL COORDINATES

$x_S = x_1 + x_2$

$x_A = x_1 - x_2$

The key observation is that the two equations of motion are coupled. That is, the dynamics of the first mass depends upon the dynamics of the second mass and vice versa.

Though there are more sophisticated methods to solve these equations in the case where there are more than two masses, for our purposes we resort to a simple trick that allows us to see the solutions more directly. In particular, by adding equations 11.2 and 11.3, we obtain

$$m\frac{d^2x_s}{dt^2} = -kx_s \tag{11.4}$$

where we have defined

$$x_s = x_1 + x_2. \tag{11.5}$$

Similarly, if we subtract equation 11.3 from equation 11.2, we find

$$m\frac{d^2x_a}{dt^2} = -3kx_a, \tag{11.6}$$

where we have defined the asymmetric normal coordinate as

$$x_a = x_1 - x_2. \tag{11.7}$$

We have discovered the existence of the "natural coordinates" of the problem, the so-called normal modes. The equations of motion in this form have now been simplified to the standard harmonic oscillator problem with solutions

$$x_s(t) = A \cos \omega_s t + B \sin \omega_s t \tag{11.8}$$

and

$$x_a(t) = C \cos \omega_a t + D \sin \omega_a t, \tag{11.9}$$

with the symmetric mode vibrating at a frequency $\omega_s = \sqrt{k/m}$ and the asymmetric mode vibrating at a frequency $\omega_a = \sqrt{3k/m}$.

Though the details are beyond the scope of this book, our analysis for the two-atom system can be generalized to include as many atoms as we like. In this case, the equation of motion for the ith coordinate (for example, the x-position of atom 12) is given by

$$m\frac{d^2 x_i}{dt^2} = -\sum K_{ij}x_j, \tag{11.10}$$

where K_{ij} is the stiffness matrix and captures the effect of all the springs that connect the different atoms within the molecule. Just as we found for the two-atom case, there are different vibrational frequencies for each mode of vibration, and each such mode involves different linear combinations of the atomic displacements.

The analysis of the vibrational modes of allosteric molecules has yielded interesting mechanistic insights and hypotheses into the inner workings of allostery. Specifically, a key mechanistic question that has been addressed in the language of modes of vibration is that of action at a distance, namely, how does the binding of a ligand at one part of a protein lead to conformational changes somewhere quite distant within the molecule? The simplest answer is that vibrational modes are extended objects. For example, if we consider a chain of atoms connected by springs, if we work out the modes of vibration, for each vibrational frequency (the eigenvalues of the so-called stiffness matrix), there will be a pattern of displacements that involves *all* the atoms. Having now seen how a given allosteric molecule has many distinct modes of vibration, we turn to the question of ensemble allostery and how the contribution of those vibrational modes can be integrated out of the problem.

11.1.2   Integrating Out Degrees of Freedom

Our starting point for the analysis of the energetic consequences of the hidden internal dynamics of an allosteric molecule follows our standard statistical mechanics protocol: identify the microscopic states and their corresponding energies. To be concrete about hidden internal dynamics, we focus on the different vibrational states of a molecule. Figure 11.4 shows a rendering of this idea for an allosteric molecule, using a quantum mechanical version of the energy of internal vibrations. For simplicity, the figure uses what is often referred to as an Einstein model, simplifying the problem by imagining that all the vibrations have the same frequency. This simplifying assumption can easily be replaced with an analysis based on the entire vibrational spectrum, as we will show later in this chapter.

The energy spectrum for a quantum harmonic oscillator is of the form $E_n = (n + \frac{1}{2})\hbar\omega$, where ω is the vibrational frequency of the oscillator, and n is the number of quanta of that vibrational mode that are excited. As noted earlier, for now we consider an Einstein model in which each atom making up the molecule jiggles around with precisely the same frequency. This idea is depicted graphically in Figure 11.5(A). In general, if there are N atoms in our molecule of interest, it will have $3N$ degrees of freedom, corresponding to the (x, y, z) positions of each atom. Associated with these $3N$ conformational degrees of freedom, there will be $3N - 6$ modes of vibration, each of which in principle has its own distinct frequency. The reason we subtract 6 from the

Figure 11.4

Hidden vibrational states of an allosteric molecule. Each of the two macrostates has its own vibrational spectrum. The different microstates are labeled by an integer that denotes how many quanta of vibration are present. We can integrate out the entirety of these hidden degrees of freedom to obtain an effective two-state model in which the statistical weight of the inactive state is Z_I, and the statistical weight of the active state is Z_A.

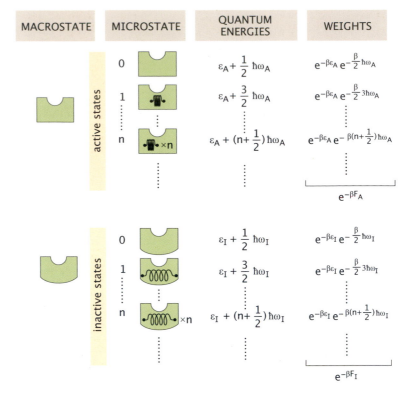

Figure 11.5

Models for vibrational spectrum of an allosteric molecule. (A) Einstein model of vibrational spectra of the inactive and active states of the molecule. All vibrational frequencies of the molecule are assumed to be the same. (B) Rectangular band model of a vibrational spectrum in which the modes are spread out uniformly over a range of different frequencies of width $\Delta\omega_I$ and $\Delta\omega_A$ for the inactive and active states, respectively.

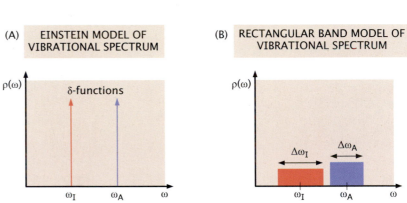

number of degrees of freedom is that three of those degrees of freedom correspond to the translation of the center of mass and three to the overall rotation of the molecule. The way we capture the spectrum of different vibrational states conceptually is through the so-called vibrational density of states, a plot that enumerates the number of different vibrational states between frequency ω and frequency $\omega + \Delta\omega$. For the case of the Einstein model that serves as our first toy model, this vibrational spectrum is all localized at a single frequency, as shown in Figure 11.5(A). Note that the key argument in this model is that the vibrational frequency for our Einstein model is different for the inactive and active states.

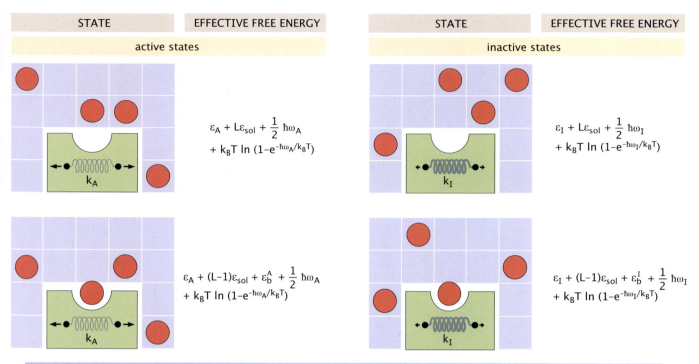

STATE	EFFECTIVE FREE ENERGY	STATE	EFFECTIVE FREE ENERGY
active states		inactive states	

Active states (left column):

$$\varepsilon_A + L\varepsilon_{sol} + \frac{1}{2}\hbar\omega_A + k_BT\ln(1-e^{-\hbar\omega_A/k_BT})$$

$$\varepsilon_A + (L-1)\varepsilon_{sol} + \varepsilon_b^A + \frac{1}{2}\hbar\omega_A + k_BT\ln(1-e^{-\hbar\omega_A/k_BT})$$

Inactive states (right column):

$$\varepsilon_I + L\varepsilon_{sol} + \frac{1}{2}\hbar\omega_I + k_BT\ln(1-e^{-\hbar\omega_I/k_BT})$$

$$\varepsilon_I + (L-1)\varepsilon_{sol} + \varepsilon_b^I + \frac{1}{2}\hbar\omega_I + k_BT\ln(1-e^{-\hbar\omega_I/k_BT})$$

Figure 11.6

States and effective energies of the active and inactive conformations as a result of thermal vibrations. Note that the "structure" of the molecule is not different between the active state (left column) and inactive state (right column). Their differences are in their vibrational spectra, as characterized by their different "spring constant" values k_A and k_I and associated vibrational frequencies ω_A and ω_I.

For the purposes of analyzing the free energy of the inactive and active states, we appeal to the quantum model of Figure 11.4. The figure shows that the vibrations are quantized and that each additional quantum of vibration brings an energy $\hbar\omega$. In light of this observation, we can write the partition function as

$$Z = \sum_{n=0}^{\infty} e^{-\beta(n+\frac{1}{2})\hbar\omega}, \tag{11.11}$$

where for now, we don't specialize to either the active (ω_A) or inactive (ω_I) states. If we define $x = e^{-\beta\hbar\omega}$, we see that the partition function is a geometric series and can be summed as

$$Z = e^{-\frac{\beta\hbar\omega}{2}} \sum_{n=0}^{\infty} x^n = e^{-\frac{\beta\hbar\omega}{2}}(1 - e^{-\beta\hbar\omega})^{-1}. \tag{11.12}$$

Since the free energy can be written as $F = -k_BT\ln Z$, we have

$$F = \frac{1}{2}\hbar\omega + k_BT\ln(1 - e^{-\beta\hbar\omega}). \tag{11.13}$$

Given this result, as seen in Figure 11.6, we can now restore the original states and weights used to describe an allosteric molecule. The primary outcome of the analysis of the ensemble of vibrational states is that we replace the energies ε_A and ε_I of the active and inactive states with effective energies

$$\varepsilon_A^{eff} = \varepsilon_A + \frac{1}{2}\hbar\omega_A + k_B T \ln(1 - e^{-\beta\hbar\omega_A}) \tag{11.14}$$

and

$$\varepsilon_I^{eff} = \varepsilon_I + \frac{1}{2}\hbar\omega_I + k_B T \ln(1 - e^{-\beta\hbar\omega_I}). \tag{11.15}$$

Figure 11.6 shows us that whereas in the original states and weights for an allosteric molecule of Figure 2.10 (p. 54) we had the energies ε_A and ε_I to describe the energy difference between active and inactive states, we really should be thinking of those energies as effective variables that mask a variety of hidden degrees of freedom, such as the changes to the energy and entropy of solution upon ligand binding, the changes to the vibrational spectrum of the MWC molecule when it goes from inactive to active, and so on.

By performing these operations, we have constructed an effective theory of the energy difference between the inactive and active states. This free energy accounts not only for the "bare" energy differences between these configurations but also for the hidden degrees of freedom associated with the thermal vibrations of the molecule. Of course, we have written a highly oversimplified toy model of the phenomenon, but the point is to illustrate how the entropy insinuates itself into the parameters characterizing the probabilities of the inactive and active states. Even more important, note that the mapping from the continuum of states to the two discrete inactive and active states is rigorous. It is not based upon wishful thinking but, rather, emerges naturally as a result of the determination of the free energy of the inactive and active states.

Our Einstein model of the vibrational spectrum of our allosteric molecule as shown in Figure 11.5(A) pretends that all the normal modes of vibration of the molecule have the same frequency. We can improve the realism of our model, as shown in Figure 11.5(B), by accounting for the distribution of vibrational frequencies. Here again we make simplifying assumptions with this highly schematized vibrational density of states to reveal the structure of the calculations and leave the further generalization of this model to more realistic vibrational spectra as an exercise for the reader. To obtain the total free energy of our allosteric molecule as amended by thermal vibrations, we compute

$$F_{tot} = \int_0^\infty F(\omega)\rho(\omega)d\omega, \tag{11.16}$$

where $\rho(\omega)$ is the density of states (see Figure 11.5(B) for an example of a "rectangular band" density of states), and $F(\omega)$ was calculated in equation 11.13.

Though we are interested in the biological consequences of the hidden degrees of freedom this section has explored, I still believe the most important message of this chapter and one of the most important of the whole book is the demonstration that it is formally possible to construct theories which rigorously integrate out degrees of freedom. The reason this is so important for biology is that the molecular biology revolution has left in its wake a strong tendency to believe that the only valid notion of mechanism is one based on microscopic degrees of freedom. What our calculations here show, practiced with enthusiasm in physics for centuries, is that one can construct higher-level effective theories in which microscopic degrees of freedom have been removed. I believe this is one of the central challenges that needs to be taken up in the wake of the molecular/structural biology revolution.

11.2   Ensemble Allostery through Tethering

The main point of this chapter is to explore generalizations of the allostery concept. One of the principal insights of recent years is the recognition that allosteric proteins have huge numbers of hidden degrees of freedom, such as are illustrated in Figure 11.1. In this section, we explore an explicit example of ensemble allostery in the form of the ensemble of conformational states available to disordered tethers.

11.2.1   Biochemistry on a Leash

One of the most fundamental features of living organisms is movement. As noted in our discussion of chemotaxis, cells make "decisions" about where to go, and these decisions in eukaryotes are implemented in the form of polymerization of actin filaments, as shown in Figure 1.5 (p. 7). What chains of events link the detection of some external cue and the formation of new actin filaments in a motile cell? As seen in the figure, nucleation of new actin filaments is mediated by signaling molecules such as N-WASP and Arp2/3. As seen in Figure 11.7, N-WASP is subject to combinatorial control by Cdc42 and PIP_2. When both these molecules are present at sufficiently high concentrations, they will bind N-WASP and mediate the conformational change to the active state.

The study of molecules such as N-WASP and its synthetic analogues will allow us to flex several sets of muscles we have developed throughout the book by considering molecules with the interesting feature that they include a tethered ligand and receptor pair that compete with free ligands. When the tethered ligand–receptor pair are bound, the signaling molecule is inactive. When the free ligands compete away the tethered ligand, the signaling molecule then switches into its active state, as shown in Figure 11.7. These tethering motifs are a common feature of signaling molecules and give us a new example of how to implement allosteric conformational changes.

One simple way to see why tethered receptor–ligand pairs work as allosteric switches is illustrated in Figure 11.8. The idea is that the tethered ligand is confined to a volume dictated by the length of the tether. In particular, if the tether has a length L resulting in a radius of gyration R_G, the length scale that characterizes the size of the polymer conformations, then the effective concentration of the tethered ligand can be estimated as

$$\text{effective concentration} = \frac{1}{\frac{4}{3}\pi R_G^3}. \qquad (11.17)$$

To develop an intuitive sense of the significance of this tethering, this estimate can be used to roughly determine the concentration at which the free

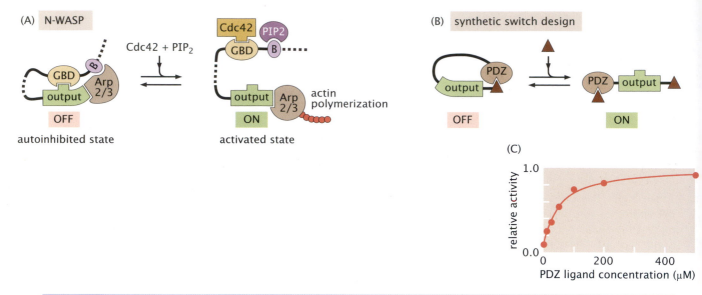

Figure 11.7

Schematic of the signaling process leading to actin polymerization. (A) Activation of Arp2/3 by ligands Cdc42 and PIP$_2$. (B) Synthetic switch constructed to activate Arp2/3 as a result of the presence of an alternative ligand. (C) Activity of the synthetic switch as a function of the concentration of PDZ ligand. Readout of the activity of the synthetic switch was measured using an in vitro pyrene-actin polymerization assay. The experiment measured actin polymerization in the presence of synthetic switch and Arp2/3. Adapted from Dueber et al. (2003).

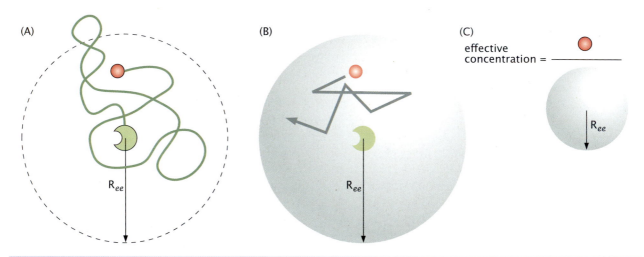

Figure 11.8

Tethering and effective concentration. (A) As a result of tethering, the ligand can explore only a limited region of space. (B) The concentration of the tethered ligand can be estimated by considering a sphere with a radius given by the radius of gyration of the tether. (C) To compute the effective concentration due to tethering, consider one ligand per volume given by a sphere with a radius equal to that of the radius of gyration.

ligands compete with the tethered ligand. Specifically, we have the approximate condition

$$c_{free} = \frac{1}{\frac{4}{3}\pi R_G^3}$$
(11.18)

that determines what concentration of free ligand will outcompete the tethered ligand–receptor pair. Clearly, at high enough concentrations, the free ligands will dominate the binding.

From the standpoint of the cell signaling in actin polymerization shown in Figure 11.7, a small signaling molecule can relay information to N-WASP, a protein that can interface with a complex of proteins called the Arp2/3 complex to create new actin filaments. This mechanism is shown in Figure 11.7(A). In particular, the presence of two ligands, Cdc42 and PIP$_2$, activate N-WASP by binding to this protein and inducing the active conformation that then makes it possible to activate Arp2/3. The presence of Cdc42 and PIP$_2$ leads to the unbinding of GDB and B domains from the C domain and Arp2/3, and N-WASP begins to stimulate actin polymerization by recruiting (and perhaps appropriately orienting) actin monomers to the proximity of the Arp2/3. With the help of activated N-WASP, Arp2/3 promotes actin polymerization by providing heterogeneous nucleation sites. Here, our aim is to study the rate of this process as a function of the concentration of effectors such as Cdc42 and PIP$_2$.

As with the analysis of genetic networks, one exciting way in which signaling pathways have been dissected is by rewiring such pathways to form various synthetic signaling networks. Figure 11.7(B) shows a synthetic activator of Arp2/3 in which a domain known as PDZ is attached to the output domain that activates Arp2/3. On the other end of the construct is a peptide sequence that binds to PDZ. This synthetic protein mimics N-WASP and can be activated by soluble ligands that bind to the PDZ domain. Specifically, these synthetic molecules made it possible to measure the activation of actin polymerization as a function of the PDZ ligand concentration, as shown in Figure 11.7(C). Our interest is in how the allostery is dictated by the ensemble of conformational states of the tethers.

11.2.2 Random-Walk Models of Tethers

To analyze the function of this signaling process, we invoke statistical mechanics in the same spirit as we have throughout the book, but this time accounting for a polymer physics treatment of the disordered part of the allosteric protein. The goal of our statistical mechanical model of the synthetic switch is to work out the probability that the molecule is in the active state. In particular, the active state corresponds to the case in which the tethered receptor is not bound to the tethered ligand. That is, the tethered ligand and receptor are separately flopping around freely. As usual, we resort to a states-and-weights diagram to work out the probability of the active state. As shown in Figure 11.9, there are three classes of states, each with its own corresponding statistical weight: (i) the switch is in the autoinhibitory state, and the tethered ligand and receptor are bound to each other (top panel); (ii) the tethered ligand and receptor are both flopping around freely, and the receptor has no bound free ligands (middle panel); (iii) the tethered ligand and receptor are both flopping around freely, and the receptor has bound one of the free ligands (bottom panel).

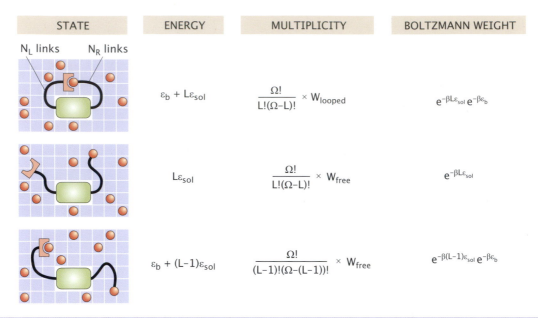

STATE	ENERGY	MULTIPLICITY	BOLTZMANN WEIGHT
N_L links N_R links	$\varepsilon_b + L\varepsilon_{sol}$	$\dfrac{\Omega!}{L!(\Omega-L)!} \times W_{looped}$	$e^{-\beta L\varepsilon_{sol}}\, e^{-\beta\varepsilon_b}$
	$L\varepsilon_{sol}$	$\dfrac{\Omega!}{L!(\Omega-L)!} \times W_{free}$	$e^{-\beta L\varepsilon_{sol}}$
	$\varepsilon_b + (L-1)\varepsilon_{sol}$	$\dfrac{\Omega!}{(L-1)!(\Omega-(L-1))!} \times W_{free}$	$e^{-\beta(L-1)\varepsilon_{sol}}\, e^{-\beta\varepsilon_b}$

Figure 11.9

States and weights for the synthetic signaling problem. The first state corresponds to the tethered ligand and tethered receptor being bound. The multiplicity factor W_{looped} characterizes the number of configurations available to the tethered ligand-receptor pair when they are bound to each other. The second state is the free receptor state, and the third state is the state in which the receptor is occupied by a free ligand. The multiplicity factor W_{free} characterizes the number of configurations available to the tethered ligand and tethered receptor when they are not bound to each other. Computing the multiplicity factors requires polymer physics ideas that are beyond the scope of this book.

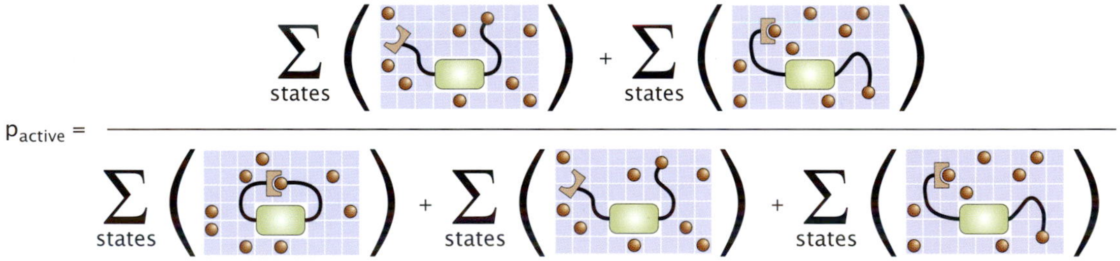

$$p_{active} = \frac{\displaystyle\sum_{states}\left(\; \right) + \sum_{states}\left(\; \right)}{\displaystyle\sum_{states}\left(\; \right) + \sum_{states}\left(\; \right) + \sum_{states}\left(\; \right)}$$

Figure 11.10

Probability of activation of Arp2/3. The numerator is the sum of the statistical weights of the active states.

To develop an intuitive sense of how this situation plays out, the probability of finding the switch in the active state is represented schematically in Figure 11.10. The essence of the situation is that as the concentration of free ligand is increased, the probability that the receptor will be bound by one of the free ligands will increase until this outcome dominates the probability. From the standpoint of testing our understanding of such systems, one of the other design parameters that can be varied is the length of the flexible tethers. As will be shown explicitly when we demonstrate the contributions of the autoinhibitory

STATE	ENERGY	MULTIPLICITY	BOLTZMANN WEIGHT
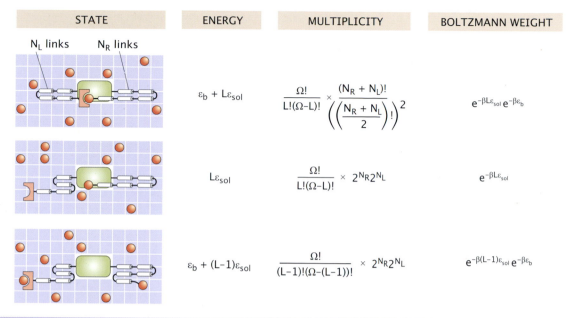 $\varepsilon_b + L\varepsilon_{sol}$	$\dfrac{\Omega!}{L!(\Omega-L)!} \times \dfrac{(N_R+N_L)!}{\left(\left(\dfrac{N_R+N_L}{2}\right)!\right)^2}$	$e^{-\beta L\varepsilon_{sol}} e^{-\beta\varepsilon_b}$
	$L\varepsilon_{sol}$	$\dfrac{\Omega!}{L!(\Omega-L)!} \times 2^{N_R}2^{N_L}$	$e^{-\beta L\varepsilon_{sol}}$
	$\varepsilon_b + (L-1)\varepsilon_{sol}$	$\dfrac{\Omega!}{(L-1)!(\Omega-(L-1))!} \times 2^{N_R}2^{N_L}$	$e^{-\beta(L-1)\varepsilon_{sol}} e^{-\beta\varepsilon_b}$

Figure 11.11

States and weights for the synthetic signaling problem with a one-dimensional tether. The first state corresponds to the tethered ligand and tethered receptor being bound. The second state is the free receptor state, and the third state is the state in which the receptor is occupied by a free ligand.

state to the overall partition function, the length of the tether is a significant part of the overall free-energy budget.

To make this calculation concrete, we resort here to simple one-dimensional ideas on the random walk, as shown in Figure 11.11, and show how the calculation generalizes to three dimensions but leave the details for the reader. Our strategy will be to break down the *total* partition function for this system into three parts as reflected in Figure 11.11, where the sum can be written as

$$Z_{tot}(L, N_R, N_L) = \underbrace{Z_1(L, N_R, N_L)}_{\text{autoinhibitory state}} + \underbrace{Z_2(L, N_R, N_L)}_{\text{free tether}} + \underbrace{Z_3(L, N_R, N_L)}_{\text{tether with ligand}}. \quad (11.19)$$

The parameter L is the number of ligands in the system, N_R is the number of Kuhn segments (the lengths of the individual links seen in Figure 11.11 equal to twice the persistence length) in the polymer tether that has the tethered receptor, and N_L is the number of Kuhn segments in the polymer tether that has the tethered ligand. Given these decompositions, we can then write the probability that the switch will be in the active state as

$$p_{active} = \frac{Z_2 + Z_3}{Z_1 + Z_2 + Z_3}, \quad (11.20)$$

as shown in Figure 11.10.

The separate contributions to the total partition function can be worked out in much the way we have done similar problems throughout the book. The key point is that each class of state has a number of microscopically equivalent configurations (the "ensemble" in the ensemble allostery), and to find their contribution to the overall partition function, we need to multiply the Boltzmann weight for each class of state by its corresponding microscopic degeneracy (obtained by adding up all the different ways of arranging the system). For example, the contribution from the states in which the tethers are flopping around freely and there is no free ligand bound is given by

$$Z_2 = \underbrace{\frac{N!}{L!(N-L)!}}_{\text{solution ligands}} \times \underbrace{2^{N_R}2^{N_L}}_{\text{tether configs.}} \times \underbrace{e^{-\beta L \varepsilon_{sol}}e^{-\beta \varepsilon_{sol}^{lig}}}_{\text{Boltzmann weight}}. \qquad (11.21)$$

The treatment of the tether degrees of freedom is based on the simplest one-dimensional random walk, in which we imagine that every segment in the tether can point either to the left or right, and we do not worry about self-avoidance. It is straightforward to use a more robust model of the tethers, but we use this one for simplicity. What this means precisely is that each tether can be in one of 2^N different configurations, where N is the number of Kuhn segments in the tether of interest. We have also introduced the energy ε_{sol} for the energy of the ligands when they are free in solution, and the parameter ε_{sol}^{lig} for the energy of the tethered ligand when it is in solution. The most interesting class of states is that in which the states are associated with the autoinhibition of the switch and involve the tethering ligand and receptor being linked. In this case, the contribution to the partition function is

$$Z_1 = \underbrace{\frac{N!}{L!(N-L)!}}_{\text{solution ligands}} \times \underbrace{(N_R+N_L)!/((\tfrac{1}{2}(N_R+N_L))!)^2}_{\text{tether closure}} \times \underbrace{e^{-\beta L \varepsilon_{sol}}e^{-\beta \varepsilon_B}}_{\text{Boltzmann weight}},$$

$$(11.22)$$

where we have used the binomial distribution to work out the probability of tether closure. The contribution from tether closure is the number of ways of making a closed loop out of a polymer of length $N_R + N_L$ Kuhn segments and in this case requires that the number of left steps equals the number of right steps. The last contribution to the total partition function arises from those microstates in which one of the free ligands attaches to the tethered receptor. This means that the solution contribution to the partition function will involve only $L-1$ ligands, or

$$Z_3 = \underbrace{\frac{N!}{(L-1)!(N-(L-1))!}}_{\text{solution ligands}} \times \underbrace{2^{N_R}2^{N_L}}_{\text{tether configs.}} \times \underbrace{e^{-\beta(L-1)\varepsilon_{sol}}e^{-\beta \varepsilon_B}}_{\text{Boltzmann weight}}. \qquad (11.23)$$

We can now obtain p_{active} by substituting the values for Z_1, Z_2, and Z_3 obtained above into equation 11.20 resulting in

$$p_{active} = \frac{1+(Z_3/Z_2)}{1+(Z_1/Z_2)+(Z_3/Z_2)}. \qquad (11.24)$$

This leads to an expression for p_{active} of the form

$$p_{active} = \frac{1+(c/c_0)e^{-\beta \Delta \varepsilon_1}}{1+p_{loop}e^{-\beta \Delta \varepsilon_2}+(c/c_0)e^{-\beta \Delta \varepsilon_1}}, \qquad (11.25)$$

(A)

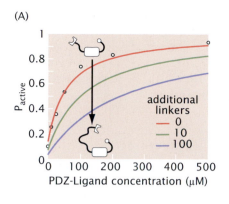

(B)

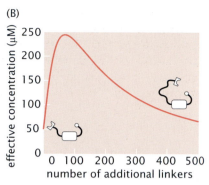

Figure 11.12

Prediction of dependence of activation on effective tail length. (A) p_{active} as a function of ligand concentration for different tether lengths. Experimental data are shown as small circles. (B) The effective concentration of tethered ligand as seen by the tethered PDZ domain as a function of tether length. Data from Dueber et al. (2003).

where we have introduced $c = L/(Nv)$, $c_0 = 1/v$, and p_{loop}, which is the probability of forming a loop, $\Delta \varepsilon_1$ is the binding energy for a free ligand, and $\Delta \varepsilon_2$ is the binding energy for the tethered ligand–receptor pair. For the one-dimensional model considered earlier, we have

$$p_{loop} = \frac{(N_R + N_L)! / (((N_R + N_L)/2)!)^2}{2^{N_R + N_L}}, \qquad (11.26)$$

which amounts to the ratio of the number of closed configurations for the polymer of length $N_R + N_L$ to the *total* number of configurations. However, the one-dimensional model has outlived its usefulness and we could instead use the result of a full three-dimensional analysis of p_{loop} using the Gaussian model of a polymer, for example. This calculation is left as an exercise for the reader.

The outcome of this kind of analysis is shown in Figure 11.12. There are several subtleties that were not accounted for in the calculation just described. First, as shown in the figure, the tethers do not emanate from the same point, resulting in a fundamental difference in the behavior of p_{loop} as a function of tether length, as shown in Figure 11.12(B). Second, in the figure, we used a three-dimensional Gaussian model for the tethers rather than the one-dimensional example we worked out. This fun example gives a look at the way that internal disorder within a protein can be exploited to set up a conformational switch.

11.3 Irreversible Allostery

This chapter has focused on the ways in which the original allostery concept can, and sometimes should, be extended to account for hidden degrees of freedom or alternative mechanisms of inducing conformational change. We already got a flavor for these ideas several times throughout the book, first in the context of light as a ligand in section 5.3.3 (p. 187) and, second, in the context of posttranslational modifications and nucleosomes in section 10.5 (p. 334). In this final section, we provide a qualitative discussion of the other routes to conformational change as a preface to the next chapter, which focuses on one particular class of applications of this idea to kinetic proofreading.

Figure 11.13 is a gallery of examples in which the activity of a protein is modified (often irreversibly) as a result of some external signal but with mechanisms that are outside the purview of the original Monod-Wyman-Changeux model.

Figure 11.13

Generalization of the allostery concept. Proteins can exist in states of different activity through a variety of mechanisms. Ligand binding can induce oligomerization, which converts a molecule into an active form. Cleavage of a molecule can change its activity. In the example shown here, a cleaved protein acts as a transcription factor. Covalent modification of a protein through the addition of a chemical group can change its activity. Work can be performed, resulting in a change of conformation.

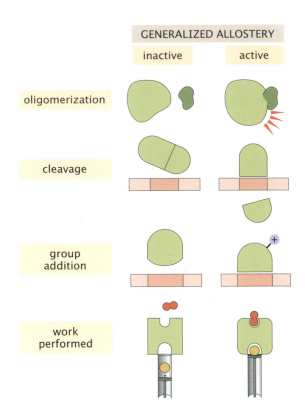

The first example is ligand-mediated oligomerization, in which a ligand serves to invite several subunits to associate to form a complex, which in the oligomerized state is now active. The opposite effect of the oligomerization process is protein degradation, in which as a result of some signal, a protein is cleaved at one or many sites, thus changing its activity. This mechanism was already revealed in our discussion of Janus factors in section 8.5 (p. 299) in the context of transcription factors that behave as activators or repressors depending upon whether they have suffered a cleavage reaction or not. For example, there we recalled the properties of Notch-Delta signaling with an emphasis on how the presence of the Delta ligand leads to cleavage of the intracellular domain of Notch, which is then phosphorylated and translocated to the nucleus, where it performs its role as a transcription factor. The Notch-Delta example serves to remind us of one of the most common mechanisms for inducing a conformational change in a way that goes beyond traditional allostery, namely, through posttranslational modifications. Finally, as we will take up in the next chapter, we can also imagine situations in which an allosteric transition is undertaken because of the expenditure of energy.

11.4   Summary

The notion of allostery has outgrown its early roots as formulated in the 1960s to now include a variety of other mechanisms and models. This chapter explored several examples in which the number of distinct conformational states has

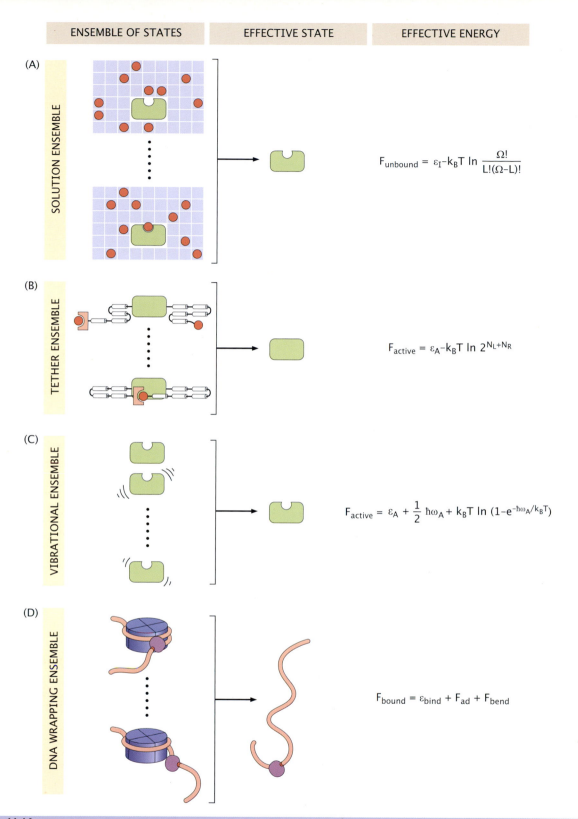

| ENSEMBLE OF STATES | EFFECTIVE STATE | EFFECTIVE ENERGY |

(A) SOLUTION ENSEMBLE

$$F_{unbound} = \varepsilon_I - k_B T \ln \frac{\Omega!}{L!(\Omega - L)!}$$

(B) TETHER ENSEMBLE

$$F_{active} = \varepsilon_A - k_B T \ln 2^{N_L + N_R}$$

(C) VIBRATIONAL ENSEMBLE

$$F_{active} = \varepsilon_A + \frac{1}{2}\hbar\omega_A + k_B T \ln (1 - e^{-\hbar\omega_A / k_B T})$$

(D) DNA WRAPPING ENSEMBLE

$$F_{bound} = \varepsilon_{bind} + F_{ad} + F_{bend}$$

Figure 11.14

Ensemble allostery and effective states. For each example we show the ensemble of states, the corresponding effective single state, and its energy. (A) Ensemble allostery due to ligands in solution. There are a huge number of states corresponding to the ligands in solution. Using a lattice model of solution we can compute the number of states. (B) Ensemble allostery due to configurations of disordered tethers with tethered ligand and receptor. Using a one-dimensional random-walk model, we can compute the total number of states associated with the tether degrees of freedom. (C) Ensemble allostery due to vibrational degrees of freedom. The free energy of this ensemble of states can be used as the effective energy for a single state. (D) Ensemble allostery of nucleosomal accessibility.

exploded. The notion of ensemble allostery makes this proliferation of states explicit by acknowledging there are many microscopic states that might attend a given state of macromolecular activity, as shown in Figure 11.14. Here we show explicitly and rigorously how such states can be "integrated out" to yield a low-dimensional effective theory once again. One of the key structural realizations of the last few decades is the importance of unstructured or disordered regions of proteins. To illustrate how statistical mechanics might be used to think about these systems, we examined tethered ligand–receptor pairs and computed the contribution of the many disordered states to the statistical weights of the inactive and active states of the molecule. All told, this chapter illustrated some of the deep nuance found in real-world allosteric molecules and how statistical mechanics can be tailored to greet these nuances in a way that leads to a tractable, predictive model.

11.5 Further Reading

Cooper, A., and D.T.F. Dryden (1984) "Allostery without conformational change: A plausible model." *Eur. Biophys. J.* 11:103–109. This paper introduced many of the key ideas of ensemble allostery in appropriate mathematical form.

Motlagh, H. N., J. O. Wrab, J. Li, and V. J. Hilser (2014) "The ensemble nature of allostery." *Nature* 508:331. This paper reflects on how our understanding of the allostery phenomenon has evolved with the emergence of new experimental techniques, such as NMR, which have substantially generalized the mechanistic underpinnings of how molecules can behave allosterically.

Nussinov, R., C.-J. Tsai, and J. Liu (2014) "Principles of allosteric interactions in cell signaling." *J. Am. Chem. Soc.* 136:17692–17701. Nussinov and coworkers have been exploring the idea of replacing the original two-state conception of allostery with conformational ensembles, as explored in this chapter. This paper gives an introduction to the way they have been approaching the problem.

Phillips, R., J. Kondev, J. Theriot, and H. G. Garcia (2012) *Physical Biology of the Cell.* New York: Garland Science. Chapters 8 and 10 of this book describe the polymer physics needed to work out the multiplicity factors for the three-dimensional tethered ligand-receptor pairs considered in this chapter.

11.6 REFERENCES

Dueber, J. E., B. J. Yeh, K. Chak, and W. A. Lim (2003) "Reprogramming control of an allosteric signaling switch through modular recombination," *Science* 301:1904.

Popovych, N., S. Sun, R. H. Ebright, and Kalodimos, C. G., (2006) "Dynamically driven protein allostery." *Nat. Struc. Mol. Biol.* 13:831–838.

MAXWELL DEMONS, PROOFREADING, AND ALLOSTERY

12

When you come to a fork in the road, take it.

—Yogi Berra

12.1 Demonic Biology

We are all familiar with the story of Humpty Dumpty, the ill-fated "anthropomorphic egg" whose short biography reads:

"Humpty Dumpty sat on a wall,
Humpty Dumpty had a great fall.
All the king's horses and all the king's men
Couldn't put Humpty together again."

My own fascination with Humpty Dumpty centers on the fact that if we run a movie of his sad life history backward, it is immediately obvious that there is something amiss. That is, we know instantly that the movie does not represent an acceptable physical process. Certain processes are known to be irreversible. By way of contrast, if we follow the trajectories of the molecules I am breathing as I write these words, the forward and reverse time histories appear equally reasonable. Indeed, even molecular processes that are clearly irreversible, such as the formation of convective rolls in the clouds or vortices on the leeward side of an island, as shown in Figure 12.1, when viewed at the level of individual molecular trajectories will look remarkably similar whether run forward or backward.

A celebrated example of apparently running the movie of molecular processes backward is provided by Maxwell's demon, illustrated in Figure 12.2. In a letter written in December 1867 to his friend Peter Guthrie Tait, James Clerk Maxwell noted: "if we conceive of a being whose faculties are so sharpened that he can follow every molecule in its course, such a being, whose attributes are as essentially finite as our own, would be able to do what is impossible to us. For we have seen that molecules in a vessel full of air at uniform temperature are moving with velocities by no means uniform, though the mean velocity of any great number of them, arbitrarily selected, is almost exactly uniform. Now let us suppose that such a vessel is divided into two portions, A and B, by a division

Figure 12.1

Satellite image of vortices on the leeward side of island off the coast of Baja California. From NASA.

Figure 12.2

Running the movie of the world backward. In the first sequence, the concentration gradient dissipates over time, consistent with our daily experience. The second sequence shows what happens when the first sequence is run backward. In the third sequence, we see that instead of running backward, the second sequence could just as well have been a forward-running movie but secretly presided over by the demon who opens the trap door at opportune moments so that molecules go to the "wrong" side of the box.

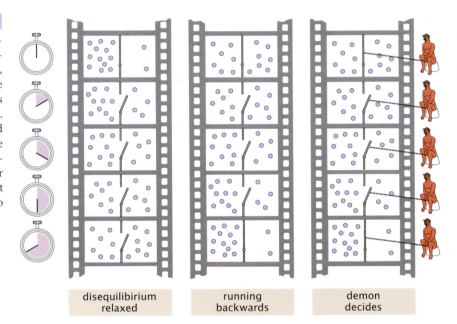

disequilibirium relaxed   running backwards   demon decides

in which there is a small hole, and that a being, who can see the individual molecules, opens and closes this hole, so as to allow only the swifter molecules to pass from A to B, and only the slower molecules to pass from B to A. He will thus, without expenditure of work, raise the temperature of B and lower that of A, in contradiction to the second law of thermodynamics."

12.2 A Panoply of Demonic Behaviors in the Living World

How many different examples can we think of in which a biological process takes place in such a way that if we ran a movie of that process forward and backward, we would be able to tell that one of the directions required the consumption of free energy? Figure 12.3 shows four such examples, all of which are biologically commonplace and yet appear to take place in a direction that is contrary to the spontaneous direction of natural processes. Specifically, the top left panel of Figure 12.3 shows how the ribosome is bathed in a constant supply of tRNAs harboring the incorrect amino acid and yet achieves an error

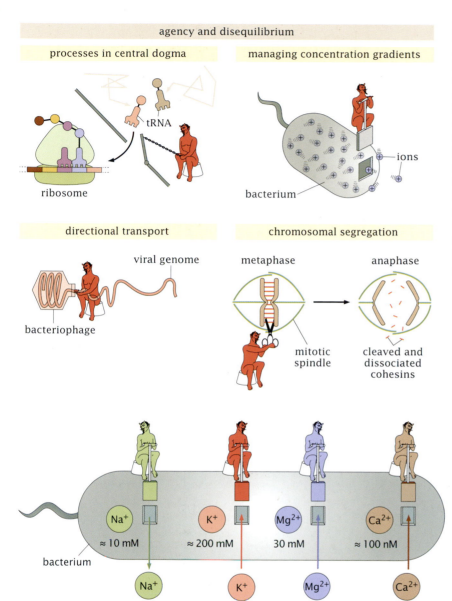

Figure 12.3

A variety of examples illustrate how biological systems exploit the consumption of free energy to drive states of disequilibrium. The process of translation involves error-correcting mechanisms to ensure that the right amino acid is added to a nascent polypeptide chain. Concentration gradients across cell membranes are maintained by molecular pumps that serve as Maxwell demons. The packing of DNA into viruses by the phage-packaging motor drives the system against both electrostatic and mechanical forces that resist such packaging. Chromosomal segregation has error-checking mechanisms to avoid premature segregation.

Figure 12.4

Concentration gradients in cells. Each species of ion has a different concentration on the inside and outside of the cell. These differences are maintained by energy-consuming membrane proteins that act as Maxwell demons. The ion concentrations shown are meant to give a feeling for the numbers and do not represent the outcome of any single experimental situation.

rate that is much lower than can be explained by codon-anticodon recognition alone. Like the process mediated by the demon in Figure 12.2, this serves as a precise example of *enriching* the concentration of the low-abundance correct tRNAs relative to the much larger background of incorrect tRNAs. Even more commonplace is the example shown in the upper right, which reminds us that nearly every living cell has enormous gradients in their ion concentrations, as illustrated in Figure 12.4. As we will show further, every 10-fold difference in concentration across the cell membrane incurs a free-energy cost of roughly $k_B T \ln 10 \approx 2.3\, k_B T$. Hence, the accumulation of such concentration gradients is energetically costly. The third example in Figure 12.3, shown in the lower left corner, is a specific example of the more general phenomenon of nucleic acid partitioning. In cells, both DNA and RNA molecules are targeted to specific locations, again incurring a free-energy cost. The case shown here shows the parallel process of packing of viral DNA into the viral capsid, which again incurs a large free-energy cost. The final frame in Figure 12.3 shows how a kind of proofreading in the process of chromosome segregation ensures that all chromosomes are properly attached in the mitotic spindle.

Each of the examples presented in Figure 12.3 shows the way in which free-energy consumption is used to drive molecular processes in a direction such that the final state of the system of interest is at a higher free energy that could, in principle, be used to extract work. My reason for including this story in a book that is ostensibly dedicated to allostery is that I will argue that in many of these cases, the demonic actions are driven by conformational changes between some inactive and active state. In that sense, we broaden the idea of allostery to include irreversible conformational changes, now no longer mediated by the binding of some ligand but, rather, for example, by attachment of some chemical group, the cleavage of some protein, or through the hydrolysis of a nucleotide.

As we have just seen, biological processes exploit free energy for a number of purposes, including (1) increasing the specificity of biochemical reactions, (2) manipulating dynamics so that processes take place in a specific order, (3) reducing variability in molecular processes, (4) amplifying weak signals, (5) erasing memory, and (6) constructing particular structures such as the ribosome or infectious viruses. For further details, see Mehta et al. (2016). To be concrete, we now consider a number of beautiful examples that illustrate these different kinds of processes, establishing the phenomenology that any theory of demonic biology must own up to.

12.2.1   The Demon and Biological Specificity

One of the signature examples of demonic behavior in biology is shown in Figure 12.5, which provides a description of error rates in the various processes of the central dogma. This plot shows that the process of replication is characterized by an error rate of 1 mistake in roughly every 10^{10} nucleotides added onto the replicating genome. The process of protein synthesis is the most sloppy of the processes of the central dogma, with an error rate of order 1 incorrect amino acid incorporated in nascent polypeptide chains out of every 10,000 newly added amino acids. We will show that even this error rate is surprisingly low. For several of the measured values reported in Figure 12.5, such as *E. coli* translation, we show the average of results of several different types of measurement.

ESTIMATE

Estimate: DNA Replication Timing in *Drosophila melanogaster*
As a case study in the wonders of DNA replication, we consider the replication of the fruit fly *Drosophila melanogaster* genome during early embryonic development, which is characterized by fast cycles of synchronized mitosis. The first 11 nuclear cycles take less than 10 minutes each. Molecular synthesis must keep up with the developmental program that is unfolding. In each one of these short 10 minute cycles, nuclei need to have enough time to double their DNA and to create enough mRNA for the synthesis of the proteins that will unleash the gene regulatory network dictating the

adult body plan, and all of this with the kinds of low error rates show-cased in Figure 12.5. To accomplish this speedy 3.4 minute replication in *Drosophila melanogaster*, the copying of the genetic material takes place in parallel, as shown in Figure 12.6. The electron microscopy image reveals a number of distinct "replication forks," shown explicitly in the schematic at the bottom of the figure. The DNA replication machinery moves along the genome at a speed of about $v_{replisome} \approx 30$ bp/s, which means that a single complex would take more than 45 days to copy the whole genome! However, given the parallel processing, as seen in Figure 12.6, we can count the bubbles in the image to identify roughly $N_{origins} \approx 14$ replication origins. Each one of these bubbles will have a replisome moving in opposite directions at each end. Thus, each bubble replicates DNA at a rate of 60 bp/s. We estimate the total DNA length in the image to be around $L_{DNA} \approx 35$ kb. As a result, the time it will take to replicate this stretch of DNA is

$$T_{replication} = \frac{L_{DNA}}{2 \times v_{replisome} \times N_{origins}} \approx 40 \text{ s.} \qquad (12.1)$$

Replication at this rate leaves time to spare in the repeated cell divisions of the fly early embryo.

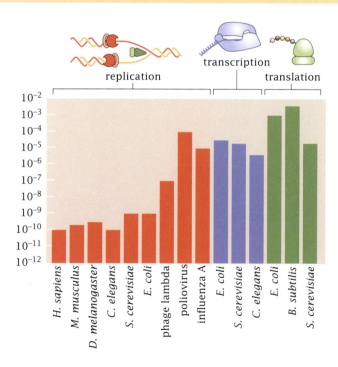

Figure 12.5

Biological specificity beyond equilibrium thermodynamics. The histograms show the error rates in the key processes of the central dogma. Data can be found on the BioNumbers website, bionumbers.hms.harvard.edu.

Despite the rapidity with which the entire 120 Mbp fly genome is replicated as described in our estimate, it is done so with a fidelity of roughly 1 part in 10^{10}. To put these numbers in everyday terms, Tania Baker advanced a wonderful analogy which asks us to think of DNA as having the diameter of a typical

Figure 12.6

Replication in the fruit fly genome. The schematic at the bottom shows the replication forks revealed in the electron microscopy image at the top. Adapted from Kriegstein and Hogness (1974); courtesy of Henry J. Kriegstein.

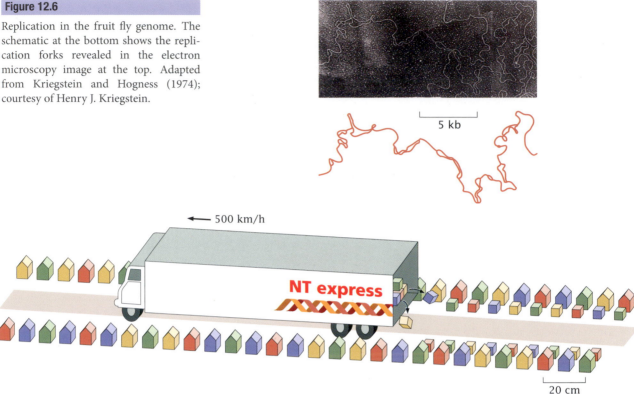

Figure 12.7

High-fidelity delivery service by the nucleotide express. DNA replication viewed as the process of delivery of packages by an NT express truck. The truck is traveling at high speed while delivering packages on both sides of the street every 10 cm. With four colors of box to deliver the possibility of delivery error is high.

sewer pipe (i.e., 1 m), implying in turn that the size of the replisome would be the same as that of a delivery truck, as shown in Figure 12.7. During its replication journey, the macroscopic replisome travels at a speed of 500 km/h, making a delivery of one of four colored boxes on both sides of the street once every 10 cm, completing its daily journey (for the case of bacterial replication) in 40 minutes. In this highly efficient delivery process, the truck would deliver a wrong package only once every 3 years! It sounds impressive, and it is. Similar analogies can be made for the process of translation, and the reader is invited to do so.

Figure 12.5 showed us the error rates in the processes of the central dogma. Because of the critical importance of the polymerization processes of the central dogma, it comes as no surprise that a big effort has been made to build specific and concrete models of these processes to account for their fidelity. Perhaps the most challenging issue is revealed by the highest-fidelity process in the central dogma, namely, DNA replication, with error rates of order 10^{-10}.

Figure 12.8 helps us develop intuition for how to think conceptually and quantitatively about errors in DNA replication (or other templated copying processes). The mechanism of copying of the DNA, as announced so dramatically with the words "It has not escaped our notice that the specific

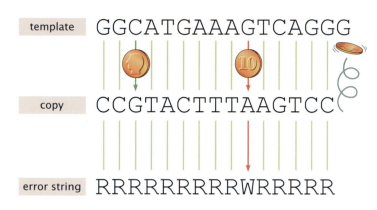

pairing we have postulated immediately suggests a possible copying mechanism for the genetic material," in James Watson and Francis Crick's 1953 paper is based upon a templating mechanism in which one strand serves as the basis of the complementary nucleotides that will be synthesized on the newly replicated strand. As shown in the figure, we can think of each addition of a new nucleotide as a very unfair coin flip with the probability of a heads more than 10^9 greater than the probability of a tails. Given ideas developed throughout the book, we recognize such coin-flipping processes as being described by the binomial distribution in the form

$$p(n_m, N_{bp}) = \frac{N_{bp}!}{n_m!(N_{bp} - n_m)!} p^{n_m} (1 - p)^{N_{bp} - n_m}, \qquad (12.2)$$

where p is the probability of a mutation during the replication of a given base, and n_m is the number of mutations in a genome of length N_{bp}. Though this result tells us how to develop our expectations about the error string shown in Figure 12.8, if we invoke the extremely small chance of a mutation, we can recast this distribution in the simpler form of a Poisson distribution.

Estimate: The Distribution of Mutations during Genome Replication

To that end, we make a succession of well-controlled approximations. First, we note that we can make the approximation

$$\frac{N_{bp}!}{(N_{bp} - n_m)!} \approx N_{bp}^{n_m}. \qquad (12.3)$$

To see where this approximation comes from, we appeal to a simple everyday scenario. Consider a large stadium such as the famed Rose Bowl in Pasadena, California. This stadium has more than 90,000 seats. If we now ask how many ways we can arrange 100 identical spectators in those seats, the exact answer is

$$\text{number of seating arrangements} = \frac{90,000!}{100!(90,000 - 100)!}. \qquad (12.4)$$

ESTIMATE

More generally, we consider a stadium with N_{seats} seats and $n_{spectators}$ but with the condition that $n_{spectators} << N_{seats}$. In this case, the number of seating arrangements is

$$\text{number of seating arrangements} = \frac{N_{seats}!}{n_{spectators}!(N_{seats} - n_{spectators})!}. \quad (12.5)$$

However, let's reimagine the seating procedure, since perhaps the mathematics is not transparent as is. If we have a hat with 90,000 numbers labeling those seats, and we ask our small number of spectators to choose those seats by reaching into the hat, the first spectator will have 90,000 choices. The second spectator will have 89,999 choices, and so on. We should have an intuitive sense that if the first spectator put his or her number back into the hat after making a seat choice, we would not be worried about one of the other spectators choosing that same seat, since there are so few spectators sampling the seats. As a result, each spectator has N_{seats} seats to choose from, and hence there are $N_{seats}^{n_{spectators}}$ ways of arranging them.

By invoking this approximation, we now have

$$p(n_m, N_{bp}) \approx \frac{(N_{bp}p)^{n_m}}{n_m!}(1-p)^{N_{bp}}(1-p)^{-n_m}. \quad (12.6)$$

We can make the approximation $(1-p)^{-n_m} \approx 1$, since n_m is small, meaning that we have a number that is nearly 1 (i.e., $(1-p)$) raised to a modest power, leaving us with this approximation. Note also that $N_{bp}p$ is the average number of mutations, λ. The last element in our analysis is to recognize that

$$(1-p)^{N_{bp}} = (1 - \frac{\lambda}{N_{bp}})^{N_{bp}} \approx e^{-\lambda}. \quad (12.7)$$

To see this, we appeal to Figure 12.9, which shows us the result of some "experimental mathematics" (i.e., playing around with numbers). We see that if we repeatedly evaluate the quantity $(1 + \frac{1}{N})^N$ for increasing N, it tends to the limit

$$\lim_{N \to \infty}(1 + \frac{1}{N})^N = e^1, \quad (12.8)$$

where e is the base of natural logarithms. We can generalize this expression by noting that if we have $(1 + \frac{x}{N})^N$, we can rewrite it if we define $M = N/x$, resulting in

$$\lim_{N \to \infty}(1 + \frac{x}{N})^N = \lim_{M \to \infty}[(1 + \frac{1}{M})^M]^x = e^x. \quad (12.9)$$

The net result of all this analysis is that the probability of getting n mutations in our genome is given by the Poisson distribution, namely,

$$p(n) = \frac{\lambda^n e^{-\lambda}}{n!}, \quad (12.10)$$

where $\lambda = N_{bp}p$ is the average number of mutations. We plot this distribution in Figure 12.10. This result demands that we circle back to the proofreading process itself, since such a low number of copying mistakes per genome replication owes its fidelity to proofreading.

(A)

N	$\left(1 + \frac{1}{N}\right)^N$
1	2
10	2.5937
10^2	2.7048
10^3	2.7169
10^4	2.7181
10^5	2.7183

(B)

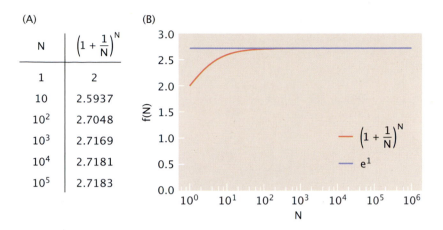

Figure 12.9

Taking the limit of the $(1 + 1/N)^N$ sequence. (A) Table of values showing the value of $(1 + 1/N)^N$ for different choices of N. (B) Plot comparing the sequence $(1 + 1/N)^N$ for increasing N and the value of e^1.

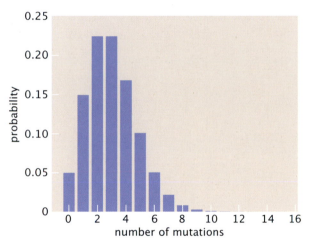

Figure 12.10

Poisson distribution for mutations. Probability of finding n_m mutations in a genome given an average of three mutations.

The process of translation is also a high-fidelity process. Figure 12.11 reminds us that out of the hundreds of thousands of tRNAs (375,000 total tRNA; BNID 108611) jiggling around within the *E. coli* cytoplasm, only a small fraction of those tRNAs correspond to the right ones for the codon waiting at the A site of the ribosome. To be more precise, given that there are between 30 and 40 unique species of tRNA, only a small percentage of these tRNAs are correct partners for the codon on the mRNA of interest. To overcome the insufficiency of equilibrium codon-anticodon recognition, there are energy-consuming steps in ribosome-mediated peptide bond formation, as shown in Figure 12.12. The consumption of energy illustrated schematically in step (B) is the molecular implementation of the Maxwell demon depicted in Figure 12.3.

12.2.2   Making Stuff Happen in the Right Order

Another fascinating example of the use of free energy to achieve processes in living matter that are outside the province of equilibrium behaviors (i.e., demonic behaviors) is the nature of specific ordering of kinetic processes. Figure 12.13 contrasts different pictures of macromolecular assembly relevant in biology. Perhaps the most familiar kind of assembly is that due to thermodynamic

Figure 12.11

Fidelity in protein synthesis. Many different species of tRNA are competing for the same site within the ribosome. Only one of those species is the correct one.

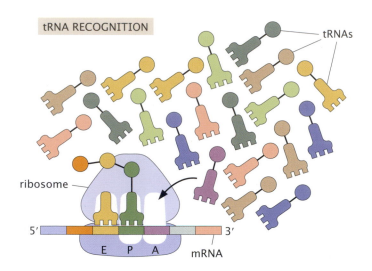

equilibrium, where, depending upon concentrations and interaction energies, the system adopts a distribution of all products and reactants, as shown in Figure 12.13(A). The second example shown in Figure 12.13(B) is very familiar from biological cartoons in which a specific linear order of processes unfolds. The distinction between these two examples is revealed by an everyday analogy of what has been called "socks before shoes" kinetics, as shown in Figure 12.14. The final example in Figure 12.13(C) is branched assembly, such as the assembly of viruses, shown in Figure 12.15. In this case, several parts must ultimately be put together in some order, and each of the pieces to that eventual assembly themselves assemble by some linear pathway.

In processes such as the assembly of the ribosome or viruses, there is a definite and required progression. Recall that the ribosome is an enormous macromolecular complex with a mass in excess of 2 MDa and consisting of both RNA and proteins. The two primary rRNAs are complemented by on the order of 50 distinct proteins that together make the protein factory responsible for polypeptide bond formation. Interestingly, as the result of the constant labors of generations of researchers and techniques, the ribosome is known to assemble according to a strict molecular ballet.

Perhaps even more amazing is the sequential assembly of the bacterial viruses known as bacteriophages, shown in Figure 12.15. For tailed bacteriophages like that illustrated, we see that the capsid and the tail structure are assembled independently, and only after the DNA genome has been packaged into the capsid does the tail get added to the virion.

Many of the conformational changes that drive cellular responses like those described in the previous few sections of this chapter are effectively irreversible. We have seen other examples throughout the book, as well. For example, the Janus transcription factors described in chapter 8 (p. 299) behave as activators or repressors depending upon the cleavage (essentially irreversible) of part of the protein. Similarly, as we saw in our discussion of nucleosomes, their free-energy landscape can be altered through the action of energy-consuming enzymes that impose posttranslational modifications that send the protein into some "permanent" different conformational state that can be undone only through the action of some other energy-consuming process.

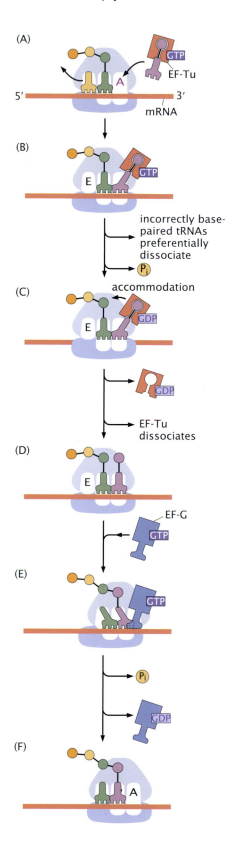

Figure 12.12

Steps in protein synthesis resulting in high fidelity. Note that in step (B), a hydrolysis event is associated with rejection of the incorrect tRNA.

Figure 12.13

The kinetics of ordered assembly. (A) Equilibrium assembly of molecular complexes results in a collection of many intermediates and no time ordering. (B) In a linear assembly pathway, dynamics takes place in a socks-before-shoes fashion that we are used to from everyday life. (C) In a branched assembly pathway, such as exhibited by the assembly of bacteriophages, several steps happen first, leading to an intermediate assembly, and then pieces are added onto that intermediate to make a next one, and so on, until the entire assembly is completed. (B) and (C) courtesy of Nigel Orme, as published in Steven et al. (2016).

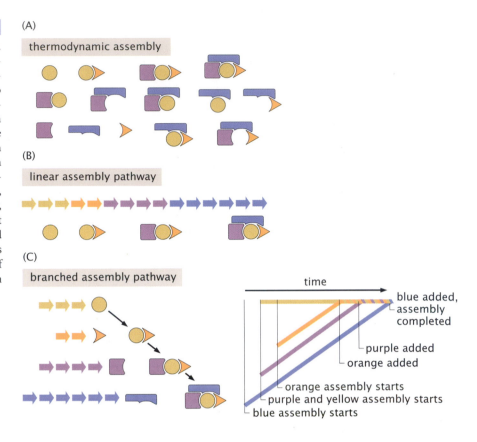

The present chapter is in many ways the most exciting and unusual of the book because it invites us to look forward and expand our definition of allostery by holding onto only its most essential element, namely, the presence of different conformational states that exhibit different states of activity. To that end, throughout the chapter, we use the pedagogical device of appealing repeatedly to energy-consuming demons as a way of illustrating that there will be irreversible allostery.

12.2.3 The Free-Energy Cost of Demonic Behavior

The general themes of this chapter can take us much farther afield than the limited province of allostery. Indeed, the subject of this chapter provides one of the most important vistas onto the great biological landscape. In his classic book *What Is Life?* Erwin Schrödinger (1944) considered some of the biggest questions from then and now about living organisms. Perhaps the more well known parts of his book are the early chapters, in which he asks, what would be the character of the molecules that carry the genetic information? But in fact, in later chapters he broaches the topic of biological organization and the appearance that biological systems are running the movie of cellular processes backward, as already discussed in Figure 12.2. Perhaps the simplest way of stating Schrödinger's observation is to note that it is often through nonequilibrium phenomena that the life force makes itself known.

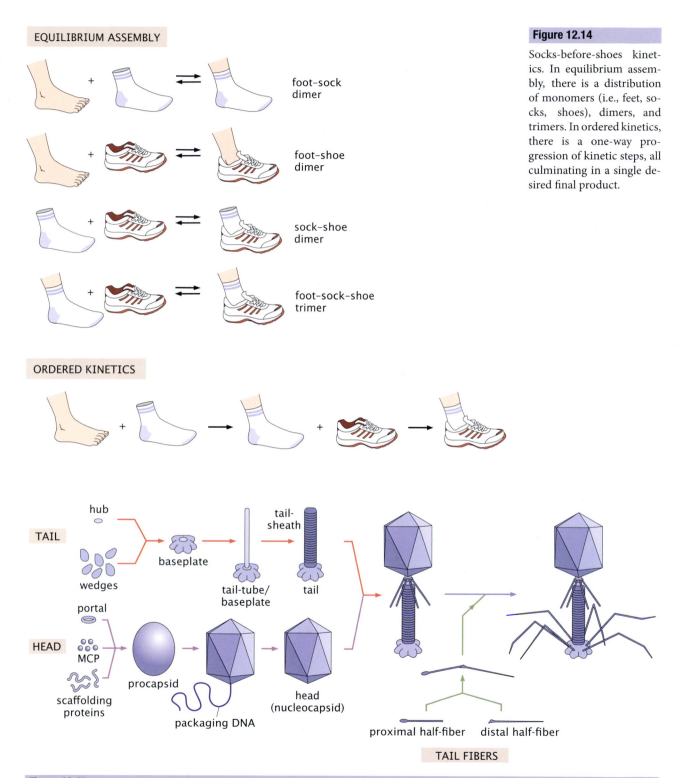

EQUILIBRIUM ASSEMBLY

foot–sock dimer

foot–shoe dimer

sock–shoe dimer

foot–sock–shoe trimer

ORDERED KINETICS

Figure 12.14

Socks-before-shoes kinetics. In equilibrium assembly, there is a distribution of monomers (i.e., feet, socks, shoes), dimers, and trimers. In ordered kinetics, there is a one-way progression of kinetic steps, all culminating in a single desired final product.

TAIL

hub

wedges

baseplate

tail-tube/baseplate

tail-sheath

tail

HEAD

portal

MCP

scaffolding proteins

procapsid

packaging DNA

head (nucleocapsid)

proximal half-fiber   distal half-fiber

TAIL FIBERS

Figure 12.15

Biological "socks before shoes" kinetics. Sequential kinetics in viral assembly. The bacteriophage tail, head, and tail fibers are assembled on three distinct pathways. Courtesy of Nigel Orme, as published in Steven et al. (2016).

Figure 12.16

Figure 12.16

Maxwell's demon as a metaphor for biological specificity. Partitioning molecules to one side of a membrane is a very common biological process.

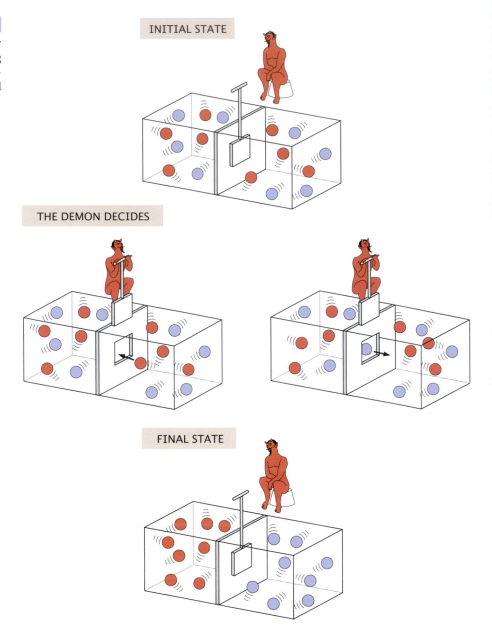

To get a sense of the quantitative implications of what Schrödinger was talking about we find it convenient to hark back to Maxwell's demon as a metaphor for biological specificity, as shown in Figure 12.16. We have already seen repeatedly that there are many biological circumstances in which it is functionally critical to partition molecules to particular places. For example, we saw that the entry site to the ribosome is a kind of demonic trapdoor that serves as a gate for the correct tRNAs to preferentially enter the confines of the ribosome itself. Nuclear export of mRNAs is another example of molecular partitioning. Concentrating ions on one side of a membrane is perhaps the simplest example to understand quantitatively. Let's examine the energetics of such molecular partitioning.

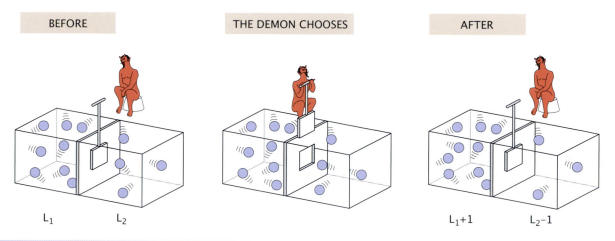

BEFORE   **THE DEMON CHOOSES**   **AFTER**

L_1 L_2 L_1+1 L_2-1

Figure 12.17

The free-energy cost of demonic behavior. The entropy change when the demon moves a single molecule from one side of the box to the other is computed by comparing the number of microscopic states available before and after the molecular transfer event.

To get a feel for the free energies involved in demonic partitioning, we consider the simpler scenario laid out in Figure 12.17. The idea of our analysis is that we will compare the free energy of the system before and after the demon makes his choice, allowing a single molecule to cross from one side to the other. We interest ourselves in the change in free energy

$$\Delta F = F_{final} - F_{initial} = (U_{final} - TS_{final}) - (U_{initial} - TS_{initial}). \quad (12.11)$$

For simplicity, we focus only on the entropic contribution to the energy change, with the knowledge that ideas such as the proton motive force also ask us to consider the change in internal energy due to electrostatic effects.

To compute the entropy change, we need to compute the entropy of the gases on both sides of the partition, both before and after the demon has allowed a single molecule to pass through the trapdoor. Specifically, we need to compute the total entropy, which is given by

$$S_{tot} = S_1 + S_2, \quad (12.12)$$

where the subscripts refer to the two compartments. This expression implies that the free energy in this state is given by

$$F = -TS_{tot}. \quad (12.13)$$

We recall that the entropy can be reckoned by counting the number of microscopic states. For our case of interest, we have

$$S_{tot}^{(final)} = k_B \ln W_1^{(final)} + k_B \ln W_2^{(final)}, \quad (12.14)$$

with a similar expression for the entropy in the initial state. Note that by the properties of logarithms, we can simplify this to

$$S_{tot}^{(final)} = k_B \ln(W_1^{(final)} W_2^{(final)}), \quad (12.15)$$

which is eminently sensible, given that the total number of microscopic states is equal to the product of the number of states in the first box and the number of states in the second box. If we adopt our usual lattice model, as introduced in Figure 2.2, (p. 37) then we note that

$$W_i(L) = \frac{\Omega^{L_i}}{L_i!}, \tag{12.16}$$

where Ω is the number of lattice sites on our lattice, and L_i is the number of ligands occupying that lattice.

We can now write the change in free energy as

$$\Delta F = -k_B T \left(\ln \frac{\Omega^{L_1+1}}{(L_1+1)!} \frac{\Omega^{L_2-1}}{(L_2-1)!} - \ln \frac{\Omega^{L_1}}{L_1!} \frac{\Omega^{L_2}}{L_2!} \right), \tag{12.17}$$

which we can simplify to

$$\Delta F = -k_B T \ln \frac{L_1!}{(L_1+1)!} \frac{L_2!}{(L_2-1)!} \approx -k_B T \ln \frac{L_2}{L_1}. \tag{12.18}$$

We can perform a little intellectual surgery to now beat this into a much more familiar form using the language of concentrations. If we multiple numerator and denominator within the logarithm by Ωv, where v is the volume of a single lattice site in our lattice model, then $\Omega v = V_{tot}$, and hence $c_1 = L_1/V_{tot}$ and $c_2 = L_2/V_{tot}$, permitting us to write

$$\Delta F = k_B T \ln \frac{c_2}{c_1}. \tag{12.19}$$

This simple result allows us to develop a rule of thumb that gives us a sense of the cost of setting up concentration gradients. If we imagine the case in which $c_2 = 10^n c_1$, this implies that

$$\Delta F = n k_B T \ln 10 \approx 2.3 n k_B T. \tag{12.20}$$

In other words, physiological gradients across membranes have a characteristic energy scale of ≈ 10 k_BT. For comparison, note that the hydrolysis of a single ATP is worth roughly 20 $k_B T$ in free energy.

As alluded to in Maxwell's letter to Tait, in which he noted of the demon: "He will thus, without expenditure of work, raise the temperature of B and lower that of A, in contradiction to the second law of thermodynamics," we can restate the quandary posed by the demon more concretely, as shown in Figure 12.18. Alternatively, we can think of the demon's partitioning of molecules as shown in Figure 12.17 as setting up a pressure difference that can be used to lift a weight (i.e., do work). Figure 12.18 shows a protocol whereby the demon equilibrates a single-molecule gas and then inserts a partition. Once he has determined on which side of the partition the molecule is found, the demon flips a switch connected to a movable pulley and weight that tethers it to that side of the partition. Thus, when the gas undergoes free expansion it does work. This protocol was originally conceived by Leo Szilard as a response to Maxwell's conundrum.

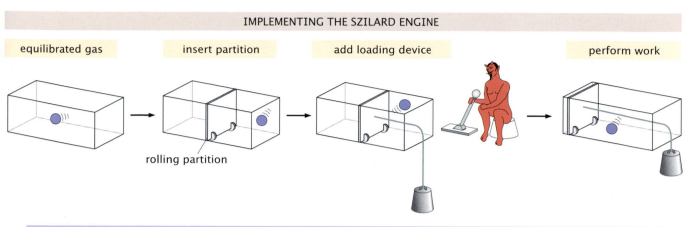

Figure 12.18

Harnessing the efforts of the demon. Once the demon has determined the position of the molecule, he flips a switch to tether the weight to the correct side of the partition so that when the single-molecule gas undergoes its expansion a weight can be lifted. Adapted from Sagawa and Ueda in Klages et al. (2013).

The history of science has been colored by a constant search for "mechanism," or, said differently, for ways to provide mechanistic insights into various natural processes. The illustration of Maxwell's demon harnessing information to do mechanical work provides an example of such a mechanism. In some ways, this search for mechanical mechanisms is an inheritance of the mechanical worldview as embodied, for example, in the famous paper of Rudolf Clausius (1857) entitled "On the nature of the motion which we call heat," in which Clausius sought an interpretation of the macroscopic properties of temperature and heat in terms of the underlying motions of the atoms.

Contemporaneously with these theoretical investigations, James Prescott Joule undertook experiments that permitted a direct determination of the so-called mechanical equivalent of heat, using the apparatus shown in Figure 12.19(A). In this device, Joule was able to perform a known quantity of work by lowering a weight connected to a mechanism with spinning plates in an insulated container of water and measuring the temperature change in the stirred water. This experiment not only demonstrated the equivalence of mechanical work and heat but also allowed Joule to assign it a specific numerical value.

Modern physical biology has carried these ideas forward in the form of devices such as optical traps and electrophysiology pipettes, as shown in Figure 12.19(B), where we see that the force of an optical trap or the membrane tension applied by a pipette can both be thought of as equivalent to the lowering of a weight. For example, when we pull on a piece of DNA with an optical trap we have to account for the free energy not only of the stretched DNA but of the loading device (i.e., the optical trap), as well. If we label the stretch of the DNA with the coordinate x, then we can write the total free energy as

$$G_{tot}(x) = G_{DNA}(x) + G_{load}(x). \qquad (12.21)$$

In the experiment the bead attached to the DNA is displaced from its equilibrium position in the trap, thus increasing its energy, captured by the term

(A) MECHANICAL EQUIVALENT OF HEAT

(B) EQUIVALENCE TO LOWERING A WEIGHT

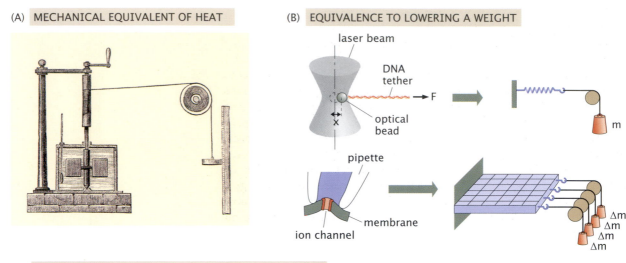

(C) EQUIVALENCE OF HYDROLYSIS TO MECHANICAL WORK

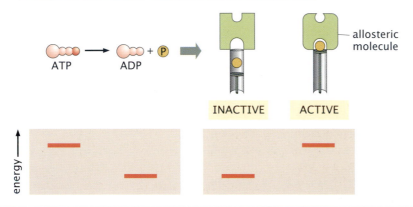

Figure 12.19

The mechanical equivalence of natural processes. (A) Experimental apparatus of James Prescott Joule used to measure the mechanical equivalent of heat. (B) Mechanical analogies for loading devices in biology. Both optical traps and pipettes apply mechanical loads, which can be thought of as equivalent to the lowering of weights. (C) The ambition of replacing chemical hydrolysis by corresponding mechanical processes.

$G_{load}(x)$. The motion of the bead in the trap is mechanically equivalent to the process of lowering a weight, as shown in Figure 12.19(B).

One of the key ideas we undertake in this chapter is to reconsider the way in which free energy is transmitted through macromolecules such as molecular motors from the standpoint of mechanical equivalences, as shown in Figure 12.19(C). In many molecular processes in biology, the free energy is delivered as a result of the hydrolysis of nucleotides. Here, we will examine how mechanical analogues such as the lowering of weights and the compression of pistons can be used to tune the amount of free energy delivered to molecular systems and how the amount of free energy delivered alters the underlying biological processes.

To lay down the groundwork on the treatment of energy transduction that will follow, let's revisit the quantitative underpinnings of what it means to have nucleotide hydrolysis take place concomitantly with some process of interest, such as the stepping of a molecular motor or the addition of an amino acid to a nascent polypeptide chain. Our analysis of the motive power of hydrolysis will be predicated upon imagining a reservoir in which ATP, ADP, and P_i are all held at a fixed chemical potential. Another way of saying this is that we imagine our tiny molecular system is placed in a giant bath, in much the same way that we could put an Eppendorf tube containing some molecular reactants of interest in the chilly Pacific Ocean with the consequence that the temperature of those reactants would very quickly adopt that of the surrounding ocean without affecting the temperature of the ocean itself in any way. Our nucleotide reservoir is similarly enormous, so that as a given molecular reaction such as the addition of the next base pair in DNA replication takes place, the free energy of hydrolysis is delivered without any impact on the reservoir itself. In the context of such a reservoir, each hydrolysis reaction will deliver a free energy that can be written formally as

$$\text{free-energy release due to hydrolysis} = \mu_{ATP} - \mu_{ADP} - \mu_{P_i}. \quad (12.22)$$

Recalling that the chemical potential can be written in the form $\mu_{ATP} = \mu^0_{ATP} + k_B T \ln \frac{c_{ATP}}{c_0}$ and similarly for ADP and P_i, we can rewrite the free-energy release due to hydrolysis from our nucleotide reservoir as

$$\text{free-energy release due to hydrolysis} = \mu^0_{ATP} - \mu^0_{ADP} - \mu^0_{P_i} + k_B T \ln \frac{c_{ATP}c_0}{c_{ADP}c_{P_i}}. \quad (12.23)$$

As shown in Figure 12.19, we will find it convenient to replace free-energy sources such as the ATP hydrolysis of equation 12.23 with mechanical equivalents. Again and again, whether in Joule's famed experiments on the mechanical equivalent of heat (Figure 12.19(A)), Szilard's classic treatment of Maxwell's demon lifting a weight (Figure 12.18), the use of tools such as optical traps or pipette aspiration (Figure 12.19(B)), or possibly even in the contemplation of nucleotide hydrolysis, mechanical equivalents help us understand the underlying physics of a broad array of processes.

   The question addressed in the remainder of this chapter is the way in which irreversible processes are harnessed in biology to ensure higher fidelity than is possible on the basis of equilibrium processes alone. Specifically, we will argue that in many of these instances, some molecule has to undergo an allosteric reaction to achieve this specificity, uniting the topic of this chapter with the conceptual thread that runs through the book as a whole.

12.3   Overcoming Thermodynamics in Biology: Kinetic Proofreading

12.3.1   Equilibrium Discrimination Is Not Enough

The fidelity of biological polymerization such as that depicted in Figure 12.5 is astonishing. Of course, to use a word like *astonishing*. we must first have some prejudice (i.e., a null model) which informs our expectations about this

Figure 12.20

States and weights for tRNA/mRNA interaction. The mRNA can either be empty, bound by the right tRNA, or bound by the wrong tRNA.

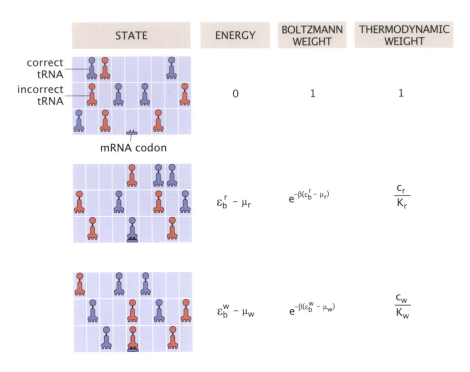

STATE	ENERGY	BOLTZMANN WEIGHT	THERMODYNAMIC WEIGHT
	0	1	1
	$\varepsilon_b^r - \mu_r$	$e^{-\beta(\varepsilon_b^r - \mu_r)}$	$\dfrac{c_r}{K_r}$
	$\varepsilon_b^w - \mu_w$	$e^{-\beta(\varepsilon_b^w - \mu_w)}$	$\dfrac{c_w}{K_w}$

process, and in the case of biological polymerization, the simplest such expectations come from the use of equilibrium statistical mechanics. In the discussion that follows, we begin by examining just how much specificity can be gotten out of equilibrium processes. The simple estimates that come from this analysis immediately reveal that in the processes of the central dogma, for example, something more is in play, since the observed specificities are much higher than those attainable on the basis of equilibrium alone. On the basis of that insight, we then examine the class of arguments given by Hopfield and Ninio that go under the heading of kinetic proofreading. What we will see there is that the essential element of such schemes is the consumption of free energy to drive some irreversible step in a reaction, thus improving the fidelity of the process by a multiplicative factor.

To be concrete, we will explore the example of protein synthesis by the ribosome. Note that in the translation process as revealed in Figure 12.5, roughly 1 in every 10,000 amino acids added during protein synthesis is incorrect. If we oversimplify the process of translation to think of it as a problem in molecular recognition in which cognate and noncognate tRNAs compete for the same "receptor" (i.e., mRNA codon), we can make a simple estimate of the error rate. The idea is that we think of the codon as a receptor that can exist in three distinct states: empty, bound by correct ligand (the right tRNA), and bound by the incorrect ligand (the wrong tRNA), as shown in Figure 12.20. Given these competing states, the probability of binding the cognate tRNA is

$$p_r = \frac{\frac{c_r}{K_r}}{1 + \frac{c_r}{K_r} + \frac{c_w}{K_w}}, \qquad (12.24)$$

while the corresponding probability for binding the noncognate tRNA is

$$p_w = \frac{\frac{c_w}{K_w}}{1 + \frac{c_r}{K_r} + \frac{c_w}{K_w}}, \tag{12.25}$$

where we have used the subscripts r and w to signify the right and wrong tRNAs. If we assume that the concentration of right and wrong tRNAs is equal, this immediately implies that the error is given by

$$\frac{p_r}{p_w} = \frac{K_w}{K_r} = e^{-\beta \Delta \epsilon}, \tag{12.26}$$

where the K_ds are the dissociation constants for binding of the right and the wrong tRNAs, and $\Delta \epsilon$ is the free-energy difference between the correct and incorrect base pairing of the codons and anticodons. Given that the relevant energy differences are several $k_B T$, and using the rule of thumb that every 2.3 $k_B T$ of energy difference is worth a 10-fold error discrimination (i.e., $e^{2.3} \approx$ 10), we would need energy differences of order 10 $k_B T$ to get to 10,000-fold discrimination, and that is only between two competing tRNAs rather than the many, many more that are actually present and competing for the attentions of the codon in the real process of translation.

To see how the calculation generalizes in the case where we imagine a huge excess of wrong tRNAs to right ones (and on the simplifying assumption that all wrong tRNAs bind with the same dissociation constant), we have

$$\frac{p_r}{p_w} = \frac{K_w}{K_r} \frac{c_r}{c_w} = \frac{c_r}{c_w} e^{-\beta \Delta \epsilon}. \tag{12.27}$$

Given the relative numbers of tRNAs (375,000 total tRNA with $\approx$10,000 of each species, BNID 108611), we can take $c_r/c_w \approx 1/30$, resulting in the provocative form

$$\frac{p_r}{p_w} = e^{-\beta(\Delta \epsilon - k_B T \ln \frac{c_r}{c_w})} = e^{-\beta(\Delta \epsilon + k_B T \ln 30)}. \tag{12.28}$$

This result tells us that we need to further compensate this additional $k_B T \ln 30$ of free energy to drive down the error rate. But as we saw earlier, even the less stringent error rate of equation 12.26 requires an unrealistically high discriminatory energy for equilibrium alone.

12.3.2 The Hopfield-Ninio Mechanism

The conceptual answer to the failure of equilibrium thinking was provided more than 40 years ago in the work of Hopfield and Ninio and was christened *kinetic proofreading* in the title of Hopfield's (1974) elegant paper entitled "Kinetic proofreading: A new mechanism for reducing errors in biosynthetic processes requiring high specificity."

Proofreading has been a critical part of human activity for as long as we have adopted the written word. As many of us have transitioned in our reading from charming, durable, and comforting hard copies of our books to e-Readers of all kinds, the quality of our manuscripts has sunk well below the level demanded

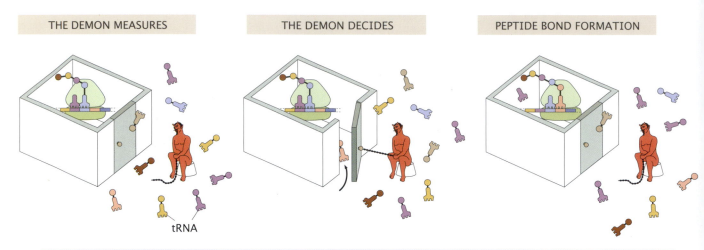

THE DEMON MEASURES   THE DEMON DECIDES   PEPTIDE BOND FORMATION

tRNA

Figure 12.21

The demon decides in translational fidelity. In the translation process, there is a many 10-fold difference in the number of right and wrong tRNAs competing for the same site in the ribosome. Here we schematize the proofreading process as a Maxwell demon that monitors the A site of the ribosome.

by the medieval monks who were responsible for bringing great texts such as Euclid's *Elements* to the modern world. As I write these words I am in the midst of reading Nathaniel Hawthorne's *The Scarlet Letter* in Kindle format and an amazed at error rates of multiple errors per chapter, with many of the errors not only point mutations but, worse yet, full recombination events involving wholesale losses of text. It is clear that the standards for proofreading human writings have plummeted in our electronic age. Nature has not been so stingy. In Hopfield's original work, he considered the specific case of protein translation, investing that process in the language of this chapter with a ribosomal demon, as shown in Figure 12.21. To illustrate the idea of kinetic proofreading, we leave the complex process of translation to consider a simple enzymatic reaction that cleaves a substrate and examine the relative rates of cleavage for the right and wrong substrates.

12.3.3 Proofreading Goes Steampunk: Building Proofreading Engines

William Gibson and Bruce Sterling's (1990) *The Difference Engine* describes a different emergence of the information age. The idea of the book is that steam-engine technology is merged with Charles Babbage's analytical engine to give rise to a version of the information age that parallels our own but instead of being based on semiconductors and other marvels of the age of quantum mechanics is based upon the classical technologies of the nineteenth century. This science fiction novel is a classic in the steampunk genre. In this section, we develop a steampunk approach to kinetic proofreading with the description of several playful thought experiments aimed at using mechanical devices to explore the interplay between work expended, speed, and accuracy.

Figure 12.22 provides a highly schematized diagram of the key conceptual elements in the kinetic proofreading framework, though we alert the reader that

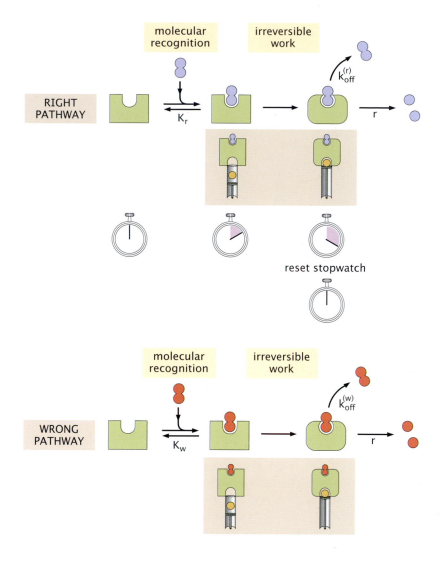

Figure 12.22

The concept of kinetic proofreading. The two panels show the fate of the right (blue) and wrong (red) substrate after binding to the enzyme, labeled in green. After the preliminary molecular recognition step, the enzyme undergoes an irreversible conformational change that poises it in the active state, where it is now competent to cleave the substrate. The molecular stopwatch is reset at this point, and the off rate of the right ("small" off rate) and the wrong ("large" off rate) substrate are different.

many nuances and clever generalizations are to be found in the literature. As seen in the figure, discrimination takes place in several steps, the first of which is traditional molecular recognition and can be thought of as a problem in the statistical mechanics of competitive binding, where for simplicity we imagine that the concentrations c_r and c_w of the right and wrong substrates are equal. In this case, we already worked out the probability of binding the right and wrong substrates in equations 12.24 and 12.25. By taking their ratio, we found that the first step in error correction is this molecular recognition step and results in

$$\frac{p_r}{p_w} = \frac{K_w}{K_r} = e^{-\beta \Delta \epsilon}. \tag{12.29}$$

After the initial molecular recognition step, as shown in Figure 12.22, work is done on the system, inducing an irreversible conformational change that puts the enzyme in the state in which it is now competent to perform its enzymatic function. In the figure, we make the performance of work explicit by showing

Figure 12.23

A piston demon. Schematic of the various states in a protocol for increasing the specificity of an allosteric enzyme that has to distinguish between a correct and an incorrect substrate. Pushing on a piston increases the concentration of the regulatory ligand (yellow). After the conformational change, the enzyme is active and can carry out the cleavage reaction of the right (blue) and wrong (red) substrates.

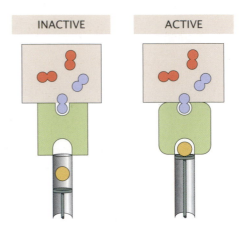

the compression of a piston. Figure 12.23 shows how a piston can be used to do work on a solution containing the regulatory ligand that induces the conformational change from the inactive to the active state. Pushing in the piston increases the concentration of the allosteric effector. Because the probability of being in the active state depends upon the concentration of this effector, doing work shifts the probability of being active from low to high. With this step, we mimic the conformational change induced by hydrolysis envisioned in the original Hopfield and Ninio proofreading schemes.

The key point is that once this "irreversible" reaction takes place, the stopwatch timing the overall reaction is reset. From the moment the irreversible conformational change is made, every substrate molecule has only one of two fates: the substrate molecule can either fall off the enzyme or it can be cleaved by the enzyme, forming the product molecules, as shown in Figure 12.24. Proofreading occurs because the wrong substrate has a higher chance of falling off before it is cleaved than does the right substrate.

As already described earlier in the chapter, in Figure 12.24 (p. 389), once the enzyme enters the active state, the stopwatch timing the dynamics of the system is reset. At this point, as shown in Figure 12.25, the ligand can suffer one of two fates. It can either fall off the enzyme with a rate k_{off}, or it can undergo the cleavage interaction with rate r. The fraction of the time that the system undergoes the cleavage reaction is

$$\text{fraction of reactions resulting in cleavage} = \frac{r}{r + k_{off}}. \tag{12.30}$$

Given this result, we can then compute the error rate. Specifically, the substrate can either undergo the product reaction with a rate r, or it can fall off the complex with a rate $k_{off}^{(r)}$ or $k_{off}^{(w)}$, respectively. Given these rates, we can work out the fraction of the time for both the right and wrong substrates that they will undergo the cleavage reactions, mediated by the enzyme. The probability of the right substrate being cleaved is given by

$$p_r = \frac{r}{r + k_{off}^{(r)}} \tag{12.31}$$

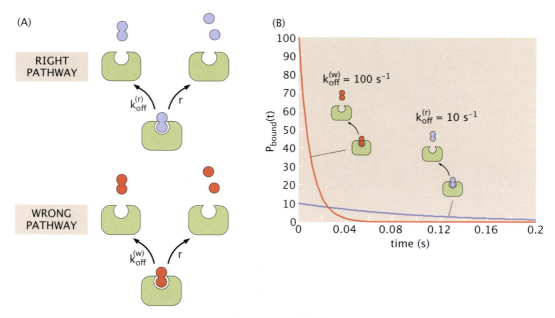

Figure 12.24

The proofreading step. (A) Once the enzyme is in the active state, the substrate can either fall off or be cleaved. The off rates for the right and wrong substrate are different. (B) The waiting-time distribution for substrate occupancy on the enzyme. The wrong substrate has a shorter lifetime on the enzyme before falling off.

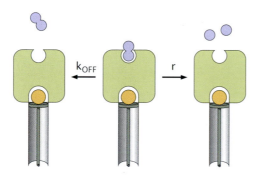

Figure 12.25

Fate of a substrate molecule after the proofreading piston has been compressed. The binding of the effector molecule (yellow) to the enzyme drives the allosteric change of the enzyme to its active state. Once the enzyme is active, there are two fates for the substrate: it falls off with a rate constant k_{off}, or it is cleaved with a rate constant r.

and similarly for the wrong substrate, the probability of being cleaved is

$$p_W = \frac{r}{r + k_{off}^{(w)}}. \qquad (12.32)$$

Given this result, the relative discrimination during the proofreading step is given by

$$\frac{p_W}{p_r} = \frac{\frac{r}{r + k_{off}^{(w)}}}{\frac{r}{r + k_{off}^{(r)}}} = \frac{r + k_{off}^{(r)}}{r + k_{off}^{(w)}}. \qquad (12.33)$$

We can make this intuitive assessment of the error rate more formal by appealing to the waiting-time distributions. Specifically, after the irreversible

Figure 12.26

Deriving the waiting-time distribution. (A) To survive a time t, the ligand must first *not* fall off during any of the previous time steps. The figure shows the possible microscopic transitions the system can suffer during each time step of length Δt. Either the ligand stays bound with probability P_- or it falls off with probability P_+. (B) To survive a total time t, the ligand must string together a succession of N steps in which it doesn't fall off, followed by falling off on the $N + 1^{\text{th}}$ step.

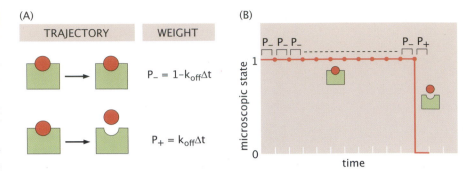

step into the proofreading state, we imagine a stopwatch starts and that there are two eventualities for both the right and the wrong substrates, as shown in Figure 12.24(A). As shown in Figure 12.24(B), the waiting-time distributions for the right and wrong substrates are different. Specifically, we see that the wrong substrate (red) has a higher off rate, and hence the probability that it will still be bound to the enzyme at longer times is lower than that for the right substrate.

But where does this waiting-time distribution come from? Figure 12.26 shows how to compute the probability that the substrate will stay bound to the enzyme until a time between $t = N\Delta t$ and $t + \Delta t$. The key conceptual point is the realization that the probability the substrate will fall off during any time interval of length Δt is $k_{off}\Delta t$, and hence, the probability of *not* falling off during that time interval is $1 - k_{off}\Delta t$. The probability of not falling off during N such intervals each of length Δt is $(1 - k_{off}\Delta t)^N$.

We can now assemble all these insights to obtain the waiting-time distribution itself. According to the preceding calculations, the probability of the ligand staying bound until the time interval between t and $t + \Delta t$ is given by

$$p_{bound}(t)\Delta t = (1 - k_{off}\Delta t)^N k_{off}\Delta t. \tag{12.34}$$

We can rewrite this as

$$p_{bound}(t)\Delta t = (1 - \frac{k_{off}t}{N})^N k_{off}\Delta t. \tag{12.35}$$

The probability distribution $p_{bound}(t)$ is itself a probability density, since it has units of 1/time, as indicated by the factor k_{off} in the distribution. We can take this further by exploiting the limit shown in Figure 12.9. Specifically, we note that

$$p_{bound}(t)\Delta t = \lim_{N \to \infty} (1 - \frac{k_{off}t}{N})^N k_{off}\Delta t = k_{off}e^{-k_{off}t}\Delta t. \tag{12.36}$$

With the waiting-time distribution in hand, we can now compute the probability that the substrate will undergo the enzymatic cleavage reaction for both the wrong and the right substrates. Conceptually, the question we are asking is whether the substrate will undergo a reaction between times t and $t + \Delta t$. We can write that probability as a product of the probability that it will be bound at all at time t times the probability that if it is bound at that time, it will react in the next time increment Δt, an idea we express in pseudo-equation form as

$$p(\text{bound \& react})(t) = p(\text{react} \mid \text{bound})(t)p(\text{bound})(t). \tag{12.37}$$

To be more precise, the probability that the enzymatic reaction will take place between t and $t + \Delta t$ is given by

$$p_{react}(t)\Delta t = re^{-rt}p_{bound}(t)\Delta t. \tag{12.38}$$

The probability $p_{bound}(t)$ that the ligand is still bound at time t can be found by evaluating

$$p_{bound}(t) = \int_{t}^{\infty} dt' k_{off} e^{-k_{off}t'} = e^{-k_{off}t}, \tag{12.39}$$

which amounts to summing up over all the waiting times greater than the time t. Note that the variable t' is a dummy variable over which we are integrating. To find the probability that the enzyme reaction will happen at any time, we need to take the result from equation 12.38 and integrate over all reaction times, with the result

$$p_{react} = \int_{0}^{\infty} dt' re^{-rt'} p_{bound}(t'). \tag{12.40}$$

If we plug the result for $p_{survive}(t)$ into our expression for the probability of a product formation, we finally have the result of interest, namely,

$$p_{react} = \int_{t}^{\infty} dt' re^{-k_{off}t'} e^{-rt'} = \frac{r}{r + k_{off}}. \tag{12.41}$$

We are now prepared to examine the fidelity that can be expected from this simple model of proofreading. We can think of the process of production of final product in several steps, as indicated in the reaction scheme of Figure 12.22. During the first stage prior to performing work by compressing the piston, the probability of having the right versus the wrong substrate bound to the enzyme is given by

$$\left(\frac{p_r}{p_w}\right)_{\text{step 1}} = \frac{K_w}{K_r}. \tag{12.42}$$

Another way of stating this result is to note that when we now compress the piston, we have the right versus the wrong substrate present with this ratio. But now, in the next step, once the system is in the state in which the enzyme is competent to perform the cleavage reaction, the ratio of probabilities for performing the enzymatic step on the right versus the wrong substrates is given by

$$\left(\frac{p_r}{p_w}\right)_{\text{step 2}} = \frac{\dfrac{r}{r+k_{off}^{(r)}}}{\dfrac{r}{r+k_{off}^{(w)}}}, \tag{12.43}$$

This means that the overall error rate is given by the product of the two steps as

$$\frac{p_r}{p_w} = \left(\frac{p_r}{p_w}\right)_{\text{step 1}} \times \left(\frac{p_r}{p_w}\right)_{\text{step 2}} = \frac{K_w}{K_r} \times \frac{r + k_{off}^{(w)}}{r + k_{off}^{(r)}}. \tag{12.44}$$

Note that in the limit when $k_{off} >> r$ for both off rates, this implies that the overall error ratio is given by

$$\frac{p_r}{p_w} = (\frac{K_w}{K_r})^2 \qquad (12.45)$$

and that in the opposite extreme, where $r >> k_{off}$ for both off rates, then the second step brings no extra fidelity. Another way of writing the exact result is

$$\frac{p_r}{p_w} = \frac{K_w}{K_r} \times \frac{K_m^{(w)}}{K_m^{(r)}}, \qquad (12.46)$$

where we note that the subtle distinction is that the second correction term is given by the ratio of the Michaelis constants.

The treatment of proofreading offered here is really a streetfighter's version in that we have used rough intuitive arguments rather than a strict kinetic scheme that writes the rates of all states of the system and solves the resulting kinetic equations. The conceit of the present chapter in a book on allostery is that in many of these cases involving biological fidelity, a conformational change takes a protein from the inactive to the active state as part of the proofreading process. The connection of allostery to proofreading is a fertile arena for a careful theory-experiment dialog.

12.4 Summary

The fidelity of biological processes is astounding. Tania Baker's macroscopic analogy for the fidelity of DNA replication makes it clear how low the error rates really are. In this chapter, I have linked two of the great themes of biology, allostery, and organization through energy expenditure in an analysis of biological fidelity. We used the original ideas of Hopfield and Ninio, but viewed through the prism of allostery, as a way of describing a kinetic scheme in which an enzyme that distinguishes "right" and "wrong" substrates is ushered through a (nearly) irreversible conformational change into an active state that can perform the enzyme action. We saw how by exploiting this energetically costly conformational change, the system could increase its fidelity, recovering the fidelity limits prescribed by Hopfield and Ninio. We also explored a gallery of different examples and generalizations of the proofreading concept.

12.5 Further Reading

One of my big disappointments in writing this book is that I fell short of my desire to provide a "Proofreader's Gallery" that showcases much of the beautiful and inspiring recent work that explores the connection between free-energy expenditure and fidelity. One of the key points is that the kind of close-to-equilibrium mentality portrayed throughout much of the book needs to be amended to explicitly treat the nonequilibrium effects that are in play. In this short list, I point the reader to several of my favorite recent papers that explore these concepts.

Banerjee, K., A. B. Kolomeisky, and O. A. Igoshin (2017) "Accuracy of substrate selection by enzymes is controlled by kinetic discrimination." *J. Phys. Chem. Lett.* 8:1552–1556; Banerjee, K., A. B. Kolomeisky, and O. A. Igoshin (2017) "Elucidating interplay of speed and accuracy in biological error correction." *Proc. Natl. Acad. Sci.* 114:5183–5188 (2017). These two papers provide a very interesting discussion of the kinetics of multistep fidelity enhancement in a way that allows for the full generality of different rate constants. The calculations are framed in the language of first-passage times.

Cui, W., and P. Mehta (2018) "Identifying feasible operating regimes for early T-cell recognition: The speed, energy, accuracy trade-off in kinetic proof-reading and adaptive sorting." *PLoS ONE* 13 (8): e0202331. This interesting paper describes generalized approaches for thinking about proofreading in the context of the immune system.

Lan, G., P. Sartori, S. Neumann, V. Sourjik, and Y. Tu (2012) "The energy-speed-accuracy trade-off in sensory adaptation." *Nat. Phys.* 8:422; Tu, Y. (2008) "The nonequilibrium mechanism for ultrasensitivity in a biological switch: Sensing by Maxwell's demons." *Proc. Natl. Acad. Sci.* 105:11737–11741. These two papers illustrate the ways in which nonequilibrium effects are critical in explaining the chemotaxis behavior of bacterial cells and provide a template for thinking about going beyond the equilibrium framework explored throughout the book.

Mehta, P., A. H. Lang, and D. J. Schwab (2016) "Landauer in the age of synthetic biology: Energy consumption and information processing in biochemical networks." *J Stat. Phys.* 162:1153–1166. This exceptional article showcases what theory can and should be like in biology. This paper codifies the different ways in which free energy is expended to drive key biological processes that I used in the introduction to this chapter.

Sagawa, T., and M. Ueda (2013) "Information Thermodynamics: Maxwell's Demon in Nonequilibrium Dynamics." In *Nonequiliibrium Statistical Physics of Small Systems*, edited by R. Klages, W. Just, and C. Jarzynski. Hoboken, NJ: Wiley-VCH Verlag. Though this paper is not explicitly biological, it is full of provocative ideas that I have a hunch will pay nice dividends in the context of biological problems.

12.6   REFERENCES

Alberts, B., A. D. Johnson, J. Lewis, D. Morgan, M. Raff, K. Roberts, and P. Walter (2016) *Molecular Biology of the Cell*, 6th ed. Newyork: W. W. Norton.

Clausius, R. (1857) "On the nature of the motion which we call heat." *Philos. Mag.* 14(91):108–127.

Gibson, W., and B. Sterling (1990) *The Difference Engine*. London: Victor Gollancz.

Hopfield, J. J. (1974) "Kinetic proofreading: A new mechanism for reducing errors in biosynthetic processes requiring high specificity." *Proc. Natl. Acad. Sci. USA* 71(10):4135–4139.

Kriegstein, H. J., and D. S. Hogness (1974) "Mechanism of DNA replication in *Drosophila* chromosomes: Structure of replication forks and evidence for bidirectionality." *Proc. Natl. Acad. Sci.* 71:135–139.

Maxwell, J. C. (1867) Letter to P. G. Tait. Quoted in C. G. Knott, *Life and Scientific Work of Peter Guthrie Tait.* London: Cambridge University Press (1911), 213–214; reproduced in *The Scientific Letters and Papers of James Clerk Maxwell*: Volume 2, 1862–1873, edited by P. M. Harman. Cambridge: Cambridge University Press (1995), 331–332.

Schrödinger, E. (1944) *What Is Life? The Physical Aspect of the Living Cell.* Cambridge: Cambridge University Press.

Steven, A. S., W. Baumester, L. N. Johnson, and R. N. Perlham (2016) *Molecular Biology of Assemblies and Machines.* New York: Garland Science.

Watson, J. D., and F.H.C. Crick (1953) "Molecular structure of nucleic acids: A structure for deoxyribose nucleic acid." *Nature* 171:737—738.

A FAREWELL TO ALLOSTERY

13

The sun has gone
To bed and so must I
So long, farewell
Auf wiedersehen, goodbye.

—Richard Rodgers, *The Sound of Music*

One of the central pillars upon which signaling, regulation, and feedback is built is allostery, and this book has been a reflection on the ways in which allostery makes it possible for living organisms to respond to their environments by exploiting these mechanisms. In this light, we see allostery as a substrate for both physiological and evolutionary adaptation. On physiological time scales, organisms often respond to their environments by ligand-induced changes in the structure and, in turn, function of their macromolecules. On evolutionary time scales, the key phenotypic properties of allosteric molecules such as leakiness, dynamic range, EC_{50}, and the effective Hill coefficient are shaped by repeated mutation of these molecules. The tools of statistical physics afford us the opportunity to examine the biophysical underpinnings of both the physiological and evolutionary implications of the allostery phenomenon.

In addition to its obvious biological importance, allostery is also an interesting and important opportunity for statistical physics. In his Nobel Lecture, twentieth-century physicist Richard Feynman noted: "Theories of the known, which are described by different physical ideas may be equivalent in all their predictions and are hence scientifically indistinguishable. However, they are not psychologically identical when trying to move from that base into the unknown. For different views suggest different kinds of modifications which might be made and hence are not equivalent in the hypotheses one generates from them in ones attempt to understand what is not yet understood. I, therefore, think that a good theoretical physicist today might find it useful to have a wide range of physical viewpoints and mathematical expressions of the same theory available." (December 11, 1965) I argue that the approach developed in this book has provided us with ways of thinking about signaling and regulation that are psychologically inequivalent to the views that have emerged from genetics and biochemistry, for example. As a result of this psychological inequivalence, the statistical physics of allostery might offer unusual and productive ways of viewing connections between age-old problems in biological adaptation. One example explored in this book is the hypothesized connection between the Bohr effect in hemoglobin, and histone modifications in nucleosomal DNA.

Part of the intent of this book is a tentative attempt at seeing how far we can go with a worldview that self-consciously eliminates molecular details. The thinking presented here attempts to intentionally coarse-grain away things which we know to be true about the underlying molecules but for which we might not have to account. As an example from other provinces of science, think about the pressure we feel on our ears during the ascent or descent of an airplane. We all know that the pressure is the result of a nearly unthinkably huge number of microscopic collisions of molecules with the membrane in our ear. But happily, that knowledge does not inhibit us from defining a single variable called pressure that subsumes all that molecular complexity. The use of pressure is not a matter of naivete or negligence but, rather, a reflection of sophistication, revealing what Philip Anderson canonized as "more is different." As such, a central theme of this book that I argue is critical for the future of biology is to explicitly search for the higher-level theories of biological phenomena through coarse graining as the basis of theory construction. Our battle cry might be that more is different, and less is better. Throughout the book, we have showcased examples of removing molecular details, for example, eliminating any reference to ligands in solution through the chemical potential, or using *effective* parameters that capture the action of effector molecules. Indeed, I would argue that eliminating reference to subsets of degrees of freedom in the world around us is precisely what the great ensembles of statistical physics are designed to do.

13.1   Diversity and Unity: Diverging and Converging Views of Biology

One of the joys of observing, studying, and thinking about the living world is the incredible diversity of living organisms. Figure 13.1 gives several of my favorite examples of this diversity and what scientists such as Darwin have had to say about it. At first blush, the organisms shown in this figure are so distinct as to nearly defy the imagination. And yet, all of them exploit allosteric mechanisms as but one example of many of their shared biological heritage. Though embracing the diversity of the living world is one of the joys of its study, there is also great joy in seeing the unity in apparently distinct biological phenomena.

Indeed, part of what is deceiving about the astounding outward diversity of living organisms is an equally astonishing list of ways in which these organisms are the same. Figure 13.2 makes this point by illustrating the ways in which foundational concepts from biology, chemistry, and physics are shared, giving rise to a hidden unity of biological phenomena. Ultimately, much of our understanding of this unity can be traced to the descent with modification highlighted in Figure 13.1(E). For example, the processes of the central dogma, whether the shared genetic code or the use of ribosomes for protein synthesis, reflect this shared history. The chemistry of life (acknowledging that distinguishing the biology and chemistry of life is perhaps attempting to put too fine a point on things), as evidenced by the fact that nucleotide hydrolysis powers cellular processes in organisms ranging from *E. coli* to *Stella humosa*, from *Pyrococcus*

Figure 13.1

Gallery of biological diversity. (A) The coccolithophore *Emiliania huxleyi,* an important presence in ocean ecosystems. (B) The net-casting spider *Deinopis subrufa.* (C) The blue whale! Nothing more needs to be said. (D) The giant trees of the American West. (E) Darwin comes to terms with the diversity of life from the point of view of descent with modification. (A) courtesy of Jeremy Young. (B) from Mike Gray © Australian Museum. (C) From Chase Dekker / Shutterstock.com. (D) from Bernt Rostad, https://www.flickr.com/photos/67975030N00/2878920184 (CC BY 2.0). (E) from Darwin, C., *On the Origin of Species,* John Murray, 1859. Courtesy of the American Museum of Natural History.

furiosus to *Pan troglodytes,* similarly reveals a unity in molecular mechanisms, as does the chemistry of many metabolic pathways such as glycolysis.

But the unity that is the main watchword of this book is the kind introduced in Figures 1.22 (p. 23) and 1.25 (p. 26), where we saw that the physics of resonance and of interference is a kind of physical unity that presides over a dazzling variety of different phenomena. In our book *Physical Biology of the Cell,* we talked of this unity as being an example of physical proximity, the idea that biological topics can be seen as "similar" from a physical perspective that juxtaposes apparently very different biological phenomena. For example, the same physical limits that determine integration time of a ligand-receptor interaction for a

Figure 13.2

Gallery of biological unity. The living world exhibits biological, chemical, and physical unity. The biological unity is illustrated through the processes of the central dogma and the shared heritage of the proteins that make up cells. Chemical unity is displayed by the use of hydrolysis reactions and by the chemistry of carbon metabolism (and many other examples to boot). Our main interest here is in the physical unity offered by concepts such as the allostery concept, which has a specific implementation in the language of statistical physics embodied in overarching concepts such as the Boltzmann distribution and the linear transport laws that relate fluxes and degree of disequilibrium.

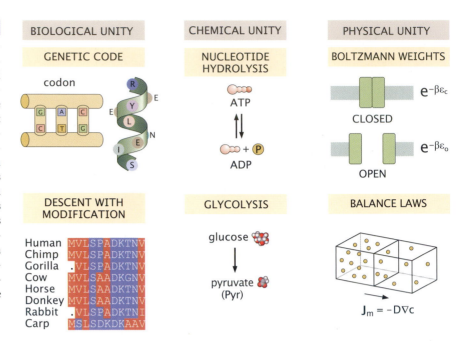

bacterium to "know" the magnitude of the concentration gradient in which it is swimming dictate the rules of how a morphogen gradient is read out by cells in a developing embryo. In the latter case, the adjacent cells have to "decide" whether to express a given gene or not, and this ultimately becomes a ligand-receptor time-integration problem once again, the laws of which are sometimes known as the Berg-Purcell limit. The view I have offered here is that the allostery concept and its mathematical incarnation offers us a view from statistical physics of one of broadest kinds of molecular unity in the living world. This unity allows us to classify different biological phenomena by a compelling kind of physical proximity.

As a reminder of the unity offered by statistical physics in the context of allostery, consider the key phenotypic parameters of allosteric molecules such as the leakiness, the EC_{50}, the dynamic range, and the effective Hill coefficient. Though the terminology may differ from one field to the next, the example of leakiness is played out repeatedly in the book, with ion channels exhibiting spontaneous openings even in the absence of ligand, genetic circuits exhibiting nonzero expression even in the absence of inducer, and nucleosomes spontaneously opening even in the absence of transcription factors.

In the preface, I noted that my fascination with allostery and the unity it brings to our view of the molecules of life really took its fully formed shape at a meeting at the Institut Pasteur to celebrate the 50th anniversary of the allostery concept. Some of the work presented there can be found in the special volume of the Journal of Molecular Biology, "Allosteric Interactions and Biological Regulation (Part I)" (Kalodimos and Edelstein 2013). From my own point of view, the aspect of these talks that I found most interesting was that all these seemingly disparate examples were to a first approximation described by the same

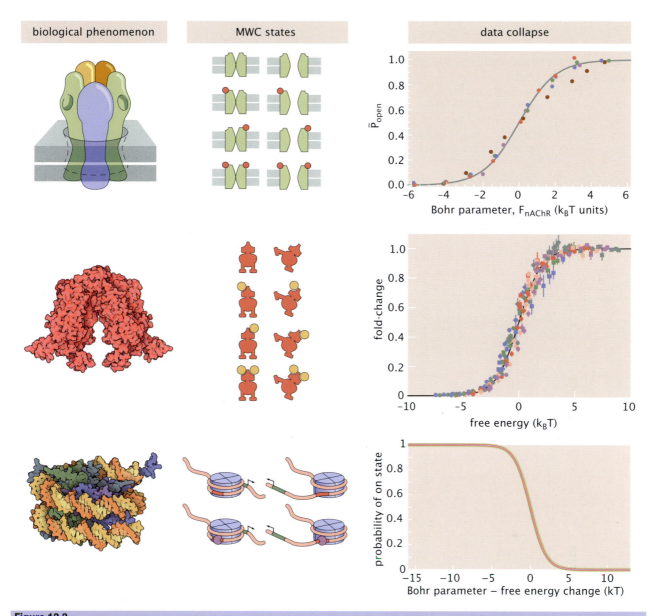

Figure 13.3

Broad reach of the MWC concept. Three distinct examples (ligand-gated ion channel, inducible transcription factor, nucleosome) of biological phenomena that can be analyzed within the MWC framework, showing their MWC states and the data collapse that emerges from using statistical mechanics to think about these problems.

fundamental equation that serves as the central equation of allostery,

$$p_{active}(c) = \frac{(1 + \frac{c}{K_A})^n}{(1 + \frac{c}{K_A})^n + e^{-\beta\varepsilon}(1 + \frac{c}{K_I})^n}. \qquad (13.1)$$

that we have now invoked over and over again throughout the book. What I find so appealing about that is shown in Figure 13.3, where we see the essential

Allosteric transcription factors in *E. coli*. A mechanistic census of the transcription factors of *E. coli* reveals that many of them are ligand dependent, in that the activity of the factor is tunable by its interactions with regulatory ligands. Adapted from Schmidt et al. (2016).

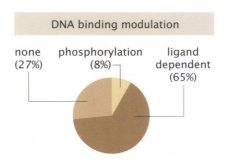

sameness of problems such as ion-channel gating, transcriptional regulation, and DNA packing. Even though each allosteric molecule might have its own peculiarities, there is more to be gained than to be lost in thinking of these molecules in terms of their sameness. Figure 13.3 also gives us the chance to consider the natural variables of allosteric problems as opposed to what one might call "pipettor's variables," which are the tunable knobs manipulated in our experiments. The method of data collapse is one way to illustrate that we have found these natural variables.

Each of our vignettes has related a beautiful story from modern biology as viewed through the prism of allostery. We began with the world of ion channels, with a specific focus on ligand-gated ion channels. From there we traveled to the bacterial membrane, where cells take stock of the chemicals around them, deciding where to go for food and whether they are surrounded by friend or foe. GPCRs then gave us an opportunity to see eukaryotic signaling at work in the fascinating case of the interaction between light and allosteric molecules. Enzymes gave us our next exercises in allostery, with classic examples like ATCase leading the way. Hemoglobin has status as an honorary enzyme, and this one molecule is so important it commanded an entire chapter of its own. Transcriptional regulation is often mediated by transcription factors, and, as shown in Figure 13.4, a large fraction of these proteins in *E. coli* are inducible by the binding of ligands. The packing of DNA in the eukaryotic nucleus gave us an opportunity for an unexpected use of the allostery concept in the context of DNA that exists in inactive and active conformations. With our final example we might have saved the best for last. One of the signature achievements of living organisms is their ability to consume energy in a way that achieves specificities manifold higher than can be achieved by equilibrium alone. I view the connection between allostery and energy-consuming pathways one of the exciting challenges for the field in the coming years.

I review all these examples so that the reader and I can together reflect on both the depth and the superficiality of what we have done. Our ability to consider each of the case studies considered throughout the book is the result of the patient and thoughtful labors of generations of researchers, many of whom devoted entire careers to the results described in only brief sections. A philosophical implication to this permits a bidirectional critical regard, both toward this book for repeatedly making caricatures of different fields and toward the experts in those fields themselves who are sometimes resistant to the importance of such caricatures. In section 3.8 (p. 112) we encountered this bidirectional critique concretely, as the MWC treatment of ligand-gated ion channels gave us a first chance to see this model in action at the level of its biophysical phenotypic parameter. This biophysical treatment foreshadowed the broad applicability of the "one equation that rules them all" and its implications for understanding physiological and evolutionary adaptation in a dizzying variety of contexts. But also, there we saw that beautiful experiments by Ruiz and Karpen made it possible to precisely control the number of ligands bound to a cyclic nucleotide–gated channel and to measure the resulting currents, thus revealing the existence of conductance substates. In this case, we learn that the all-or-nothing picture of Monod-Wyman-Changeux provides a caricature that is not fully respected by the molecules themselves. The ion-channel example gives us a powerful opportunity to see both the extreme power and the limits of our models.

13.2   Shortcomings of the Approach

In my view, the scientific literature has lost something since previous eras, in which it was more common for works to be self-critical. I have always been particularly inspired by Chapter 6 of *On the Origin of Species* by Charles Darwin, simply but elegantly entitled "Difficulties on Theory." Darwin's words can help us cultivate a habit of mind in which we assert our own complaints about the status of our subject and our imperfect approaches to those subjects. In this section, the goal is to produce a list of objections to the developments highlighted throughout the book. My hope is that this short list can serve as a map of the ways in which future research might unfold in the important arena of signaling and allostery.

Whence cometh the allosterome? Which molecules are allosteric and what small molecules serve as effectors and inhibitors of that allostery? Writ large, we don't know the answer to these questions. This absence of factual knowledge is to my mind one of the biggest blemishes on our subject. Preliminary hints as to how one might go about addressing this challenge were shown in Figure 1.20 (p. 21). An example of the damage inflicted by this ignorance is the ubiquitous appeal of "network" diagrams that show how batteries of signaling molecules or genes are linked together in complex arrays of interactions and feedbacks. But often the linkages that connect nodes in these networks are individual proteins such as transcription factors, and in many cases, these transcription factors are themselves controlled by some small molecule of which for now we remain ignorant. In the absence of the allosterome, it is very hard to know what cells really "care about," since it is their interactions with these small molecules that change the activity of those allosteric targets that is really "the biology" of these problems. Until the allosterome becomes a priority like all the other "omes," we will remain handcuffed in our ability to understand the complex set of feedbacks and interactions that make physiological and evolutionary adaptation possible.

Missing theory-experiment dialogue. On several occasions throughout the book, I have noted that there are very strong and contradictory opinions in the scientific community often framed with the skepticism-inducing words "we all know." In particular, there is one camp that argues we already know that the MWC model works and hence there is no further need to dig into the details. The other camp argues that everyone knows that the model is "too simple" to handle the real biological complexity. In writing this book, I felt no allegiance to either of these camps and tried to occupy an agnostic middle ground. My reason is that after a decade of evaluating the literature, my sense is that the number of examples in which anyone set out from the beginning to carefully make specifically crafted predictions from the model and then to carefully design exactly the experiment allowing those predictions to be tested is very limited. Even in the exhaustively studied example of hemoglobin, there are attempts to fit data in which the statistical mechanics of other effectors besides oxygen are never featured in the states and weights. One of my hopes for the future is that you, the reader, will have been inspired to do the kinds of experiments needed to test some of the critical case studies described here, such as the analogy between the Bohr effect in hemoglobin and posttranslational modifications to histones. More important, and especially in this era with disturbing labels such as "alternative facts" and "fake news," never has it been more important to respect science as a knowledge engine, and the only way to really construct knowledge is by digging deep on repeated, careful iterations between theory and experiment. In many cases, the experiment-theory dialogue provides the answers to many of the "shortcomings" on the remainder of my list.

Not in equilibrium. An enduring objection to the kinds of approaches used often throughout this book is that biological systems are out of equilibrium, and hence, one cannot use equilibrium ideas. Stated in this form, this argument is flimsy, since there are many, many examples of problems in which a separation of time scales allows us to pretend as though certain steps are effectively in equilibrium. That said, there is great merit to this objection, both in needing to dig into the dynamics of allosteric molecules and in carefully articulating how to recognize out-of-equilibrium effects when we see them. A more rigorous way to state the objection is to note that we are lacking a clear and definitive treatment of the conditions under which equilibrium thinking is justified. The switching behavior of the bacterial flagellar motor described briefly in chapter 4 gave us an example of the inability of equilibrium ideas to account for the duration time of counterclockwise rotations of the motor and should serve as a template for the kinds of experiment and theory we should look to in the future. One of my strongest hopes is that one of the many gifts biology can give to physics is to help us understand how to treat nonequilibrium problems. The great mathematician Stanislaw Ulam once noted that "using a term like nonlinear science is like referring to the bulk of zoology as the study of non-elephant animals." (Campbell 2004) More recently Jullien Taileur restated this at a lecture at college de France as "nonequilibrium physics is like non-elephant biology." The time is now: what are the limits and validity of equilibrium thinking, and what general ideas can be marshaled to treat nonequilibrium in all its glory?

It's about the dynamics! A serious shortcoming of the approach primarily espoused in this book is its reliance on dose-response curves. By focusing on these, we have pigeonholed our thinking into an equilibrium mindset. Further, as we saw with our treatment of the transcription factor CRP interacting with cAMP, both the MWC and KNF models are largely consistent with the existing data. And perhaps for many ways in which we use our understanding of macromolecular function, this will suffice. But if we are interested in digging even more deeply into our study of allosteric molecules and their mechanisms, we must invariably go beyond the dose-response curve. Often, a very powerful approach is to appeal to dynamics. For example, in the context of ion channels, measuring the current passing through channels as a function of time tells us more than the dose-response curves alone. Similarly, FRET experiments tell us about the dynamical transitions of GPCRs as they perform their functions in the presence of signal.

Too simplistic—there are more than two states. As we saw in section 3.8, (p. 112), even the "simple" case of ligand-gated ion channels forced us to consider a proliferation of states, as evidenced by the Eigen model shown in Figure 3.36 (p. 119). The idea of generalizing the MWC and KNF models to molecular situations with more states is, from a statistical mechanical perspective, straightforward and really demands that we do nothing more than follow the statistical mechanical protocol of Figure 2.1 (p. 36). From this perspective, it is my hope that readers of the book now see how to implement this protocol for situations that go beyond the simplest MWC or KNF treatment. However, even if we see how to implement the statistical mechanics, a big challenge is how to identify what states belong in our description of a given macromolecule. A further compelling challenge is that once we have identified such states, all the questions about dynamics and nonequilibrium effects remain.

One equation does not rule them all. Throughout the book we have returned to the refrain "one equation to rule them all" multiple times. However, as we saw in the case of ion channels, there are clearly circumstances in which the MWC framework is far too simplistic. For a fun and pithy description of the absurdity of this framework working as well as it does, the reader is referred to the article by Miller in Further Reading. Beyond its shortcomings as an effective theory, the MWC framework in and of itself does not tell us how to understand the ways in which specific base pairs in the gene or amino acids in the protein dictate either molecular or organismal phenotype. In his classic book *What Is Life?*, Erwin Schrödinger argued that the macroscopic order seen in biology is different from that offered up in conventional thermodynamics. Schrödinger called the traditional macroscopic order, "order from disorder" to showcase how many kinds of order we see in macroscopic systems are the result of Gibbs's statement of the second law of thermodynamics, namely, that isolated systems reach a terminal privileged state of maximum entropy. By way of contrast, macroscopic biological order is of a different character, what Schrödinger called "order from order." The key point is that we may well need a different kind of theory to describe these states of matter, since the emergence of systems in which these single, tiny microscopic changes in genomes can manifest themselves all the way up to the macroscopic phenotype of the organism is not present in the traditional uses of statistical physics described throughout the book. While the MWC framework offers a comforting and familiar statistical physics embrace, it leaves much to be desired at the level of the genotype-phenotype mapping of such great interest in biology.

Too many parameters. One of the challenges in biological "model building" is the proliferation of unknown parameters as the model complexity increases. I noted earlier that many of the case studies considered throughout the book forced upon us a proliferation of the number of states needed to account for kinetic data. With each additional state, our equilibrium models require further energies to determine Boltzmann weights, and our kinetic diagrams accumulate further arrows, with each such arrow bringing along its own associated parameter. The BioNumbers database attempts to take stock in a curated fashion of the enterprise in biological numeracy required to meet the challenges of describing allosteric molecules and many, many other problems. That said, the quest for intuition demands that we go beyond high-dimensional models. My own taste leans toward a desire to see this field adopt the same "self-conscious" attitude revealed in statistical physics in the context of the renormalization group, in which the idea is to formally and rigorously remove degrees of freedom to obtain an effective theory of the remaining degrees of freedom.

Degeneracy of parameters. Interestingly, even in the context of the simplest three-parameter MWC model entertained through much of this book, it has been realized in contexts ranging from ligand-gated ion channels to hemoglobin to inducible transcription factors, that often the data are insufficient to distinguish between different parameter sets that are all roughly of the same quality in fitting the data. Specifically, one of the inheritances of the thinking based on dose-response curves that appeared a number of times throughout the book is that the resulting statistical mechanical mindset leads to a degenerate set of parameters, all of which are consistent with the data. One way out of this quandary is to design and perform new classes of experiments whose ambition is to remove these degeneracies. For example, measuring the leakiness directly provides a window on the energy difference between inactive and active states in the absence of ligand, thus determining this parameter in a way independent of the dose-response curve.

A lack of microscopic detail. The notion of mechanism is a complicated one. Some people are satisfied only when there is an underlying description of the phenomenon which appeals to molecular degrees of freedom. This tradition of hard-won victories in molecular biology has led to a sense that mechanism is inextricably linked with a microscopic description of what the atoms are doing. Structure reigns supreme. Given that the central repository for protein structures (Protein Data Bank) now boasts more than 100,000 structures, it is clear that the research enterprise has devoted a huge amount of time and money to advance this vision. One interesting perspective that has come to the fore recently is the use of the kinds of elastic network models hinted at in chapter 11. Such models provide a way to reconcile atomic-level structures and the mechanics of action at a distance. That said, the "more is different" view of Philip Anderson has much to recommend it. Indeed, I don't share this same enthusiasm for the mechanistic explanatory power of structure and ultimately embrace the aim of finding effective theories in which microscopic details are "integrated out."

Nonprotein allostery. Our analysis of nucleosomes in chapter 10 gave us one example of how our definition of allostery could be expanded to account for molecular conformations of molecules other than proteins. In that case, we described the conformations of the DNA wrapped around histones in terms of open and closed states analogous to more traditional protein allostery. However, other examples abound. One of the most exciting current research areas is RNA biology. The discovery of RNAs with enzymatic and regulatory roles hints at the promise of undiscovered allosteric transitions not only in RNAs but in the other macromolecules of the cell as well. We have shortchanged the reader by not exploring these other case studies.

Focusing on fitting the data. The stories we have told about allostery throughout the book often culminated in some description of the experiments done on some allosteric molecule of interest and how well the models from statistical physics respond to those experiments retrospectively. Elsewhere, I have referred to this kind of theory as Figure 7 theory. An alternative to the Figure 7 theory perspective is offered by the Figure 1 perspective, in which theory is not focused on fitting some data but, rather, on making polarizing predictions about experiments that have not yet been carried out. A general challenge for the models we have written here, but for theory more generally in biology, is to be ahead of the experiments. Ultimately, we want to suggest exciting and revealing experiments that have not yet been conceived or undertaken. One of the critical frontiers in this area is to design experiments that showcase the uniquely nonequilibrium features of living systems, providing an impetus for new kinds of statistical physics.

Evolution of allostery. One of my biggest regrets as my time with this book comes to a close is my failure to report on the exciting and impressive advances taking place in thinking about the evolution of allostery. The interested reader is encouraged to explore this fascinating topic in physical biology, in which a number of thoughtful researchers are advancing and testing hypotheses for the evolutionary history of the second secret of life.

13.3   Beyond Allostery

Ultimately, science remains the greatest engine of knowledge generation known to humanity. It is the formal tool used to answer those questions delivered up by our sense of wonder about the world around us. For millennia, using that formal tool, humans have pushed back the frontiers of our ignorance. Interestingly, there are many complementary pieces of scientific knowledge that remind me of the story about several blind people feeling the elephant and not even realizing that they are all talking about the same thing. From the point of view of the allostery phenomenon that has been the central focus of this book, an army of researchers including biochemists, molecular biologists, bioinformaticians, evolutionary biologists, and physicists have labored to each reveal their part of the allostery elephant. Sometimes the focus is on structure, sometimes on the atomic-level details that provide the "mechanism" of allosteric behavior. Other times, the focus is on the character of the input-output function or the kinetics of these molecules as they switch between their different allowed states. In the end, all these scientists are carrying out different parts of the broader goal of knowledge generation.

This book is the first in a series entitled *Studies in Physical Biology*, whose ambition is to describe the state of the art for some of the key topics in modern quantitative biology. The next book in this series continues the broad themes of signaling, regulation, and feedback by exploring the problem of how the genes in genomes are regulated. In this book, we have focused primarily on the behavior of individual allosteric molecules as nodes in signaling pathways of all kinds, with many of the examples considered here relating to the way these molecules are wired as parts of pathways. In the next book, we drill down into the question of how multiple molecules that are themselves often allosteric such as polymerases and transcription factors conspire together to use the information in genomes. We will again appeal to the tools of statistical physics as a way to sharpen the questions we can ask about biological processes. This journey will take us into an exploration of the processes of replication, transcription, and translation from a predictive point of view. We hope the reader will continue with us in these adventures in biological numeracy.

13.4   Further Reading

Anderson, P. W. (1972) "More is different." Science, 177:393–396. This paper is one of the great gems of modern science. Noted physicist Phil Anderson takes on his high-energy physics colleagues who argue that a knowledge of the fundamental particles and the forces between them provides the foundation for understanding everything else. Anderson's counterpoint is: "But this hierarchy does not imply that science X is 'just applied Y.'" Anderson argues that the different levels of phenomena will have their own laws, and these laws are not necessarily derived or inferred from a knowledge of the phenomena/laws at a different scale.

Lindsley, J. E., and J. Rutter (2006) "Whence cometh the allosterome?" *Proc. Natl. Acad. Sci.* 103 (28): 10533–10535. This short but profound article clearly articulates a big gap in our knowledge of the regulatory landscape of molecular biology, namely, our ignorance of which proteins are controlled by effectors, and in such cases, which effectors perform that function.

Miller, C. (1997) "Cuddling up to channel activation." *Nature* 389:328. This two-pager gives one of the most thorough and thoughtful descriptions of the allosteric framework and its shortcomings that I am aware of.

Phillips, R. (2017) "Musings on mechanism: Quest for a quark theory of proteins." *FASEB J.* 31:4207–4215. The argument I make in this paper is that modern biology is faced with precisely the same kind of thinking about what is fundamental that physics faced half a century ago. Specifically, the consensus view on fundamental "mechanism" in biology is that it is a statement about molecules: their structures and interactions. My piece argues facetiously that clearly we can't take any such biological mechanisms seriously, since they ignore quarks. My argument is that biology needs and will be better served by building theories which explicitly and intentionally *avoid* reference to molecules in the way that hydrodynamics and elasticity have for centuries in physics.

Tu, Y. (2008) "The nonequilibrium mechanism for ultrasensitivity in a biological switch: Sensing by Maxwell's demons." *Proc. Natl. Acad. Sci.* 105:11737–11741. This paper shows how nonequilibrium effects have to be invoked to explain the temporal behavior of the bacterial flagellar motor, thus illustrating the shortcomings of the strictly equilibrium MWC framework.

13.5 REFERENCES

Campbell, David K. (2004) "Nonlinear physics: Fresh breather." *Nature* 432 (7016): 455–456.

Kalodimos, C., and S. Edelstein, eds. (2013) "Allosteric interactions and biological regulation (Part I)." *J. Mol. Biol.* 425 (9): 1391–1592.

Schmidt, A., K. Kochanowski, S. Vedelaar, E. Ahrné, B. Volkmer, L. Callipo, K. Knoops, M. Bauer, R. Aebersold, and M. Heinemann (2016) "The quantitative and condition-dependent Escherichia coli proteome." *Nat. Biotechnol.* 34 (1): 104–110.

INDEX